QUALITY OF SERVICE IN OPTICAL PACKET SWITCHED NETWORKS

QUALITY OF SERVICE IN OPTICAL PACKET SWITCHED NETWORKS

Akbar Ghaffarpour Rahbar
Sahand University of Technology

IEEE Press Series on
Information & Communication
Networks Security
Stamatios Kartalopoulos, Series Editor

IEEE Press

For general information on our other products and services or for technical support, please contact our Customer Care Department within the United States at (800) 762-2974, outside the United States at (317) 572-3993 or fax (317) 572-4002.

Wiley also publishes its books in a variety of electronic formats. Some content that appears in print may not be available in electronic format. For information about Wiley products, visit our web site at www.wiley.com.

Library of Congress Cataloging-in-Publication is available.

ISBN 978-1-118-89118-6

Printed in the United States of America.

To my parents
and my family

CONTENTS IN BRIEF

CONTENTS

LIST OF FIGURES

LIST OF TABLES

PREFACE

Welcome to the era of unlimited communications, video-centric applications, and Internet! Internet applications require both bandwidth and Quality of Service (QoS) because of a huge number of Internet users and growing number of real-time applications (such as 3D TV, ultrahigh-definition TV, video-on demand, Internet Protocol TeleVision (IPTV), video-conferencing, Internet gaming, voice over IP, etc.) that need different levels of QoS. IP networks consist of core networks and access networks. By increasing IP traffic, access networks can grow in both size and count [1]. For example, traffic of broadband access networks such as ADSL and Fiber To The Home (FTTH) is continually increasing every year. To transport the huge traffic offered by IP networks, the core networks capabilities must be increased to avoid them from becoming bottleneck for IP traffic. This could be a problem when the network bandwidth is limited, the network supports only the best effort traffic, and the Internet traffic does not have a uniform characteristic.

The need for more and more bandwidth forces us to think of more granularity. The best promising solution is to use Wavelength Division Multiplexing (WDM) all-optical networks in core networks. Note that an optical network that uses optical transmission and keeps optical data paths through the nodes from source to destination is called all-optical network. Due to the fact that all-optical networks use photonic technology for the implementation of both switching and transmission functions, signals in these networks can be maintained in optical form without any conversion to the electronic domain resulting in much high transmission rates. All-optical networking with deployment of Dense Wavelength Division Multiplexing (DWDM) appears to be the sole approach to transport the huge network traffic in future backbone networks. The DWDM technology provides the multiplexing of many wavelength channels in a single optical fiber, resulting in several Tbits/s bandwidth capacity.

Similar to the electronic domain in which packet switching is the most granular method of switching, the most promising technique for optical core networks could be Optical Packet Switching (OPS) due to its high throughput and very good granularity and scalability. In an OPS edge node, a header is attached to each client packet received from a legacy network, where the header includes the information about source edge node, destination edge node, and content of packet payload such as its length. The packet is then transmitted in the optical domain, called an optical packet, toward the OPS network. In OPS, an optical packet stays in the optical domain inside the core network and switched optically. The optical packet can only be converted to the electronic domain in its destination edge node. Packet switching provides connectionless transmission of packets. Thus, there is no need to establish a path (i.e., a circuit) between source–destination nodes like in circuit switching. However, contention of optical packets in the core network is the major problem in OPS networks.

Since different applications need different levels of QoS, service differentiation must be considered in optical networks as well. Under the best-effort service in which no guarantees are given to any packet regarding loss rate, delay, and delay jitter, all traffic in the network is equally treated. This will, in turn, degrade the QoS requirements for real-time traffic. Thus, having a QoS-capable optical backbone network will be a requirement in which low latency, low jitter, low loss, and bandwidth guarantees must be provided for real-time traffic.

For providing QoS in OBS networks, [2] details (a) the basic mechanisms developed for improving end-to-end QoS and (b) relative and absolute QoS differentiation among multiple service classes. On the other hand, for OCS networks, the work in [3] focuses on the methods developed for service-differentiated and constraint-based wavelength routing and allocation in multi-service WDM networks. However, there is no comprehensive work on QoS in OPS networks.

In future, OPS networks must be setup for worldwide communications in order to transport the huge traffic generated by Internet users and applications. In addition, research and development on optical communication networking have been matured significantly during the last decade to the extent that some of these principles have moved from the optical research laboratories to formal graduate courses. Moreover, there are a large number of experts working on designing optical devices and physical-layer of optics that are interested in learning more about OPS network architectures, protocols, and the corresponding engineering problems in order to design new state-of-the-art OPS networking products. Finally, there are many books written for device level of optical communications, and there are even devices suitable for OPS. However, there is almost no work dedicated solely for system level of OPS (say architectures and protocols), improving quality of service, and the operation of OPS networks.

In general, there are some books published for covering optical networking such as [4–10]. However, the number of published books dedicated to the system level of OPS is limited to OPS in access networks [11], design of optical buffers for OPS [12], edge node design for contention avoidance in slotted OPS [13], scheduling in star-based OPS networks [14], and OPS for ring networks [15].

This book provides a comprehensive study on OPS networks, its architectures, and developed techniques for improving its quality of switching and managing quality of service. This book is organized in six chapters, each covering a unique topic in detail:

- Chapter 1 provides an introduction to OPS networks, its architectures, and QoS in OPS. Since many optical networking books have stated optical systems in much detail, this chapter does not include them. In addition to OPS networks, GMPLS-supported optical networks and optical networks based on Orthogonal Frequency Division Multiplexing (OOFDM) are studied in this chapter.

- Chapter 2 describes contention avoidance schemes proposed for OPS networks in which edge switches send optical packets to the OPS network in a way to reduce their collisions. Broadly, these schemes are classified as either hardware-based or software-based.

- Chapter 3 details contention resolution schemes proposed for OPS networks in which OPS switches resolve the collision of contending optical packets. In general, contention resolution schemes are classified as either hardware-based or software-based.

- Chapter 4 studies the hybrid contention resolution schemes that use a number of contention resolution schemes in the same architecture in order to reduce optical packet loss rate. In addition, hybrid contention resolution and contention avoidance schemes are reviewed that can efficiently reduce optical packet loss rate in a cost-effective manner.

- Chapter 5 describes hybrid optical switching schemes in which OPS networking is combined with another optical switching technique (say optical circuit switching) in order to improve the performance of traffic transmission in the optical domain.

- Chapter 6 states different OPS architectures designed for metro area. These networks are mainly based on ring and star topologies with active optical switches.

This book is a useful resource for students, engineers, and researchers to learn more about optical packet switched networking from system level points of view. It is intended as a textbook for graduate level and senior undergraduate level courses in electrical engineering and computer science on (advanced) optical networking. Knowledge about computer networks is a prerequisite for understanding this book. For advanced optical networks course relevant to OPS, the book can be entirely used.

Reasonable care has been taken in eliminating any types of errors. However, readers are encouraged to send their comments and suggestions to the author via e-mail. I personally hope that this book will give the reader enough information in OPS networks and motivate his/her interests to develop efficient, QoS-capable, and cost-effective OPS networks suitable for future core optical networks.

AKBAR GHAFFARPOUR RAHBAR

Sahand University of Technology
ghaffarpour@sut.ac.ir

REFERENCES

1. A. Shami, M. Maier, and C. Assi. *Broadband Access Networks: Technologies and Deployments.* Springer, 2009.

2. K. C. Chua, M. Gurusamy, Y. Liu, and M. H. Phung. *Quality of Service in Optical Burst Switched Networks.* Springer, 2007.

3. A. Jukan. *QoS-based Wavelength Routing in Multi-Service WDM Networks.* Springer, 2001.

4. B. Mukherjee. *Optical WDM Networks.* Springer, 2006.

5. R. Ramaswami, K. Sivarajan, and G. Sasaki. *Optical Networks: A Practical Perspective.* third edition, Morgan Kaufmann, 2009.

6. T. E. Stern, G. Ellinas, and K. Bala. *Multiwavelength Optical Networks: Architectures, Design, and Control.* second edition, Cambridge University Press, 2008.

7. J. M. Simmons. *Optical Network Design and Planning.* Springer, 2008.

8. V. Alwayn. *Optical Network Design and Implementation.* Cisco Press, 2004.

9. R. J. B. Bates. *Optical Switching and Networking Handbook.* McGraw-Hill, 2001.

10. M. Maier. *Optical Switching Networks.* Cambridge University Press, 2008.

11. K. Bengi. *Optical Packet Access Protocols for WDM Networks.* Springer, 2002.

12. E. H. Salas. *Design of Optical Buffer Architectures for Packet-Switched Networks: An Optical Packet Buffer Overview.* LAP Lambert Academic Publishing, 2010.

13. A. G. Rahbar and O. Yang. *OPS Networks: Bandwidth Management & QoS.* VDM Verlag, Germany, 2009.

14. N. Saberi. *Photonic Networks: Bandwidth Allocation and Scheduling.* LAP LAMBERT Academic Publishing, 2011.

15. B. Uscumlic. *Optical Packet Ring Engineering: Design and Performance Evaluation.* LAP LAMBERT Academic Publishing, 2011.

ACKNOWLEDGMENTS

To all those wonderful people I owe a deep sense of gratitude especially now that this book has been completed. To my wife and daughter for their consistent patience and encouragement. To the publisher's staff for their collaboration and project management.

Akbar Ghaffarpour Rahbar

ACRONYMS

3LIHON	3 Level Integrated Hybrid Optical Network
ACK	ACKnowledge
AF	Assured Forwarding
APTB	Aggregated Packet Transmission Buffer
AW	Additional Wavelengths
AWG	Array Waveguide Grating
BE	Best Effort
BER	Bit Error Rate
BH	Burst Header
BP	Buffer Pool
bps	bits per second
BPSK	Binary Phase-Shift Keying
BV	Bandwidth Variable
BvN	Birkhoff and von Neumann
CBR	Constant Bit Rate
COPS	Composite Optical Packet Scheduling
CoS	Class of Service
CPA	Composite Optical Packet Aggregation

CPDU	Control packet PDU
CRSA	Contention Resolution Scheduling Algorithm
CSMA/CA	Carrier Sense Multiple Access with Collision Avoidance
CWDM	Coarse WDM
DA	Distribution-based bandwidth Access
DCF	Dispersion Compensated Fiber
DiffServ	Differentiated Services
DMUX	DeMultiplexer
DR	Deflection Routing
DRwBD	Deflection Routing with Backward Deflection
DRwoBD	Deflection Routing without Backward Deflection
DSF	Dispersion Shifted Fiber
DWDM	Dense WDM
EAP	Even Assignment Problem
EBvN	Efficient BvN
EBvN_FEC	EBvN with Filing Empty Cells
EDFA	Erbium-Doped Fiber Amplifier
EDF$_{de}$	Even Density distribution through gauging Frame
EDF$_{di}$	Even Distance distribution through gauging Frame
EF	Expedited Forwarding
EON	Elastic Optical Network
ES	Edge Switch
FCFS	First-Come-First-Served
FD	Fair Dissemination distribution
FDL	Fiber Delay Line
FDM	Frequency Division Multiplexing
FEC	Forward Error Correction
FEC	Forwarding Equivalent Class (in MPLS networks)
FF	First-Fit
FIFO	First-In-First-Out
FRWC	Full Range Wavelength Converter
FSC	Fiber-Switch-Capable
FTTH	Fiber-To-The Home
FTWC	Fixed-input/Tunable-output WC
FWC	Fixed Wavelength Converter
GF	Gauging Frame
GMPLS	Generalized Multi-Protocol Label Switching
GST	Guaranteed Service Transport
HCT	High Class Transport
HOPSMAN	High-performance Optical Packet-Switched MAN

HOS	Hybrid Optical Switching
HOTARU	Hybrid Optical neTwork ARchitectUre
HP	High-Priority
HSWC	Hybrid Shared Wavelength Conversion
HTDM	Hybrid TDM
IAS	Impairment-Aware Scheduling
ID	IDentification
ILP	Integer Linear Programming
IP	Internet Protocol
IPD	Intentional Packet Dropping
IPT	Immediate Packet Transmission
IPTV	Internet Protocol TV
ISA1	Ingress Switch Architecture 1 for class-based OPS networks
ISA2	Ingress Switch Architecture 2 for class-based OPS networks
LB	Load Balanced distribution index
LCR	Local Cyclic Reservation
LCR-SD	Local Cyclic Reservation with Source-Destination
LDP	Label Distribution Protocol
LER	Label Edge Router
LGRR	Loan-Grant-based Round Robin
LP	Low-Priority
LRWC	Limited-Range Wavelength Converter
LSC	Lambda-Switch-Capable
LSP	Label Switching Path
LSR	Label Switching Router
MAC	Media Access Control
MAN	Metropolitan Area Network
MBS	Merit-Based Scheduling
MEMS	Micro-Electro-Mechanical Systems
MF	Multi Fiber
MFAW	Multi Fiber + Additional Wavelengths
MI	Minimum Interference, Multilayer Interference
MING	Minimum Gap Queue
MINL	Minimum Length Queue
MP	Mid-Priority
MPBvN	Multi-Processor BvN
MPLS	Multi-Protocol Label Switching
M_PR	Modified Prioritized Retransmission
MPR	Multi-Path Routing
MQWS	Minimum Queue length Wavelength Selection

MUX	Multiplexer
MW-OPS	Multi-Wavelength Optical Packet Switching
NACK	Negative Acknowledge
NBR	Non-Blocking Receiver
NCPA	Non-Composite Optical Packet Aggregation
NCT	Normal Class Transport
NoD	No Deflection
NR	No Retransmission
NRP	Number-Rich Policy
NRPWC	Non-Recursive Parametric Wavelength Conversion
NWB-OPS	Non-Wavelength-Blocking OPS
NZDSF	Non-Zero Dispersion-Shifted Fiber
OBM	Optical Bandwidth Manager
OBS	Optical Burst Switching
OCGRR	Output-Controlled Grant-based Round Robin
OCS	Optical Circuit Switching
O/E/O	Optical/Electrical/Optical
OOFDM	Optical Orthogonal Frequency Division Multiplexing
OP	Optical Packet
OpMiGua	Optical Migration capable network with service Guarantees
OPS	Optical Packet Switching
ORION	Overspill Routing in Optical Networks
OSNR	Optical Signal-to-Noise Ratio
OTDM	Optical Time Division Multiplexing
OVP	OVerspill Packet
OXC	all-Optical Cross-Connect switch
PA	Packet Aggregation
PAU	Packet Aggregation Unit
PDP	Preemptive Drop Policy
PDU	Protocol Data Unit
PLR	optical Packet Loss Rate
PMD	Polarization Mode Dispersion
POADM	Packet Optical Add and Drop Multiplexers
PQOC	Probabilistic Quota plus Credit
PR	Prioritized Retransmission
PS	Packet Scheduler
PSC	Packet-Switch-Capable
PSK	Phase Shift Keying
PTES	Packet Transmission based on Scheduling of Empty Time Slots

PWC	Parametric Wavelength Converter
QAM	Quadrature Amplitude Modulation
QoS	Quality of Service
QoT	Quality of Transmission
QPSK	Quadrature Phase-Shift Keying
RIB	Reservation Induced Blocking
RNENF	Random choice among Neither Empty Nor Full queues
ROADM	Reconfigurable Optical Add/Drop Multiplexer
ROB	Removing of Overdue Blocks
RPWC	Recursive Parametric Wavelength Conversion
RR	Random Retransmission
RS	Reed-Solomon
RSA	Routing and Spectrum Assignment/Allocation
RSVP	Resource Reservation Protocol
SA	Spectrum Assignment/Allocation
SBvN	Separated BvN
SDU	Service Data Unit
SFD	Smoothed Flow Decomposition
SHP	Shortest Hop-Path
SLA	Service Level Agreement
SM/BE	Statistically Multiplexed Best Effort
SMF	Single-Mode Fiber
SM/RT	Statistically Multiplexed Real Time
SOA	Semiconductor Optical Amplifier
SPC WC	Single-Per-Channel Wavelength Converter
SPIL WC	Shared-Per-Input-Link Wavelength Converter
SPIW WC	Shared-Per-Input-Wavelength Wavelength Converter
SPL WC	Shared-Per-Link Wavelength Converter
SPN WC	Shared-Per-Node Wavelength Converter
SPOL WC	Shared-Per-Output-Link Wavelength Converter
SPOW WC	Shared-Per-Output-Wavelength Wavelength Converter
SPR	Shortest-Path Routing
SSMF	Standard Single-Mode Fiber
SWING	Simple Wdm rING
TCP	Transmission Control Protocol
TDM	Time Division Multiplexing
TFWC	Tunable-input/Fixed-output WC
TLWC	Two-Layer Wavelength Conversion
Tout	Timeout

TTL	Time-To-Live
TTWC	Tunable-input/Tunable-output WC
TWC	Tunable Wavelength Converter
VRP	Variety-Rich Policy
WAN	Wide Area Network
WAR	Wavelength Access Restriction
WB-OPS	Wavelength-Blocking OPS
WC	Wavelength Converter
WDM	Wavelength Division Multiplexing
WDS	Wavelength Delay Section
WRN	Wavelength Routed Network
WSC	Waveband-Switch-Capable
WSS	Wavelength Selective Switch
WXC	Wavelength Cross-Connect

GLOSSARY

*	Notation used in switch sizes, say a switch with a inputs and b outputs is denoted by $a * b$
x mod y	Remainder of x divided by y
Asynchronous OPS	Pure (non-slotted) OPS
Client packet	The upper layer packet arriving at an ingress switch from legacy networks
Conversion ratio	Ratio of total number of WCs used in an $N * N$ OPS switch with W wavelengths to total number of wavelengths in the switch (i.e., $N \times W$)
Core switch	An optical switch that performs switching in the optical domain
Drop link	f fibers connecting a core switch to an egress switch for delivering f optical packets on each wavelength from the core to the edge switch at the same time
Drop port	One fiber connecting a core switch to an egress switch for delivering one optical packet on each wavelength from the core to the edge switch
Egress switch	An edge switch with the function of receiving traffic from an optical network
FDL bank	A bank of FDL buffers that provides delay in range 0 to $B \times D$, where D is a constant delay time and B is the buffer depth
Gbits/s	Gigabits per second
Ingress switch	An edge switch with the function of transmitting traffic to an optical network
Input port	An input fiber to a core switch from a neighbor edge/core switch in a single-fiber network
Input link	f fibers input to a core switch from a neighbor edge/core switch in a multi-fiber network
Legacy network	Any network (including an old network, an Ethernet network, a TCP/IP network, and a SONET/SDH network) connected to an edge switch
Local optical packet	The optical packet either added to an OPS switch by its local ingress switch, or dropped by the OPS switch to its local egress switch
max(x, y)	The larger value of x and y
Mbits	Megabits

min(x, y)	The smaller value of x and y
nf-slot-set	The set of $n \times f$ optical packets sent from n ingress switches on the same wavelength channel and on f fibers within a given time slot
nm	Nano meters
ns	Nano seconds
Optical packet	The packet transmitted by an ingress switch to an OPS network that may include one or more client packets
OP set	A set of optical packets transmitted at the same time slot over the available fibers/wavelengths in an ingress switch
OPS core switch	An optical switch that performs OPS switching functionality in the optical domain
Output port	An output fiber from a core switch to a neighbor edge/core switch in a single-fiber network
Output link	f fibers output from a core switch to a neighbor edge/core switch in a multi-fiber network
Slot set	Set of $f \times W$ rows in a column of a frame in frame-based scheduling
Synchronous OPS	Slotted OPS
Tbits/s	Terabits per second
Torrent	All the (class-based) traffic going to the same egress switch in an ingress switch. Therefore, Torrent-i traffic goes to egress switch i
Transit optical packet	The optical packet that passes through an OPS switch toward another OPS switch
Uniform selection	Uniform selection from a list with m items; i.e., selecting an item randomly with probability $\frac{1}{m}$.
WC bank	A bank of wavelength converters with the same type

SYMBOLS

A	Traffic demand matrix
$AgPk$	An aggregated packet from client packets
B	Buffer depth in an FDL bank
B_C	Wavelength channel bandwidth in bps
D	A constant value in delaying of optical packets in an FDL bank, where an FDL provides a delay in an integer number of D
d_c	Wavelength conversion degree in limited-range wavelength converters
$D_{P,i}$	Propagation delay from edge switch i to the core switch in overlaid-star OPS
E	Number of empty time slots in each column of a frame
f	Number of fibers on each connection link between two nodes in an OPS network
F	Frame size
G	Number of legacy networks connected to an ingress switch
H	Hurst parameter under bursty traffic
$H1$	Hybrid index 1
$H2$	Hybrid index 2
$H3$	Hybrid index 3
L	Average traffic load of client packets arriving at an ingress switch, normalized with respect to total bandwidth of each edge switch; i.e., $f \times W \times B_c$
L_a	Average length of client packets
L_{OP}	Average normalized optical packet arrival rate from an ingress switch to the OPS network, defined as the optical packet generation rate over optical packet service rate in an ingress switch
M	Number of traffic classes supported in an OPS network
N	Number of input/output links/ports of an OPS core switch

n	Number of optical switches in an OPS network
n_e	Number of egress switches in an OPS network from an ingress switch point of view which is equal to $n - 1$
$N_{d,m}$	Number of drop links in core switch m connected to edge switch m
$N_{j,c}$	Number of client packets from class c carried in optical packet j
N_{WC}	Total number of wavelength converts used in an OPS switch
OP_i	Optical packet i
$p_{j,c}$	Percentage of client packets from class c carried in optical packet j
P_k	Permutation matrix k
PLR	Optical packet lost rate in an OPS network
PLR_{req}	Desired (maximum) optical packet lost rate in an OPS network
PMD_j	PMD of optical packet j until current core switch
R	Doubly stochastic matrix
R'	Doubly sub-stochastic matrix
r_{wc}	Number of wavelength converts used in a converter bank in an OPS switch
$S_{k,i,m}$	Set of m contending optical packets indexes incoming on wavelength λ_i destined to output port k
S_O	A time-gap interval plus a header interval (for transmitting header information) in an optical packet
S_T	Time duration of a time slot in OPS networks within which traffic is carried
S_{th}	Optical packet size threshold (either in volume or number) in packet aggregation
S_{ts}	Time-slot size in seconds in slotted-OPS networks
T_{HP}	Timeout for high-priority class
T_{LP}	Timeout for low-priority class
T_{MP}	Timeout for medium-priority class
V_c	Significance parameter for client traffic class c, where (c = HP, MP, LP) or (c = EF, AF, BE)
W	Number of wavelengths on each fiber in an OPS network
W_a	Number of additional wavelengths on each fiber in an OPS network
X_{DE}	Index for EDF_{de} within a GF frame
$X_{DE,i}$	Index for EDF_{de} for Torrent-i
X_{DI}	Index for EDF_{di} within a GF frame
$X_{DI,i}$	Index for EDF_{di} for Torrent-i
X_{FD}	Index for FD
X_{LB}	Index for LB
λ_i	Wavelengths number i
ω	Total number of wavelengths required in a network on each connection link (i.e., $\omega = f \times W$)
τ	Gauging frame size in OP-sets
v_j	Rank for optical packet j

CHAPTER 1

INTRODUCTION TO OPTICAL PACKET SWITCHED (OPS) NETWORKS

The demand for high bit-rate traffic is increasing rapidly each year, especially by Internet users [1]. Among different networking mechanisms, optical networks can support this overwhelming traffic demand. In optical networking, various amounts of traffic with different bit rates can be multiplexed into a single fiber and switched in the network. Optical fiber communication can offer much higher bandwidth than conventional copper wires used extensively as the transmission medium. In addition, optical fibers can provide low communication cost per kilometer. Moreover, signals transmitted on optical fibers encounter lower bit error rate than copper wires. In this chapter, an introduction is provided for optical networking, especially Optical Packet Switched (OPS) networks. The OPS networking can solve the mismatch between very high transmission capacity of WDM optical links and the processing power of routers/switches [2].

1.1 Optical Fiber Technology

Optical fibers are essentially very thin glass cylinders filaments that carry data and control signals in the form of light (i.e., optical signals). There are two common

Quality of Service in Optical Packet Switched Networks, First Edition.
By Akbar Ghaffarpour Rahbar Copyright © 2015 IEEE. Published by John Wiley & Sons, Inc.

fibers as single-mode and multi-mode as the basics of optical transmission. Single-mode fiber has a relatively small core diameter of about 8 to 10 μm. Using single-mode fiber effectively eliminates intermodal dispersion and enables a significant increase in the bit rates and distances between nodes. Using this fiber, a network needs a typical regenerator/amplifier spacing of about 40 km and it can operate at bit rates of a few tens gigabits per second. However, the distance between regenerators/amplifiers is primarily limited by fiber loss. The standard Single-Mode Fiber (SMF) is a common single-mode fiber used in optical technology. These fibers provide low dispersion and can be appropriate for long distances such as regional and long haul networks. On the other hand, multi-mode fibers have a relatively large core diameter of about 50 to 62.5 μm. These fibers have high dispersion and are only suitable for short distances such as metro networks.

Whereas in electronic networks different multiplexing techniques such as Time Division Multiplexing (TDM) and Frequency Division Multiplexing (FDM) are employed to use transmission resources efficiently, Wavelength Division Multiplexing (WDM) is used in optical networks as an effective technique to make use of the large amount of available bandwidth in optical fibers for applications with huge bandwidth demands. The technology of using multiple optical signals on the same fiber is called WDM. In WDM, several baseband-modulated channels are transmitted along a single fiber at the same time, but each channel is located at a different wavelength. The most important components of any WDM system are optical transmitters, optical receivers, optical amplifiers, optical switches, optical add drop multiplexers, WDM multiplexers, and WDM demultiplexers. All-optical networks using WDM appear to be the sole approach for transporting huge network traffic in future core networks.

In WDM optical networks, data are carried on wavelengths. Consider that there are W different wavelengths on a fiber, where each wavelength is able to carry data at different data rates such as $B = 2.5$ Gbits/s, 10 Gbits/s, 40 Gbits/s, and even more. The aggregate system has a capacity of $W \times B$, and therefore the amount of total carried data on one optical fiber can increase up to several Tb/s. Therefore, WDM networks can be employed in Wide Area Networks (WANs) for the future backbone networks of Internet. Nowadays, WDM primarily uses the 1.55-μm wavelength region because of the inherent loss in optical fiber which is the lowest in this region; excellent optical amplifiers have been designed for this region. In Dense Wavelength-Division Multiplexing (DWDM), many wavelengths (say 160) are packed densely into a fiber with small channel spacing.

When selecting a fiber for an optical network design, two important issues must be considered: attenuation and dispersion. Attenuation is the optical signal loss when it travels through a fiber, whereas dispersion is the tendency of different wavelengths to travel with different speeds, which leads to pulse spreading [3].

Attenuation in an optical fiber depends on the wavelength an optical signal uses (see Fig. 1.1) [4]. As can be observed, not all wavelengths are suitable for transmission of an optical signal. Four principal windows of wavelength ranges with low attenuation are displayed in this figure, called O-band (displayed with first window), S-band (displayed with second window), C-band (displayed with third window), and L-band (displayed with fourth window). Optical lasers deployed in optical commu-

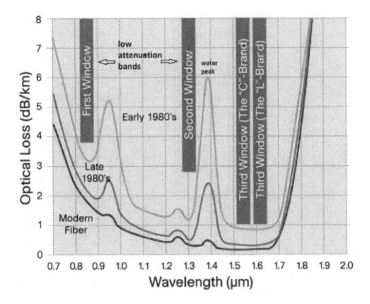

Figure 1.1: Attenuation versus wavelength curve [4]

nications typically operate at these bands. Note that in general, there are seven bands standardized for the attenuation-wavelength curve [3, 5]:

- 850 nm: wavelength range 770 nm to 910 nm. This range was used in the first generation of optical networks with opaque switches such as SONET/SDH and gigabit Ethernet networks.

- O-band (Original band) (known as 1310 nm): wavelength range 1260 nm to 1360 nm. This range is used for short range communications (say upstream wavelength in PON). The O-band provides high attenuation of 0.5 db/km, but no dispersion in standard single-mode fibers.

- E-band (Extended band): wavelength range 1360 nm to 1460 nm. This is the water-peak band that can only be used in modern fibers as they have reduced attenuation in this range. This range is only suitable for short-range communications.

- S-band (Short band): wavelength range 1460 nm to 1530 nm. This range is used for short-range communications (say downstream wavelength in PON).

- C-band (Conventional band) (known as 1550 nm): wavelength range 1530 nm to 1565 nm. This range has the lowest attenuation and was used for early WDM communications.

- L-band (Long band): wavelength range 1565 nm to 1625 nm. This range provides low attenuation and is used for WDM communications.

- U-band (Ultra-long band): wavelength range 1625 nm to 1675 nm. This range is used for WDM communications.

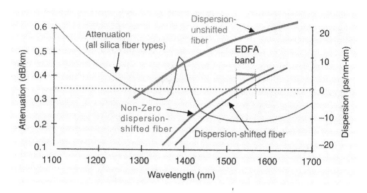

Figure 1.2: The SSMF, DSF, and NZDSF fibers [6]

Chromatic dispersion in single-mode fibers causes pulse spreading because of the fact that various wavelengths travel at different speeds. When optical signals spread too far, they overlap and cannot be correctly detected at the receiving end of the network. Similar to attenuation, dispersion is a function of wavelength (see Fig. 1.2). Since Standard Single-Mode Fibers (SSMF), displayed with dispersion unshifted fiber in Fig. 1.2, provide a zero dispersion at wavelength 1310 nm, 1310 nm transmitters are not subject to chromatic dispersion. On the other hand, at 1550 nm, CWDM and DWDM transmissions over SSMF are affected by chromatic dispersion. In short, SSMF provides high attenuation at 1310 nm compared with other bands, but zero dispersion in this range, whereas it provides the lowest attenuation at 1550 nm, but high dispersion at this range [4, 6].

In order to reduce the dispersion encountered by transmissions within the C-band and L-band windows in SSMF, other fiber types have been developed. Dispersion-Shifted Fibers (DSF) provide a zero dispersion at 1550 nm (see Fig. 1.2). DSF with low loss in the L-band range is suitable for DWDM applications. Although DSF eliminates the dispersion problem for transmissions of single wavelengths at 1550 nm, it is still not appropriate for wavelength multiplexing applications since WDM transmissions can be affected by another non-linear effect called four-wave mixing. This has led to the development of Non-Zero Dispersion Shifted Fibers (NZDSF), where the zero dispersion is shifted just outside the C-band, around 1510 nm. This can limit the chromatic dispersion (as the zero dispersion remains close enough to the transmission band) and four-wave mixing. The NZDSF fiber is optimized for long-haul applications at 1550-nm range in terms of both attenuation and dispersion. It is suitable for distances over 70 km. There are two types of NZDSF known as (−D)NZDSF (called negative NZDSF) and (+D)NZDSF (called positive NZDSF) that can provide a negative and positive slope versus wavelength, respectively. Both positive and negative NZDSF are fine over distances of up to 200 km in the C-band at data rate 10 Gbits/s. However, for 40 Gbits/s transmission rate, positive NZDSF is recommended [3, 4, 6].

To reduce the dispersion problem and improve optical signal transmission, there are various fiber types developed such as Dispersion-Compensated Fiber (DCF), Dispersion-Flattened Compensated Fiber (DFCF), Dispersion-Slope Compensated Fiber (DSCF), and Dispersion-Shift Compensated Fiber (DSCF) [6].

Optical fibers and components (such as amplifiers and filters) cause several kinds of impairments for optical signals. This results in signal quality degradation in optical receivers. To compensate signal loss, optical amplifiers must be used as essential components in transmission systems and networks. The most common optical amplifier could be the Erbium-Doped Fiber Amplifier (EDFA) operating in the C-band. In addition, L-band EDFAs and Raman amplifiers can be used. EDFAs can be used in almost all WDM networks, whereas Raman amplifiers can be used in addition to EDFA in many ultra-long-haul optical networks. The main advantage of an EDFA amplifier is that it can simultaneously amplify many WDM channels. To support high-capacity DWDM transmission, DCF fibers could be placed in between amplification stages [7, 8].

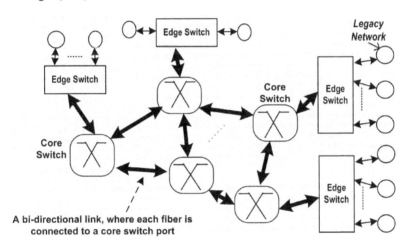

Figure 1.3: General optical network model

1.2 Why Optical Networks?

The Internet traffic demand is increasing 70 to 150 percent each year [9]. The need for high bandwidth and high data rate services are also increasing. Almost all communication applications tend to change their networks to IP; even voice applications use IP networks to carry their voice traffic. Traffic demands of various Internet applications (such as video conferencing, real-time medical imaging, emergency services, online gaming) continue to grows and this will consume more and more network bandwidth [10, 11]. In addition, current bandwidth-limited networks mostly support the best effort traffic, and therefore no differentiation can be easily made for real-time traffic. Moreover, network traffic patterns have a bursty nature, and not a uniform characteristic. Therefore, core networks in WANs are becoming bottleneck. It seems that the sole approach to overcome this huge demand and resolve the bottleneck is to deploy optical networks. Thus, the optical networking is the key technology for future communication networks.

The optical and electronic networks have essential differences in switching speed, buffering architecture, and bandwidth granularity issues. In the optical domain, switching speed is slower and building optical buffers is a more complex and a more expensive issue than electronic networks. On the other hand, an optical network can provide a higher bandwidth, a better signal quality, and a better security than an electronic network. Considering these major differences, different architectures and bandwidth management protocols must be used to utilize the huge bandwidth offered by all-optical networks [6, 12].

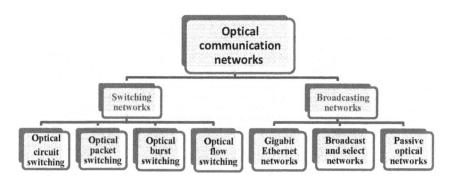

Figure 1.4: Hierarchy of optical communications networks

As Fig. 1.4 shows, optical communications networks can be divided into switching and broadcasting networking similar to electronic networks. Some common switching networks developed for data transmission over WANs are Optical Circuit Switching (OCS) networks [also called Wavelength Routed Networks (WRNs)], Optical Packet Switching (OPS) networks, Optical Burst Switching (OBS) networks, and Optical Flow Switching (OFS) networks. Similar to electronic domain in which packet switching is the most granular method of switching, the most promising technique for optical core networks could be OPS due to its high throughput and very good granularity and scalability. However, contention of optical packets in the core network is the major problem in OPS networks. On the other hand, gigabit Ethernet networks, broadcast and select networks, and Passive Optical Networks (PON) are common broadcasting networks developed for Local Area Networks (LAN) that broadcast data toward all destinations.

1.3   Optical Networking Mechanisms

There are two common mechanisms to share the bandwidth in an optical network: reservation-based and contention-based. Reservation-based mechanisms always reserve bandwidth in advance, say for a connection request, and are mostly suitable for (semi-) static traffic. The decision made in a reservation-based scheme can be either centralized or distributed. In centralized, only one central module is responsible for bandwidth provisioning, while a number of modules cooperate with each other to reserve bandwidth in the distributed manner. On the other hand, contention-based schemes perform no reservations and are appropriate for the networks with dynamic traffic. A contention-based scheme relies on a random access technique in order to access the network. Due to their lower complexity, they are easily scalable and can respond more quickly to bursty traffic than the reservation-based schemes. Contention-based schemes may suffer from collision, whereas reservation-based schemes may experience connection blocking.

An optical network is interconnected with a number of optical switches, each called a core switch (see Fig. 1.3). One edge switch operating in the electronic domain is connected to each optical switch. When an edge switch has the function of transmitting traffic to the optical network, it is called ingress switch. On the other hand, when an edge switch has the function of receiving traffic from the optical network, it is called egress switch.

The packets arriving at an ingress switch from legacy networks are called client packets. In this book, a legacy network denotes any network in the electronic domain including an old network, an Ethernet network, a SONET/SDH network, and a TCP/IP network. By buffering client packets in the electronic domain and forwarding them to the optical network, an edge switch provides interfacing between the electronic legacy networks and the optical network. The architecture of an optical network could be either single-fiber or Multi-Fiber (MF). In a single-fiber optical network, there is only one fiber used in each direction between two adjacent nodes (i.e., two fibers between two adjacent nodes). The adjacent nodes could be either (a) two optical switches or (b) an edge switch and an optical switch. On the other hand, in a multi-fiber optical network, there are say f fibers used in each direction between two adjacent nodes (i.e., $2 \times f$ fibers between two adjacent nodes), but with a small number of wavelength channels per fiber compared with the single-fiber architecture. For example, if a single-fiber network needs 48 wavelengths per fiber, then a multi-fiber architecture with $f = 3$ fibers on each connection link requires $W = 16$ wavelengths per fiber, i.e., $f \times W$ is constant.

A core switch size is displayed with say $N * N$, i.e., it has N input ports/links and N output ports/links. Note that term input/output port is used in single-fiber networks, and term input/output link is used in multi-fiber networks, where each link has f ports.

A multi-fiber architecture has a numbers of advantages. It can reduce connection blocking in circuit switching optical networks and traffic loss in optical packet/burst switching networking. In addition, the devices (such as dispersion compensator modules) required for a fiber with a large number of wavelengths are more expensive than the devices used for a fiber with a small number of wavelengths [13–16]. For example, as Fig. 1.5 illustrates, for a fiber link between any two optical switches, Distributed Raman Amplifiers (DRA) with 82-km spacing can be employed. Each amplification span includes 70 km of SMF fiber whose dispersion and dispersion slope are compensated by 12 km of Dispersion Compensating Fibers (DCF). Finally, a multi-fiber architecture can provide a better protection against a single-fiber cut than can a single-fiber architecture [14].

Core switches used in an optical network may have different sizes according to the topology of the network. For example, at $f = 2$, one core switch may need 6 input ports and 6 output ports, the other core switch may need 10 input ports and 10 output ports, and so on. However, an optical switch must have a size of $N * N$, where we have $N = 2^x$ and x is an integer number. Therefore, some ports of the core switch may be unused. These unused ports can be connected to the relevant edge switch in order to reduce the receiver contention problem, i.e., to make an almost non-blocking receiver (detailed in Chapter 3).

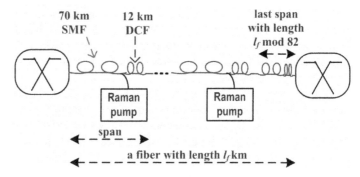

Figure 1.5: Link model in an optical network

Optical networks are categorized as all-optical (transparent) networks, opaque networks, and translucent networks [17–19]:

- **Opaque Networks**: Every optical switch, equipped with O/E/O (optical to electronic to optical) converters, converts an optical signal to the electronic domain, processes or saves it in the electronic domain, and then regenerates and converts it to the optical domain. In other words, opaque networks only apply optical fibers as their transmission medium, and processing and switching are carried out in the electronic domain. Electronic equipments used in O/E/O optical switches limit the data rate in opaque networks and high bandwidths cannot be achieved in such networks.

- **Transparent Networks**: An all-optical network uses a transparent optical signal transmission without any conversion of its data traffic into the electronic domain in its core switches, while it may process the header of data traffic in the electronic domain. Note that transparent optical switches are not only cheap but can also completely utilize the bandwidth offered by DWDM. Realizing all-optical networks needs very low-loss fibers, switches, and multiplexers/demultiplexers.

- **Translucent Networks**: In translucent optical networks, some optical switches are transparent and some other are opaque with O/E/O capability in such a way that a transmitted signal stays in optical domain in most of the core switches.

Different optical networking mechanisms have been proposed for managing network bandwidth, where each mechanism is based on a switching technique. In the following, a number of switching schemes used in synchronous (also called slotted) and asynchronous (pure) optical networks are studied. A survey of optical switching schemes can be easily found in [12, 20–24].

1.3.1 Asynchronous Optical Switching

In this section, different switching mechanisms introduced for asynchronous (pure) optical switching, where wavelength channels are not time-slotted.

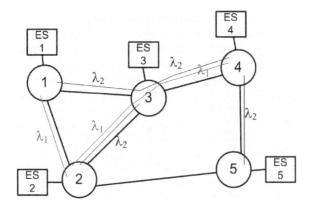

Figure 1.6: Light-path setup example in a WRN optical network

1.3.1.1 Optical Circuit Switching   To transmit data in a Wavelength Routed Network (WRN), the OCS mechanism (as a two-way reservation-based mechanism) is used in which an optical connection, called light-path, must be set up between a source–destination pair using a Routing and Wavelength Assignment (RWA) technique prior to data transmission [25–30]. A light-path is setup on a wavelength channel in a path that may span a number of fiber links. Then, the entire bandwidth on this light-path is used for a connection until its termination. Since light-path setup is the most important operation to make a connection request between a pair of nodes, its effective establishment should be taken into account. If an RWA algorithm cannot establish a light-path due to lack of network resource, the relevant connection request will be blocked. For example, Fig. 1.6 shows a network in which four light-paths have been set up: (1) a light-path from Edge Switch (ES) 1 to ES 4 on wavelength λ_2 crossing core switches 1, 3, 4; (2) a light-path from ES 1 to ES 4 on wavelength $\lambda 1$ passing through core switches 1, 2, 3, 4; (3) a light-path from ES 2 to ES 3 on wavelength $\lambda 2$ crossing core switches 2, 3; and (4) a light-path from ES 4 to ES 5 on wavelength $\lambda 2$ crossing core switches 4, 5.

Two types of traffic are considered for OCS networks. Under static traffic [31, 32], a set of fixed connection requests is given and the network should follow two main objectives in establishing these connections: (1) maximizing the number of established connections when the number of wavelengths is limited and (2) minimizing the number of wavelengths needed to set up the connections. Under dynamic traffic [33], a connection request arrives based on a random process with a random holding time, and therefore a light-path is established and then terminated after elapsing its connection holding time. Therefore, RWA decisions should be made rapidly when a connection request arrives at the network. Due to insufficient resources or unavailable wavelengths, the network may not be able to find a light-path for a given connection request, thus resulting in connection blocking. Therefore, the main objective of dynamic RWA is to find a route and choose a wavelength that maximizes the probability of establishing each connection request and at the same time attempt to minimize the blocking probability for future connections. Obviously, it is impossible to keep the resource utilization optimal in dynamic traffic.

There are three types of routing approaches used in RWA: fixed routing, fixed alternate routing, and adaptive routing. In fixed routing, there is only one fixed route (e.g., the shortest path or the shortest hop) between a pair of ingress and egress switches. The shortest path/hop for each source–destination pair is calculated in an offline manner using the Dijkstra or Belman–Ford algorithm. This is the simplest form of routing, but it may block a connection request if there is no feasible wavelength along the shortest route found for the request. In fixed-alternate routing, each ingress switch maintains a routing table that includes an ordered list of fixed routes to each egress switch (say the first-shortest-path route, the second-shortest-path route, the third-shortest-path route, and so on). The actual route for a connection can only be chosen from these fixed routes. In adaptive routing, any route between a source–destination pair can be used for a connection request provided that it has available wavelength resources. Adaptive routing techniques increase the probability of establishing a connection by using networks state information. Under adaptive routing, a routing decision is made dynamically based on the current wavelength usage information on each link (such as the shortest-cost path first or the least-congested path first). For example, under the least-congested path first adaptive routing, a path is selected that has the highest number of available wavelengths among other paths because the links that have a small number of available wavelengths are considered more congested [34].

After finding a route between a source–destination pair, a wavelength assignment algorithm should select an available wavelength that maximizes wavelength utilization. There are a number of heuristic wavelength assignment approaches used in RWA such as (1) the first-fit method that attempts to select the first available wavelength in numerical order, (2) the most used method that attempts to allocate the most utilized wavelength first, (3) the least used method that attempts to allocate the least utilized wavelength first, (4) the best-fit method that picks a wavelength on which the light-path fits best (i.e., among the wavelengths available for a connection request, a wavelength is chosen that has the least free capacity remaining after accommodating the connection request), and (5) the random method that attempts to allocate a wavelength randomly. Under static traffic, the objective of wavelength assignment is to minimize the number of wavelengths, whereas in dynamic traffic it is assumed that the number of wavelengths is fixed and the network tries to minimize connection blocking [25, 35].

Wavelength continuity is a problem in WRN networks, where a connection request may have to be rejected (i.e., blocked) even though a route is available for the request because of non-availability of the same wavelength on all the links along the route. Connection blocking is increased when the traffic load goes up, thus reducing the performance of RWA. In summary, connection blocking is more likely to occur under the following cases:

- under dynamic traffic rather than static because there is no information about future connections.

- in the networks without Wavelength Converters (WCs) because of the wavelength continuity constraint. A wavelength converter is a device that can convert the wavelength of an incoming optical signal to another wavelength.

- lack of available wavelengths in the network.

- when the number of hops (path length) is high. In a large-diameter network, it will be harder to find the same wavelength free on long paths.

- when the network global information is not used in routing.

- when optical switches are not strictly non-blocking. Different architectures of strictly non-blocking optical switches have been studied in [36] such as Clos architecture, Cantor architecture, and so on. Note that a strictly non-blocking switch is a sort of switching module in which an unused input port can always be connected to an unused output port, without having to rearrange existing connections inside the switch. It should be noted that a Clos switch becomes strictly non-blocking under a specific condition. There are three switching stages in an $N*N$ Clos switch. The first stage uses $\frac{N}{m}$ switches, each with size $m*k$. There are k switches in the intermediate stage, each with size $\frac{N}{m}*\frac{N}{m}$. Each of these k switches is connected to any switch in the first and the third stages. The third stage uses $\frac{N}{m}$ switches, each with size $k*m$. A Clos switch becomes strictly non-blocking if $k \geq 2 \times m - 1$ [37].

The wavelength continuity problem can be resolved using wavelength converters, strictly non-blocking switches, and light-path rerouting. Using wavelength conversion, a light-path may use several wavelengths along its path when traversing through different fiber links. Light-path rerouting means the action of changing the physical path and/or the wavelength(s) of an established light-path. Rerouting is a viable and cost-effective mechanism to improve the connection blocking performance [38, 39]. Bandwidth wastage and scalability issues are two drawbacks for OCS [20]. OCS suffers from low channel utilization since a single connection may not employ the whole bandwidth of a light-path. To resolve this drawback, grooming techniques are used to highly utilize the bandwidth of a light-path [40]. In grooming, a number of low bit rate connections are mixed and sent on a setup light-path. Traffic grooming performs a two-layer routing to effectively pack low-rate connections onto high-rate light-paths.

OCS networks perform two-way reservation for establishing a connection request before data transmission. The network can be thought of as a two-plane architecture consisting of a control plane and a data plane. One wavelength channel on each link is used for exchanging control information. The actual data are transmitted in the data plane.

There are two main signaling methods for light-path setup in WRN as centralized RWA and distributed RWA. In distributed RWA signaling [41, 42], each network element is in charge of light-path setup. Under distributed RWA, all connection requests are processed at different network core switches and each core switch makes its decision based on its available resources. In centralized RWA signaling [43], a central network element reachable by all network core switches is in charge of establishing all connection requests. The central element is aware of the complete network topology, physical parameters, and resource availability, thus leading to better performance results than the distributed RWA. However, the central element failure is a big concern. The centralized RWA could be suitable for static traffic and small networks, not for dynamic traffic because of much information that must be managed by the central element. Centralized RWA results in scalability and complexity problems for dynamic traffic. This is why distributed RWA is much more popular than centralized RWA. Distributed control can improve the scalability and reliability performances. A distributed RWA uses two strategies to find a suitable wavelength on a given route R between source node S and destination node D as follows [28, 44], where the purpose of both strategies is to find a continuous wavelength from source node S toward destination node D over a given physical path (i.e., no wavelength conversion is used in these algorithms):

- **Forward Reservation Method (FRM)**: In FRM, node S reserves one of the unused wavelengths over its ongoing link toward the next node. For wavelength reservation, node S sends a message toward the next node over route R. The node that receives this message repeats the process of reserving the selected wavelength in other nodes until the message arrives at node D. If this reservation is blocked due to the unavailability of the selected wavelength along route R, node S will select another free wavelength and repeat the process.

- **Backward Reservation Method (BRM)**: For a given connection request, node S initiates a wavelength set that contains all free wavelengths of its ongoing link towards the next node over route R. Every intermediate node modifies this set based on the intersection of the incoming wavelength set and the set of free wavelengths of its ongoing link toward node D. The final set that reaches node D is the set that contains unused wavelengths over the whole path from S to D. Node D then processes this incoming set for finding an appropriate free wavelength over route R. If there is no wavelength in the final set, the call will be blocked. After these processes, node D initiates the reservation process in a backward manner towards node S.

Many studies on RWA have concentrated on establishing light-paths under the assumption of an ideal physical layer. However, this assumption could only be suitable for opaque networks, where a signal is regenerated at each intermediate optical switch. As an optical signal propagates along a light-path toward its destination in a transparent wavelength-routed optical network, the signal quality degrades since there is no conversion to the electronic domain and therefore no signal regeneration. This in turn increases Bit Error Rate (BER) of the signal. However, high BER is not acceptable by users. In addition, it is not acceptable if the establishment of a light-path results in increasing the BER of other existing light-paths. Establishing light-paths with low BER can reduce the number of retransmissions by high layers, thus increasing network throughput. Therefore, the RWA techniques that consider physical layer impairments for the establishment of a light-path, called Quality of Transmission-Aware (QoT-aware) RWA, could be much more practical.

The following RWA surveys can be found in literature. A survey of pure RWA techniques (without considering transmission impairments) can be found in [25]. The work in [45] has reviewed the techniques (including pre-emption, wavelength management, and routing) that can be utilized for the differentiation of light-paths in the RWA process. The RWA techniques suitable for translucent networks have been reviewed in [46], where a translucent network employs electrical regenerators at some intermediate nodes only when it needs to improve the signal quality. A review on the management and control planes required for QoT-aware RWA techniques has been provided in [32, 47]. The surveys in [32, 48, 49] have provided a review on static (offline) physical layer impairment-aware RWA algorithms in all-optical networks. The surveys presented in [27, 32, 48] have studied in general different topics related to optical networking and physical layer impairments, including DWDM technology, physical layer impairments, optical components, service level agreements, failure recovery, static impairment-aware RWA techniques, impairment-aware control plane techniques, and some dynamic impairment-aware RWA algorithms in both translucent and all-optical networks.

1.3.1.2 Optical Packet/Burst Switching   Due to its finer granularity, a packet switching technique (as a contention-based mechanism) can obtain a higher channel utilization and can yield better bandwidth efficiency than OCS by two approaches as discussed below [21, 50]. Comparison of OPS and OBS switching techniques can be easily found in [23, 51, 52].

- **Asynchronous Optical Packet Switching (OPS)**: An OPS network could be the promising network to employ the huge bandwidth provided by optical fiber

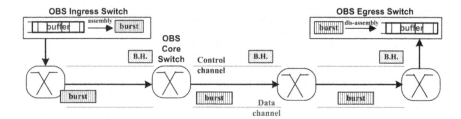

Figure 1.7: An example for OBS operation

technology. In asynchronous OPS, client packets are converted from the electronic domain to the optical domain in ingress switches (called optical packets), and then each optical packet is sent to the OPS network. An optical packet may include a single client packet or a number of client packets. In this switching, each optical packet is individually switched in a core network like its electronic counterpart while keeping payload in the optical domain and processing the optical packet header electronically [1, 2, 53]. Optical packets are switched to their desired ports as they arrive at an OPS switching node and there is no requirement for synchronization stages. In asynchronous OPS, client packets can have any size and thus are suitable for IP packets. Egress switches are the places for converting optical packets to electronic domain and sending them toward their final destination networks. The OPS network is necessary for both metro and backbone networks [21]. The OPS is also compatible with bursty nature of IP traffic and can increase throughput and efficiency of network bandwidth so that OPS needs less wavelengths and resources for handling the same traffic compared with an OCS network.

- **Asynchronous Optical Burst Switching (OBS)**: This is a switching protocol [54, 55] with a finer granularity than OCS and a coarser granularity than OPS. An OBS network aggregates a number of client packets destined to a given egress switch in a burst, sends a Burst Header (BH) on a control channel for necessary resource reservation in the intermediate core switches along the egress switch, waits for an offset time, and then transmits the burst over an existing data wavelength without waiting for an acknowledgment from the egress switch. Each OBS core switch processes the BH and provides an output channel to forward the arriving burst while resolving the possible contention of the burst. A data burst in OBS can involve client packets with various data rates. A BH includes different information for the burst including its ingress switch address, its egress switch address, the number of client packets aggregated in the burst, the class of service for the client packets, the burst duration, the data wavelength on which the burst will be transmitted, and so on. Figure 1.7 depicts an example for OBS operation in which client packets are assembled in the ingress switch and dis-assembled in the egress switch. Since processing of the BH takes some time in each intermediate OBS core switch, the header and BH are getting closer to each other while they are at the egress switch. Therefore, the offset time must take into account the processing time of the BH in all intermediate OBS core switches along the path to the egress switch.

Resource reservation in OBS can be implemented by two approaches. (1) In one-way reservation (called Tell and Go (TAG) signaling protocol), a control packet is transmitted by the ingress switch to allocate the necessary resources for a data burst. Then, the data burst follows its control packet without waiting for the reservation acknowledgment from the egress switch. This results in a high burst blocking rate, especially at high traffic load. However, the end-to-end delay is minimized. This type of reservation is ideal for delay-sensitive applications; and (2) In two-way reservation (called Tell and Wait (TAW) signaling protocol), a control packet is sent over the path of the burst in order to collect information on the availability of resources. At the egress switch, a resource assignment algorithm is executed. An acknowledgment packet is sent back to the intermediate OBS switches to reserve the necessary resources. If the reservation fails at an intermediate switch, a failure packet is sent to the egress switch. If the reply packet reaches the ingress switch, the burst data are transmitted. This eliminates the loss of data bursts, but it can also lead to high end-to-end delay.

In an ingress switch, an optical burst is created by the packet aggregation technique. The ingress switch is responsible for aggregating the client packets coming from legacy networks destined to a given egress switch into a burst. By receiving a burst, the egress switch unpacks all the client packets in the burst and then routes each packet individually toward its relevant destination. In each ingress switch, there is a dedicated queue to each egress switch. Packet aggregation can be either timer-based or threshold-based, or a combination of both timer-based and threshold-based. In the timer-based aggregation technique, a timer is started whenever a client packet arrives at the queue relevant to egress switch i and this queue is empty. The aggregation algorithm waits for an aggregation period and during this period, it collects all the client packets destined to egress switch i. A burst is created when timeout happens at the end of the aggregation period. On the other hand, in the threshold-based technique, when waiting traffic in a queue reaches a given threshold size, a burst can be formed. The aggregated traffic size in the burst must be equal to or less than the threshold size. Finally, in combined timer-based and threshold-based, a burst is created whenever there is enough traffic to fill a burst or whenever timeout occurs. Note that the timeout mechanism can limit the waiting time of packets in queues, but at the expense of creating small-size bursts [56–58].

There are two mechanisms for aggregation of class-based traffic. First, client packets of the same class of service can only be aggregated in a burst [58, 59]. Second, client packets from different classes of service can be aggregated in a burst (i.e., composite burst aggregation) [60]. Since the latter mechanism can aggregate bursts sooner than the former, it can lead to smaller end-to-end packet delay and burst loss rate than the former mechanism.

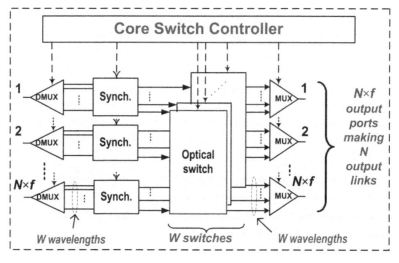

Figure 1.8: Synchronized multi-fiber OPS switch architecture

1.3.2 Synchronous (Slotted) Optical Switching

To provide the finest granularity and improve bandwidth usage, the Optical Time Division Multiplexing (OTDM) concept can be deployed in optical networks. Under OTDM, many source–destination pairs can share the network bandwidth. However, synchronization of traffic arrival at core switches is the limitation of this switching. In synchronized networks, fiber delay lines are required for synchronization issue at the input ports of core switches [53, 61].

1.3.2.1 Optical Circuit Switching In slotted OCS (as a reservation-based mechanism) [62–64], the bandwidth of a wavelength is divided into frames of fixed time slots and traffic for a given source–destination pair is periodically transmitted in pre-allocated time slot(s) in each frame. The Routing and Wavelength Assignment problem in the wavelength-routed networks is changed to the Routing, Wavelength, and Time-Slot Assignment problem in slotted OCS. Bandwidth is reserved for ingress switches by intermediate core switches in the optical network.

1.3.2.2 Optical Packet/Burst Switching There are different approaches to deploy the packet switching concept in slotted networks (as contention-based mechanisms):

- In conventional slotted OPS [53, 65], a fixed-length client packet together with a header makes an optical packet. Then, the optical packet is transmitted to the OPS network at a time slot boundary. A large-size client packet must be fragmented and transmitted in a number of time slots [2]. Optical packets are synchronized before entering an OPS switching module using switching delay lines. A slotted OPS network has a lower complexity in terms of switch control than an asynchronous OPS network [21]. In addition, it has a higher throughput than asynchronous OPS due to a smaller contention rate in slotted OPS than in asynchronous OPS [21]. This is similar to slotted ALOHA in which the vulnerable period of packets is reduced, thus leading to a lower packet contention than with pure ALOHA. The simplest technique to manage wavelengths in this technique could be based on round robin. Under this scheme, a pointer in an egress switch points to the transmission wavelength to be used by the next incoming client packet. After sending the client packet in a time slot, the pointer is incremented by 1 modulo W, where W is the total number of wavelengths available in an ingress switch. Slotted OPS has lower complexity of switch control and lower optical Packet Loss Rate (PLR) than asynchronous OPS.

In slotted OPS, each time slot includes two time intervals. The optical packet interval for transmitting optical packet payloads is denoted by S_T; and the time-gap interval plus the header interval for transmitting header information is denoted by S_O. Hence, time-slot size is equal to $S_{ts} = S_T + S_O$.

Variable length client packets cannot be efficiently utilized in slotted OPS networks because they may cause some bandwidth degradations in slotted OPS. For example, of the variable length IP packets, almost half of the packets are 40–44 bytes, almost 18% have the length of 1500 bytes, another 18% of packets are either 552 or 576 bytes, and a very small number of packets have a size larger than 1500 bytes [66]. Hence, the variance of the IP packet sizes is very high and choosing a time slot to carry 1500 bytes will result in a high bandwidth wastage. On the other hand, using a time-slot size to carry say 40 bytes requires not only a much faster optical switching speed, but also a faster processing at core switches to process a large number of packet headers in a very short time [21]. Clearly, this requirement will even be high when 40-Gbits/s or higher transmission rates are used in OPS networks. This issue also leads to the packet fragmentation problem that increases header complexity and induces cost due to the reconstruction of received packets at egress switches. The other problem with fragmentation is that when a part of an optical packet is dropped, the whole optical packet will be useless. To resolve the problems of the conventional slotted OPS network, one can use a relatively larger time slot in which a number of client packets of any type and class can be aggregated within a time slot before transmission to network.

In slotted OPS, a switch fabric can only be reconfigured at the start of each time slot for all input links. This needs alignment and synchronization of all optical packets coming from input ports according to a local clock. The reason for synchronization is due to the fact that different optical packets could arrive at an OPS switch at different point of times because of different distance of physical links, temperature fluctuations, and chromatic dispersion. Optical packets synchronization is the major disadvantage of slotted OPS.

- In slotted OBS [67], each burst is divided into multiple time slots and transmitted at fixed positions in a periodic frame structure. The control wavelengths carry burst header cells so that core switches along a path can set up a connection. This architecture may need optical buffers in core switches in order to interchange time slots. The problem with the burst division is that the whole burst can be blocked due to the blocking of a small part of the burst, thus leading to an inefficient resource utilization.

- In Photonic Slot Routing (PSR) [68, 69], simultaneous slots (each slot may carry a number of client packets) transmitted on distinct wavelengths are aggregated in one photonic slot and routed through core switches. Since this technique does not require wavelength-selective optical switches, it is cost-effective. However, this network is not too flexible and the network bandwidth may not be fully utilized because the whole traffic of a fiber can only be switched to a specific output port of a core switch. Consequently, this switching usually finds applications in ring networks.

There is also a combined approach of asynchronous OPS and slotted OPS in which the network operation is asynchronous while switches operate synchronously. In this technique, the optical packets sent to the OPS network can have variable sizes and can arrive at a core switch at any time. However, the switching module inside an OPS switch begins to operate only in the start of time slots, where a time slot is the duration of transmitting the smallest packets size (say 40 bytes in IP networks). Here, large-size optical packets are transmitted in several consequent time slots. This approach provides the flexibility of having any packet size and makes easy contention resolution in OPS switches [70].

Note that OBS and OPS could be two dominant techniques for future optical core networks. Among the aforementioned switching schemes, OPS is not only scalable but also flexible and can dynamically allocate network resources with fine granularity [21]. OPS can efficiently use the network bandwidth that enables it to support different applications and services [71]. However, there are three implementation limitations for OPS [21]: (1) its inability to save optical data in memory with random access capability; (2) the lack of sophisticated processing in the optical domain, while high processing power is a requirement in the optical switches due to the larger amount of overhead; and (3) its requirement for fast switching speed (e.g., nanoseconds). For instance, in a slotted OPS with $B_c = 10$ Gbits/s, consider $S_T = 450$ ns and time gap $S_O = 50$ ns. In this example, one user packet of size 560 bytes can be carried within a time slot as an optical packet. Therefore, an optical switch must be able to perform header processing, contention resolution, and switching within 50 ns of the time gap.

To resolve the second and the third problems, a number of client packets should first be aggregated in a large-size optical packet and then transmitted to the network. This approach allows one to use relatively larger optical packet sizes and larger time gaps between optical packets (in slotted OPS), thus alleviating the need to use a very fast optical switch. This clearly reduces the complexity of OPS switches due to the lower number of entities per unit time to be processed [3]. For example, consider $S_T = 45,000$ ns and $S_O = 5000$ ns in slotted OPS. A time slot in this case can carry up to almost 56,000 bytes (usually more than one client packet). In addition, the time for header processing, contention resolution, and switching increases to 5000 ns compared with 50 ns in the previous example.

1.4   Overview of OPS Networking

In the following, different aspects of OPS networks are detailed. Enabling technologies for OPS have been discussed in [65]. Generally speaking, an OPS core network is responsible for transferring clients' traffic inside optical packets between two edge switches. Each edge switch is in charge of collecting clients' traffic from legacy networks and delivering clients' traffic to the legacy networks.

OPS can be used in both connection-oriented and connectionless networks. In a connectionless OPS network, an optical packet can be routed toward its destination without path limitation through any path that an OPS switch decides. In connection-oriented OPS, each optical packet should follow the path determined in connection setup phase (i.e., virtual circuit). Each connection has a unique ID, and optical packets of that connection are switched in OPS switches based on this ID.

1.4.1   Network Topologies for OPS

An OPS networking was originally designed for regional/long-haul networks. However, nowadays, OPS is even desired for metro networks in order to efficiently use network resources for the following reasons [21]:

- A metro network should be able to provide more capacity to cope not only with the ever-increasing bandwidth demands of novel applications but also with the unexpected future demand growth [72, 73]. The simple reason for the huge amount of traffic in metro networks is the mass deployment of Fiber-To-The Home (FTTH) technologies in access networks [73].

- A metro network must provide different Quality of Service (QoS) levels for new applications, mostly based on the Internet.

- A metro network must provide agility in order to deploy bandwidth for traffic demands at a finer granularity since most of the traffic nowadays is Internet traffic.

- A metro network experiences high traffic dynamics due to Internet traffic.

Thus, all-optical packet-switched networks appear to be the sole approach to provide such capacity and agility. The general OPS network model has mesh topology similar to what is displayed in Fig. 1.3, suitable for regional/long-haul networks. However, for metro networks, star and ring topologies can also be used. The benefits and disadvantages of star topology over ring topology in metro networks have been stated in [74].

The metro architectures based on star topology could be based on three architectures:

- **Passive Broadcast-and-Select Star Couplers**: Although networks using passive star couplers have zero power consumption [26], they have three problems: (1) wavelength reuse problem [26], in which a wavelength cannot be used by several connections like in WRNs; (2) need for a large number of wavelengths to support even simple traffic patterns [75]; and (3) suffer from power loss problem due to splitting loss [26, 76, 77].

- **Central Passive Arrayed Waveguide Grating (AWG)**: AWG can resolve the wavelength reuse and power loss problems. Thus, a highly efficient network architecture can be realized by using all wavelengths at all ports of an AWG simultaneously [72]. However, AWG acts like a static wavelength router. In AWG, an incoming to outgoing route is determined statically, where the routing pattern is a function of the incoming wavelength and input port [72]. In addition, the maximum number of wavelengths that can be used in the network is a function of the number of input ports of an AWG [74, 78].

- **Wavelength-Selective Cross-Connect Optical Switches (Known as Active Switches)**: An active switch with the wavelength switching capability can dynamically switch each incoming wavelength from any input port to any output port. Obviously, an active switch can provide additional degree of freedom to route traffic by changing the setting of the switch module. Opposed to an AWG, the total number of wavelength channels to be used in the network based on an active switch is independent of the number of input ports of the switch and is easily scalable. This gives another freedom in designing a metro network based on an active switch. Therefore, by using an active switch, a more efficient and flexible star metro network can be realized. Using active switches, an OPS network can operate much more efficiently than an OPS network using passive switches due to the spatial wavelength reuse and splitting loss problems in passive components. Note that a power source must be present for controlling an active switch [74, 79–87].

The advantages of star topology in a metro network are: (1) simplicity in routing and configuration, (2) reduced complexity of control plane, (3) design simplicity, (4) good scalability as the network size grows physically since it is easy to add an edge switch to a star, (5) easy synchronization required in slotted OPS networks, and (6) reduced switch crosstalk because of having only one core switch along a transmission path compared to ring and mesh topologies [79, 88]. The main disadvantage of the single-hop network is its core switch failure. To provide robustness in a star network, the overlaid star topology (see Section 6.1.1) with a number of overlaid core switches can be used, where every edge switch is connected to each one of the core switches. Using this architecture, an edge switch can easily reroute its traffic to another star when a core switch fails. The bandwidth sharing mechanisms proposed for single-hop overlaid OPS networks will be detailed in Chapter 6 (see Section 6.1.3).

Different architectures have been proposed for metro networks based on the ring topology, e.g., [74, 89, 90], where optical packets are routed over a WDM ring network. Here, Media Access Control (MAC) protocols are proposed to enable the OPS switches on the WDM ring to share the bandwidth of the network while preventing optical packet collisions. These protocols will be detailed in Chapter 6.

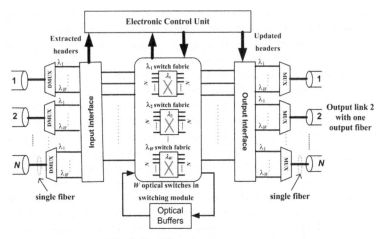

(a) Single-fiber OPS switch architecture

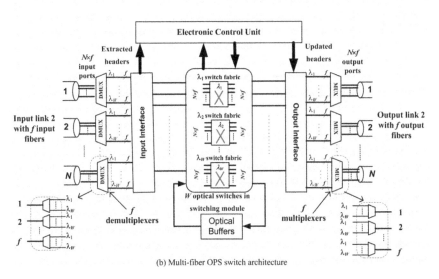

(b) Multi-fiber OPS switch architecture

Figure 1.9: General OPS switch architecture

1.4.2   Core Switch Architecture in OPS Networks

An overview on the operation of OPS networks is discussed in this section. Figure 1.9a shows a generic single-fiber OPS switch architecture with N input ports and N output ports (i.e., an $N*N$ OPS switch) and also shows W wavelength channels available per fiber [65, 70]. It includes different components as N demultiplexers,

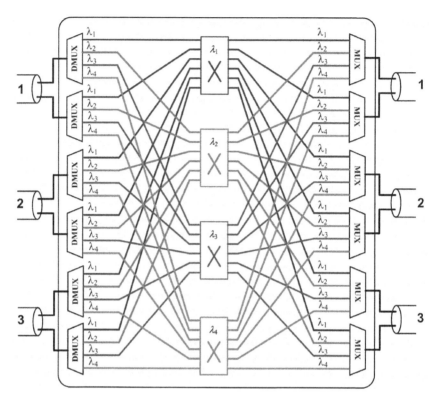

Figure 1.10: An example for switching module in an OPS core switch at $N = 3, f = 2$, and $W = 4$

an input interface, a switching module, an output interface, N multiplexers, optical buffers for delaying optical packets, and an electronic control unit. It should be mentioned that some efforts have been made to design the control unit in the optical domain [91]. The control unit keeps a routing table based on the shortest path mechanism in order to route optical packets destined to a given egress switch through the output links. In an OPS switch, the switching module includes a number of non-blocking wavelength-selective cross-connect optical switches. Here, it is considered that the control unit processes headers in the electronic domain. On the other hand, some efforts have been made to process an optical packet header in the optical domain [91–94]. An OPS switch will be referred to as a core switch from now on. On the other hand, Fig. 1.9b illustrates a generic multi-fiber OPS core switch architecture with $N \times f$ input ports and $N \times f$ output ports (i.e., an $(N \times f)*(N \times f)$ OPS core switch), along with W wavelength channels available per fiber. It includes different components such as $N \times f$ demultiplexers, an input interface, a switching module, an output interface, $N \times f$ multiplexers, optical buffers for delaying optical packets, and an electronic control unit.

Each demultiplexer separates W wavelengths of each input fiber. The input interface detects the start and end of an optical packet header and payload (of at most

$N \times W$ optical packets in the single fiber architecture and of at most $N \times f \times W$ optical packets in the multi-fiber architecture), converts the header to the electronic domain, and sends the extracted header for further processing to the electronic control unit. Synchronization, re-amplification, and wavelength conversion of input wavelengths could also be carried out in the input interface, if necessary. For example, in slotted OPS, synchronization of optical packets to the beginning of time slots should be performed in the input interface. Even re-amplification of optical packets can be carried out in the input interface. For example, EDFA amplifiers can be used inside the core switch in order to compensate the internal loss of the core switch and the loss occurred on the fiber link from the last amplifier on the link and the core switch [7, 95].

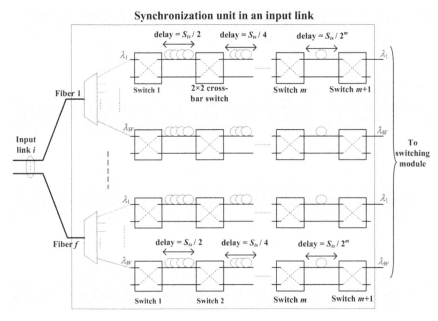

Figure 1.11: Synchronization unit for an input link of a multi-fiber OPS network

For proper operation of transmission and reception of optical packets in slotted OPS, we need to make an assumption where the fiber length between any pair of switches (edge-to-core or core-to-core) must be an integer number of time slot duration (i.e., $S_{ts} = S_T + S_O$). However, all fiber spans cannot be exactly designed as an integer number of time slot duration. In addition, the arrival of optical packets or control information at the core switches may be misaligned with each other due to the chromatic dispersion of different wavelengths, accumulative jitter of different paths inside optical switch fabrics, temperature variation, and other fiber transmission non-linearities that result in varying length of fibers. To provide synchronous switching operation, they must first be realigned by using a synchronizing unit at each input port of any core switch in order to synchronize the incoming optical packets boundaries to the local timing reference. Figure 1.8 shows a slotted multi-fiber

OPS switch architecture with f fibers in each input/output link, where synchronization modules are used to align optical packets to the beginning of time slots. This can be implemented using a finely calibrated set of optical delay lines with feedback control of the delays being provided through the system controller. Note that each core switch is operating with a reference to its own internal clock that can be derived from a network synchronization signal distributed throughout the network. Considering time-slot length of S_{ts}, Fig. 1.11 illustrates the synchronization block diagram in an OPS switch in which there are $f \times W$ synchronization modules. Each synchronization module for an input link include $m + 1$ switching elements, where each 2×2 switching element can be set either in bar state or in cross state. By controlling the switching elements, different fibers delay lines can provide different delays starting from at least 0 to at most $\sum_{j=1}^{m} \frac{S_{ts}}{2^j}$ for each wavelength channel [53, 65, 96–98].

The control unit processes the header of an optical packet to obtain the information about source and destination edge switches, and then it searches its switching table to find the suitable output port for forwarding the optical packet. Then, it sends control signals to the switching module for appropriately configuring it. In short, the control unit decides which optical packet should be switched to which output fiber and on which wavelength channel. Contention resolution is one of the most important responsibilities of the core switch. Contention happens when several optical packets must be switched to the same output fiber on the same wavelength channel at the same time.

The switching module is a wavelength-selective optical switch. This module in the single-fiber architecture includes W optical switching fabrics, where each switching fabric is used for switching a wavelength channel. Note that each switching fabric is a strictly non-blocking $N*N$ optical switch. In the multi-fiber architecture, each switching fabric is a strictly non-blocking $(N \times f)*(N \times f)$ optical switch. The switching module may use different contention resolution schemes (like wavelength conversion and optical buffering) for resolving the contention of optical packets that are going to be sent over the same fiber and wavelength channel at the same time. The optical buffering displayed in the figure is actually based on Fiber Delay Lines (FDLs) that are used for two purposes: delaying optical packets headers for contention resolution purposes and waiting for packet header processing. This switch module may also use Tunable Wavelength Converters (TWC) and other techniques for contention resolution. Figure 1.10 displays an example for switching module in a multi-fiber OPS switch with $N = 3$ input links and $N = 3$ output links, $f = 2$ fibers per link, and $W = 4$ wavelengths per fiber shown with different colors. As can be observed, four 6*6 optical switch fabrics are used in the switching module, one optical switch fabric for each wavelength channel. Each optical switch fabric is a strictly non-blocking switch. A strictly non-blocking switch can be built with different architectures such as Clos architecture, cross-bar architecture, Cantor architecture, and so on [36]. The important factor for an optical switching fabric is its reconfiguration speed which must be on the order of nanoseconds for OPS. The speed of switching fabric based on the Micro-Electro-Mechanical Systems (MEMS) technology is on the order of milliseconds, which is not appropriate for OPS. The Semiconductor Optical Amplifier (SOA) and electro-optic Lithium Niobate (LiNbO3) technologies are promising for OPS switching fabrics. SOAs have a switching speed on the order of a few nanoseconds and can be integrated to large scales, but they add some noises to optical signals. The LiNbO3 switching fabrics have sub-nanosecond switching times, but with high insertion loss, thus limiting their integration scalability for only medium-scale optical switching fabrics. Other important factors for a switching fabric could be: scalability to large port numbers, ease of manufacturing, low cost, temperature independence, and being strictly non-blocking [21, 65].

After switching of an optical packet, the output interface attaches the updated header to the packet payload. The output interface can also perform signal amplification to overcome noise and degradation of optical signals and resynchronize the optical packet to the time slots in slotted OPS. Wavelength conversion is another responsibility of the output interface. Finally, multiplexers at the output side combine wavelengths on the output fibers.

It should be mentioned that different architectures have been proposed for OPS switches. The detailed architectures will be discussed in Chapters 3 and 4.

1.4.3   Edge Switch Architecture in OPS

An important issue in an OPS network is to provide network access for traffic coming from/delivering to legacy networks. As Fig. 1.12 depicts, an edge switch is used to isolate the optical domain of the OPS switch from the electronic domain of G legacy networks. An edge switch has two main functionalities. As an ingress switch, it collects clients' traffic coming from G legacy networks, redirects traffic destined to the same egress switch in the same buffer, schedules them, converts them to optical packets, and then transmits the optical packets to the OPS core network. As an egress switch, it receives optical packets from the OPS core network, converts them to the electronic packets, saves the traffic destined to the same legacy network in the same buffer, and then delivers the electronic packets to the G legacy networks. Borrowing from [99], a "**torrent**" is defined as the whole (class-based) traffic going to the same egress switch in an ingress switch. Therefore, Torrent-i traffic goes to egress switch i.

Figure 1.12 illustrates a multi-fiber OPS network in which there are f fibers on each connection link. The OPS switch has $N \times f$ input ports and $N \times f$ output ports, where each set of f input/output ports is called a link. Among these links, there are $N - 1$ transit links (i.e., with $(N - 1) \times f$ fibers) for switching transit optical packets and one link (i.e., with f fibers) for traffic add/drop from/to the edge switch. In this network, the edge switch is connected to the OPS switch with f fibers as add ports and with f fibers as drop ports. The edge switch can use all W wavelengths provided on each fiber for traffic transmission to the OPS switch and traffic receiving from the OPS switch. In this case, the edge switch should use $f \times W$ fixed optical transmitters and $f \times W$ fixed optical receivers. The bandwidth of each wavelength channel is B_C in bps.

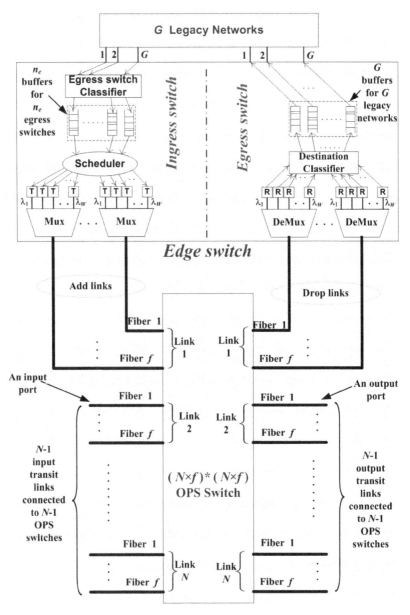

Figure 1.12: Edge switch architecture in a multi-fiber OPS network

In its simplest form, as Fig. 1.12 shows, an ingress switch contains one electronic buffer dedicated to each egress switch; $n - 1$ buffers for the OPS network with n OPS switches. When a client packet destined to a given egress switch arrives at an ingress switch, it is saved in the relevant egress switch FCFS (First-Come-First-

Served) buffer. There is a packet scheduler unit that schedules client packets from the buffers, converts each client packet to an optical packet, adds a header to the optical packet, and then sends the optical packet to the OPS network on an available wavelength channel immediately (in asynchronous OPS) or at time slot boundary (in slotted OPS).

For class-based traffic, packet differentiation can be implemented using differentiation buffers in an ingress switch. The network consists of M classes of traffic, where a class is defined according to the QoS requirements (e.g., $M = 3$ for EF, AF, and BE according to the Differentiated Services (DiffServ) model). Note that there are three general classes defined in DiffServ: Expedited Forwarding (EF) [100], Assured Forwarding (AF) [101], and Best Effort (BE). In this case, for a network supporting M classes of service, M buffers are used for each egress switch; $M \times (n - 1)$ buffers for the OPS network with n nodes. There is a QoS-based packet scheduler unit that schedules client packets from the class-based buffers, creates optical packets, and then transmits them to the network on available wavelengths.

As stated, an ingress switch can immediately send an arriving packet to the OPS network. However, this may increase traffic loss at the network due to the traffic burstiness, especially under IP traffic. There are two common mechanisms proposed for traffic shaping in ingress switches to cope with the bursty traffic and avoid contention in an OPS network:

- The first mechanism uses traffic shaping similar to electronic networks [102, 103].

- The second mechanism aggregates a number of client packets in an optical packet to reduce traffic burstiness in the OPS network [83, 104, 105], thus improving the network performance. This approach can be used in both single-class and multi-class traffic networks.

1.4.4 Signaling in OPS

In OPS, electronic control logic units are distributed in control units of OPS switches in order to process the header of each optical packet and route the optical packet toward its egress switch. Payload of an optical packet can contain one or more upper layer client packets, e.g., IP packets. It should be mentioned that the header of an optical packet may include some fields that must be updated at intermediate OPS core switches after switching the optical packet. A header of an optical packet may include different fields such as [65]

- Synchronization bits

- Ingress switch address generating the optical packet

- Egress switch address used for routing the optical packet in an OPS core network

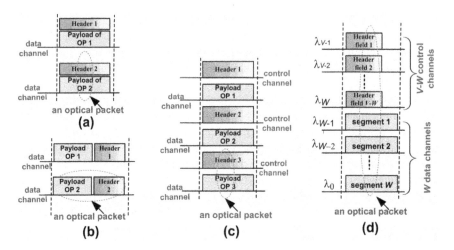

Figure 1.13: OPS signaling: (a) subcarrier multiplexing, (b) serial transmission, (c) separate wavelength, (d) header stripping

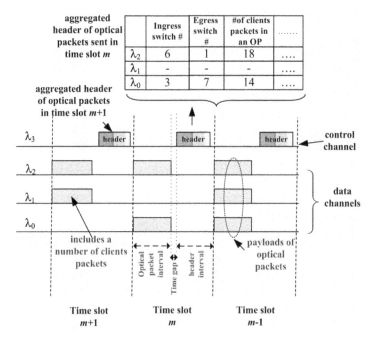

Figure 1.14: Aggregated header for three data wavelength channels

- Packet type that identifies the optical packet type and its priority for implementing QoS

- Packet sequence number to identify out-of-order or duplicated optical packets

- Operation, administration, and maintenance fields

- Header error correction field

- Number of client packets aggregated in the payload of the optical packet

- Fragmentation information for managing fragmented client packets in a number of optical packets

- Source routing information when the optical packet carries its route map

There are some time gaps (called guard bands) between the header and payload of an optical packet. Guard bands are used for waiting time of header processing in the electronic domain and for de-jittering of payloads. Different techniques have been proposed for transmitting data payload and header of an optical packet in an OPS network as follows:

- **Subcarrier Multiplexing**: In this approach (see Fig. 1.13a), an optical packet header (low-bandwidth) is placed on a subcarrier above the baseband frequencies occupied by the optical packet payload (high-bandwidth), and both are sent within the same time slot. In this technique, an optical packet includes only one client packet. Using this technique, header processing can take up to the entire data payload transmission time. However, by increasing the payload data rate, the baseband will expand and it might overlap with the header subcarrier frequency [21].

- **Serial Transmission**: In this simple technique (see Fig. 1.13b), before transmission of the data payload of an optical packet (that includes only one client packet), its header is sent serially on the same wavelength. There is a guard time between the header and the data payload to allow for the removal and the reinsertion of the new header at intermediate OPS switches [21].

- **Separate Wavelength**: Since in this approach (see Fig. 1.13c) the header and the payload of an optical packet (that carries only one client packet) are sent on two separate wavelength channels at the same time, extracting the header at the inputs of OPS switches is simpler than previous methods. There are two disadvantages for this method: crosstalk and dispersion. To compensate the dispersion effect on the delay, each intermediate OPS switch should realign the header and the payload [21].

- **Header Stripping**: In this approach (see Fig. 1.13d), optical packets are injected in an OPS network with a wavelength-striped packet format [106–108]. Different fields of an optical packet (such as packet length, quality of service, ingress switch address, and egress switch address) are separately encoded on dedicated control wavelengths, and the payload traffic is segmented and distributed over the rest of the available wavelengths. Each payload segment is modulated simultaneously at a high data rate (say at 40 Gbits/s per wavelength) to yield a high aggregate message bandwidth. The wavelength-striped optical packets are routed at each OPS switch altogether. Note that in this technique, an optical packet carries only one client packet. In this case, each OPS switch must be a simple optical cross-connect switch without wavelength selection capability that transparently switches the whole optical packet from an input port to an output port. In the wavelength-striped OPS, wavelength conversion is no longer an appropriate contention resolution scheme in OPS core switches.

- **Aggregated Header**: This approach is suitable for synchronized OPS networks (see Fig. 1.14) [104], where there is one dedicated control wavelength channel and W data wavelength channels on each fiber. A time slot includes two parts: (1) optical packet interval for transmitting optical packet payloads (with duration S_T, called slot traffic) and (2) a time-gap interval plus a header interval for transmitting header information (with duration S_O, called slot offset). Note that a time slot interval is $S_{ts} = S_T + S_O$.

For all W optical packets transmitted on W wavelength channels on a fiber within a time slot, an aggregated header is sent over the control channel during the header interval that includes traffic information (such as ingress/egress switch addresses) for each optical packet. It can be observed that there is a time gap between the transmission of a header and the transmission of relevant optical packets within a time slot. Note that in this technique, an optical packet may include an integer number of client packets. It should be mentioned that time slots are larger in this technique than the conventional synchronized OPS networks that use the serial transmission technique [104]. Figure 1.14 shows the information carried in the header for the optical packets sent within time slot m, where no optical packet is transmitted on wavelength λ_1.

Slot offset S_O should consist of a guard time to allow for timing uncertainties (T_{guard}), a processing time (T_{proc}), and a switching time (T_{sw}) at a core switch. The processing time during the S_O interval consists of the time required to evaluate potential contention, to resolve the contention, to make a new aggregated header, and finally to make the core switch ready to switch the arriving optical packets toward their desired egress switches. Therefore, we should have $S_O > T_{guard} + T_{proc} + T_{sw}$. For instance, if we need $T_{guard} = 100$ ns, $T_{proc} = 700$ ns, and $T_{sw} = 400$ ns, then we should have $S_O > 1200$ ns.

Note that S_T must be chosen in such a way that the bandwidth wastage is minimized. In general, two bandwidth overheads can be found in slotted OPS networks: (1) slot-offset overhead of $O_1 = \frac{S_O}{S_{ts}}$ and (2) client packet aggregation overhead of $O_2 = \frac{(B_C \times S_T) \bmod L_a}{B_C \times S_{ts}}$, where B_C is wavelength channel in bps, L_a is average length of client packets, and $x \bmod y$ is the remainder of x divided by y. The second overhead occurs when a number of client packets can be carried in an optical packet within a time slot. For example, consider a slotted OPS network with wavelength channel bandwidth $B_C = 40$ Gbits/sec and average client packet size $L_a = 5200$ bits. For example, if one requires $S_O = 1$ μs and wants at most 10% bandwidth overhead, then we should have $O_1 + O_2 \leq 0.1$. Hence, we obtain $36,000 + (40,000 \times S_T) \bmod 5200 - 4000 \times S_T \leq 0$. One can find a list of answers for this inequality such as $S_T = 9.11$ μs, $S_T = 9.24$ μs, and so on. However, choosing a large value for S_T would increase queuing delay in ingress switches because of high optical packet inter-departure times. Therefore, to have both a desirable bandwidth overhead and small queuing delay in ingress switches, one should select a smaller value for S_T from the list of answers.

1.4.5   Contention Problem in OPS

Packet loss in an electronic switch is mainly due to buffer overflow and bit error rate that makes the packet corrupted. Although OPS networks are promising networks for future optical core networks, they still encounter some problems. In OPS networks, an OPS switch works in cut-through mode, and there is mainly no RAM in the switch, and therefore optical packets cannot be saved in buffers like the electronic networks. Hence, the most important problem could be contention of optical packets.

In OPS networks, there is no collaboration among ingress switches for packet transmission, and traffic is sent to the optical network without any coordination among transmitters. Assume that a number of optical packets arrive at an OPS switch, where they all must be switched to the same desired output port. Contention of optical packets occurs in a single-fiber OPS switch when more than one optical packet should be switched to the same output fiber on the same wavelength at the same time, thus damaging contending all optical packets. On the other hand, in a multi-fiber OPS switch, contention occurs when more than f optical packets should be switched to the same output fiber on the same wavelength at the same time. In any architecture, some of the contending optical packets will go through, while others should be dropped.

The contention problem is different in slotted and asynchronous OPS networks. As stated, optical packets only arrive at an OPS switch at the start of time slots in slotted OPS; therefore, contention could only happen at the starts of time slots. On the other hand, in asynchronous OPS, optical packets can arrive at any time; therefore, contention occurs when the desired wavelength of an output fiber is already occupied. This is why the probability of contention in asynchronous OPS is higher than slotted OPS. Note that, compared with asynchronous OPS, slotted operation can reduce the vulnerable period of information contention and can reduce traffic loss [21]. Another sort of contention happens at an egress switch in which the number of optical receivers at the egress switch is less than the number of received optical packets [109].

When contention happens in an OPS switch and there is no contention resolution mechanism, only one of the contending optical packets in a single-fiber architecture and f contending optical packets in a multi-fiber architecture can only be switched successfully to the desired output link, and the remaining contending optical packets must be dropped. This clearly increases PLR in OPS and reduces network throughput. Therefore, contention resolution mechanisms should be implemented in each OPS switch in order to decrease PLR.

Note that the same contention problem happens on an output port of an electronic switch in an electronic network (or even an opaque switch in an opaque optical network in which optical packets are converted to electronic packets in switches). The contention in this case is simply resolved by buffering all contending electronic packets in electronic queues located at the output port and then scheduling them on the output port. However, there is no optical queue in the optical domain, and optical packets cannot be stored like electronic packets.

In general, contention is the major problem for an OPS network. To deal with the contention problem, resolution and avoidance are two common schemes. Contention resolution schemes (as reactive schemes) resolve arisen collisions in OPS switches. On the other hand, the contention avoidance schemes (as proactive schemes) send optical packets in ingress switches in such a way to reduce the number of collision events in OPS switches. Due to the bursty nature of the Internet traffic, the contention resolution schemes may not be effective to maintain a reasonable network performance under higher traffic loads. Therefore, contention avoidance schemes should also be used to regulate traffic going to an OPS network, thus reducing the traffic loss. Contention avoidance includes a different range of operations such as load balancing, using multi-fiber OPS, traffic shaping, traffic aggregation in ingress switches, even packet transmission, and so on, which will be detailed in Chapter 2.

On the other hand, contention resolution schemes (detailed in Chapter 3) try to resolve an occurred contention in an OPS switch. There are two main approaches to resolve contention of optical packets. In the first approach, received optical packets are first converted to electronic packets in an OPS switch (i.e., opaque networks), and then the OPS switch switches them as in electronic networks. However, this can be a bottleneck because of limited speed and high cost of O/E/O converters [110]. The second approach keeps optical packets in the optical domain (i.e., transparent network). When contention happens in a transparent network, an OPS switch uses some approaches to resolve the contention by changing space (by using deflection routing), switching time (using optical buffers), retransmission, converting wavelength (using wavelength converters), or a combination of them.

It seems that the contention avoidance schemes are mostly cheaper than the contention resolution schemes since most of the avoidance schemes are software level operations. On the other hand, a multi-fiber OPS architecture without using wavelength converters inside any OPS switch could also be cheaper than a single-fiber OPS when using wavelength converters in OPS switches. In addition, devices such as dispersion compensators used for a fiber with a high number of wavelengths is more expensive than the devices used for a fiber with a small number of wavelengths [13, 14]. In other words, a multi-fiber OPS with a small number of wavelength channels per fiber is cheaper than a single-fiber OPS with a large number of wavelengths per fiber.

Finally, even the size of optical packets may influence PLR in asynchronous OPS [111]. Transmission of fixed-size optical packets leads to the lowest PLR when using FDL buffering in OPS switches. On the other hand, variable-size optical packets yield the smallest PLR when the arrival process is bursty and there is no FDL buffer in OPS switches.

1.4.6   Quality of Service (QoS) in OPS

Quality of Service (QoS) is a broad term with almost 40 definitions [112]. Based on these definitions, degree of satisfaction of users could be the main objective of QoS provisioning in networks. QoS can be categorized from two points of views [113]:

- **QoS Experienced by End Users**: End user perception of QoS denotes how the quality of a particular service is received from the network. This necessitates bandwidth guarantee, minimum delay, small PLR, and controlled jitter.

- **QoS from the Network Point of View**: Network's perception of QoS denotes how the network resources and capabilities are fully and efficiently utilized by end users.

Since implementing QoS is complicated, some may believe that QoS is not needed because OPS bandwidth will be high enough to provide good QoS for all applications. However, this needs an infinite bandwidth in network components in order to support all network traffic, which is an impossible issue. Although an OPS network can provide high bandwidth, it must also improve the parameters that affect the network QoS in order to fully utilize the network bandwidth. OPS networks should be QoS-capable in order to support a variety of services and traffic types (such as data, voice, video, and multimedia streaming) generated by the applications with very different requirements of network performance or QoS such as in the Internet. This is because a variety of real-time applications (such as video on demand, voice over IP, Internet Protocol TV (IPTV), video conferencing) with different QoS requirements are executed in the Internet domain nowadays. This results in providing QoS support to (a) the large variety of service requirements and (b) the requirement to carry multiple services efficiently across both IP and optical networks [113–116].

1.4.6.1 QoS Metrics   Different QoS metrics can be stated for satisfaction of network users:

- **Network Reliability:** Reliability shows the probability that a network component will continue to perform its desired function under a given operating conditions over a predetermined period of time [117]. Different performance metrics include:

    - Network failure rate: This is the reverse of Mean Time To Failure (MTTF) [117].
    - Network repair rate: This is the reverse of Mean Time To Repair (MTTR) [117].

- **Network Security:** This metric is relevant to the strength of cryptographic algorithms for hiding information and integrity of information.

- **Network Performance:** Important performance metrics in an OPS network include:

– **Throughput**: The throughput metric is a measure of the number of packets that a network can successfully deliver. In practice, the throughput is less than the network bandwidth because of protocols overhead and network congestion. The throughput metric can be separately computed for client traffic and optical packet traffic because an optical packet may carry a number of client packets. The normalized throughput is the common performance metric in an OPS network which is computed in a network with n core switches (where one edge switch is connected to each core switch) by

$$Throughput = L \times \frac{\sum\limits_{i=1}^{n} N_{dlv,i}}{\sum\limits_{i=1}^{n} N_{sent,i}} \, , \qquad (1.1)$$

where $N_{dlv,i}$ is the number of packets successfully delivered in egress switch i, $N_{sent,i}$ is the number of sent packets to the network by ingress switch i, and L is normalized traffic load of packets arrival at the network. Clearly, when the second term tends to 1.0 (i.e., almost all transmitted packets have been successfully delivered), throughput will tend to traffic load L.

Throughput can also be defined as the volume of traffic successfully delivered within a given period of time. For example, if a traffic of volume v bits successfully arrive at a destination within τ seconds, then throughput will be $\frac{v}{\tau}$ bits/s.

– **Fairness**: Fairness determines whether network users or applications are receiving a fair share of system resources. It is usually measured by the Jain's fairness index [118]:

$$J(x_1, x_2, \ldots, x_m) = \frac{(\sum\limits_{i=1}^{m} x_i)^2}{m \times \sum\limits_{i=1}^{m} x_i^2} \, , \qquad (1.2)$$

where m is the number of users/applications and x_i is the throughput or bandwidth received by user/application i.

– **Packet Loss Rate (PLR)**: The packet loss rate can be separately computed for client traffic and optical packet traffic. This is because an optical packet may carry a number of client packets. When an optical packet carries only one client packet, these two packet loss rates will be the same. Note that it is common to consider the electronic buffers in ingress switches large enough so that no client packet is lost due to the buffer overflow in ingress switches. The general formula to compute PLR in a network with n core switches (where one edge switch is connected to each core switch) is given by

$$PLR = \frac{\sum\limits_{i=1}^{n} N_{loss,i}}{\sum\limits_{i=1}^{n} N_{sent,i}} \,, \qquad (1.3)$$

where $N_{loss,i}$ is the number of lost packets in core switch i and $N_{sent,i}$ is the number of sent packets to the network by ingress switch i.

- **Delay**: An optical packet sent by an ingress switch experiences some delay until it arrives at its destination. This delay performance can be stated in two ways as queuing delay and end-to-end delay:

  * The queuing delay is due to the waiting time of a packet in an electronic buffer of an edge switch and waiting time in the optical buffers of OPS core switches along the path to its destination. Clearly, when there is no optical buffers used in core switches, the queuing delay is limited to the waiting time in edge switches.

  * The end-to-end delay t_{e2e} for a given optical packet is composed of four parts: (1) transmission delay time to push the optical packet bits onto a wavelength in the ingress switch (t_{tx}), (2) propagation delay of the optical packet due to the speed of light on the links along its path (t_{prop}), (3) processing delay to process the header of the optical packet and make switching module ready in all core switches along its path (t_{proc}), and (4) its queuing delay in the ingress switch and all the core switches along its path (t_{queu}) as stated above. In other words, we have

  $$t_{e2e} = t_{tx} + t_{prop} + t_{proc} + t_{queu} \,. \qquad (1.4)$$

  Taking average among end-to-end delays of all successfully delivered optical packets results in average end-to-end delay in the network.

- **Jitter**: Jitter is computed from the end-to-end delay. This metric is the measure of variation of delay due to the fact that each packet in the network travels through different paths, and therefore the end-to-end delay varies. other reasons for creating jitter is network congestion and variable processing times at intermediate switches. In general, there are two methods to compute jitter. First, let d_i denote the end-to-end delay of packet i, and let m denote the number of packets from the same flow arrived at a destination. The first method to compute jitter is given by [119]

$$Jitter = E\left[|d_{i+1} - d_i|\right] = \frac{\sum\limits_{i=1}^{m-1} |d_{i+1} - d_i|}{m-1} \,. \qquad (1.5)$$

The second method to compute jitter is calculated by [120]

$$Jitter = \sqrt{\frac{\sum\limits_{i=1}^{m}(d_i - \bar{d})^2}{m-1}}, \tag{1.6}$$

where $\bar{d} = \frac{1}{m}\sum\limits_{i=1}^{m} d_i$ is average delay of m packets.

- **Bit Error Rate (BER) and Q Factor**: BER could be the most important metric that influences the end-end quality of an optical packet. BER is used to determine an optical packet signal quality. A transmitted optical packet may fail to arrive at its destination (i.e., lost due to high BER) because of signal degradation, packet corruption, faulty networking hardware, faulty network drivers, and distance between the source and destination. The Q factor is a good intermediate parameter for BER and Optical Signal-to-Noise Ratio (OSNR). For example, $BER = 10^{-11}$ is equivalent to $Q = 6.7$, and $BER = 2.86 \times 10^{-7}$ is equivalent to $Q = 5$. This factor is obtained from [7]

$$Q = \sqrt{\frac{B_o}{B_e}} \times \frac{2 \times OSNR}{1 + \sqrt{1 + 4 \times OSNR}}, \tag{1.7}$$

where B_o and B_e are optical and electrical bandwidths, respectively. We could have $\frac{B_o}{B_e} = 10$. The approximate relation between Q factor and BER is defined by the following equation when $Q \geq 3$ [121]

$$BER \approx \frac{e^{-Q^2/2}}{\sqrt{2 \times \pi \times (Q^2 + 2)}}, \tag{1.8}$$

where e ≈ 2.71828 is the Euler's number.

1.4.6.2 QoS Provisioning QoS is related to traffic control and resource management. QoS provisioning in an OPS network can be divided into three categories.

1.4.6.2.1 QoS Improvement QoS improvement mechanism is defined as any mechanism that can improve the general performance of the network through traffic control and resource management mechanisms and provide predictable or guaranteed performance for delay, packet loss, jitter, fairness, reliability, and fault tolerance parameters. A QoS improvement mechanism controls the allocation of network resources to all applications traffic in a way to improve their performance parameters in order to satisfy their users. This is suitable for a network with single class traffic of service in which all packets are treated equally. Applications like email and web browsing that are non-interactive can be counted in this category [114].

1.4.6.2.2 QoS Differentiation   There is no guarantee in PLR and delay performance parameters in today's Internet services since Internet offers only the best effort and connectionless service model. For single class BE traffic service, no guarantees can be given to any packet regarding loss rate, delay, and jitter performance metrics since all traffic in the network is equally treated [114]. However, when network resources become inadequate, the network performance parameters will be degraded. This will in turn degrade the QoS requirements for the real-time traffic. Therefore, having an optical OPS network to differentiate the optical packets generated by different network applications/users is a requirement in the near future.

Under QoS differentiation, the network must be able to differentiate between high-important traffic and low-important traffic and provide better service for the high-important traffic. Traffic differentiation is required under two cases:

- **Differentiation Based on Application Traffic:** Here, high-important traffic includes the traffic generated by highly interactive applications (such as video conferencing and online gaming that need stringent operating requirements), and low-important traffic includes the traffic generated by non-interactive or semi-interactive applications (such as e-mail, web browsing). Since different network applications need different levels of QoS, service differentiation should also be considered for future optical networks.

- **Differentiation Based on User Traffic:** Here, high-important traffic includes the traffic generated by the companies that pay high premiums to ensure network reliability in order to transmit their critical transactions. On the other hand, low-important traffic includes the traffic generated by home users only that need cheap Internet access and can tolerate a lower service level [55].

In store-and-forward electronic routers and switches, buffers are easily used to provide packet differentiation. However, this is not suitable for all-optical OPS networks due to the lack of optical buffers in OPS switches (i.e., either having a little or no optical buffers). Therefore, there are interests in providing new approaches to provide service differentiation in all-optical OPS networks without using optical buffers [122–124]. The QoS in OPS implies high throughput, low latency, and low packet loss for high-priority class traffic. It is also desirable to keep the order of optical packets because optical packet reordering increases latency at egress switches.

A number of enhancements have been introduced to offer different levels of QoS for IP networks. There are mainly two common methods for service differentiation as IntServ [125] and Differentiated Services (DiffServ) [126]. IntServ can guarantee QoS by end-to-end bandwidth reservation for traffic flows and per-flow scheduling in all intermediate routers/switches in a network. On the other hand, DiffServ can guarantee QoS differentiation for different classes of traffic aggregates. Since DiffServ is more scalable than IntServ, optical networks use the DiffServ model for QoS provisioning [59, 122, 123, 127]. Under DiffServ, edge switches classify, mark, drop, or shape client packets based on a Service Level Agreement (SLA) and prevent the DiffServ network from malicious attacks. On the other hand, core nodes perform high speed switching of differentiated packets and provide relative per-class QoS differentiation by allocating more bandwidth, lower delay, or lower loss to one class than another class. There are three general classes defined in DiffServ: Expedited Forwarding (EF) [100], Assured Forwarding (AF) [101], and Best Effort (BE). The EF service class is used for the applications that need low loss rate, low latency, low jitter, and bandwidth guarantees, while AF offers different levels of forwarding assurances to client packets. The remaining traffic is treated as BE without any QoS guarantee [128, 129].

In addition to EF, AF, and BE classes in the DiffServ model, some researches use general classes. In general, HP, MP, and LP denote high-priority, mid-priority, and low-priority traffic, respectively. Note that client packets are usually classified into two or three classes, and therefore we will have HP, MP, and LP client packets in the three-class scenario. If an optical packet carries one or more client packets from the same class c, the class of that optical packet is also considered as c. For instance, an optical packet carrying a number of LP client packets is called an LP optical packet.

The QoS provisioning can be grouped as either relative or absolute [130]:

- In absolute (hard) QoS provisioning, QoS is guaranteed to each traffic class based on the rules stated in an SLA. Note that the SLA is a contract between a user and the network provider, which defines the number of forwarding classes, level of service, and upper bounds on the values of network parameters (such as delay or loss). In other words, the absolute QoS provides the worst-case guarantee on the loss, delay, and bandwidth to applications, which is appropriate for delay and loss sensitive applications such as multimedia and mission-critical applications [131].

- The relative (soft) QoS differentiation model can provide quantitative differentiation between different service classes in such a way that a high-priority traffic is guaranteed to receive no worse service than a low-priority traffic. In other words, a high-priority optical packet should always obtain better service (such as lower loss, lower delay, and lower jitter) than a low-priority optical packet. It is clear that the amount of service received by a class depends on the current network load in each class. The main advantage of relative service differentiation over the absolute service differentiation is its simplicity and ease of deployment [116, 132].

The relative service differentiation model has been improved to the proportional differentiation QoS model in order to provide the network operators with quantitative QoS differentiation among service classes. With this QoS model, service differentiation between traffic classes can be adjusted according to pre-defined factors. The proportional differentiation model is more scalable than the relative service differentiation model. Define q_i to be the desired QoS metric for class i traffic, and define s_i to be the differentiation factor for class i traffic that is set by the network service provider. With M traffic classes in the network, we should have [132]

$$\frac{q_i}{q_j} = \frac{s_i}{s_j} , \tag{1.9}$$

for any class $i, j = 1, 2, \ldots, M$. For instance, assume q_1 and q_2 to be packet loss rates of class 1 and class 2, respectively. In addition, consider that the network service provider has set $s_1 = 2$ and $s_2 = 3$. Then, we should have $\frac{q_1}{q_2} = \frac{2}{3}$; i.e., packet loss rate of class 1 must be almost two-third of class 2. The proportional differentiation model must be held for both long time and short time periods since the long-term average is not quite meaningful under bursty traffic. For this purpose, Eq. (1.9) is changed to the following equation [132]:

$$\frac{\overline{q}_i(t, t + \tau)}{\overline{q}_j(t, t + \tau)} = \frac{s_i}{s_j} , \tag{1.10}$$

where $\overline{q}_i(t, t + \tau)$ is the average QoS metric from time t to $t + \tau$. By this general equation, the quality of differentiation between different traffic classes can be defined as a function of various QoS metrics. However, in practice, we cannot expect the equality in Eq. (1.10), and some deviations must be allowed in this equation as [132]

$$\frac{\overline{q}_i(t, t + \tau)}{\overline{q}_j(t, t + \tau)} = \frac{s_i}{s_j} \pm \Delta , \tag{1.11}$$

where Δ is a deviation from the relation determined by $\frac{s_i}{s_j}$. By changing s_i and s_j accordingly, the differentiation between them can be controlled within a bounded deviation. When τ is small enough, high-priority traffic will always receive better service than low-priority traffic independent of traffic load fluctuation [132].

As stated, electronic networks can support class-based traffic based on the DiffServ model. This model has three major classes (EF, AF, and BE) and a number of sub-classes for AF (such as AF11, AF12, AF13, AF21, ..., AF33, AF41, AF42, and AF43) [133]. Hence, the number of traffic classes in the electronic domain could be high. On the other hand, the number of classes in the optical domain must be kept as small as possible in order to reduce operational complexity of OPS switches since complex scheduling algorithms for several classes of traffic may not be applicable in the optical domain because of the limitation of buffering in the this domain [134, 135]. Some works have considered two classes of traffic in the optical domain, e.g., [114, 135], and some have considered three traffic classes, e.g., [83, 95, 104]. Therefore, the class-based traffic in the electronic domain should be mapped to the available classes in the optical domain using the method stated in [114].

1.4.6.2.3 Quality of Transmission (QoT)-Based Scheduling Many OPS studies have assumed an ideal physical layer for transmission and switching of optical packets. However, the ideal assumption is suitable for opaque networks. As an optical packet propagates toward its destination in a transparent optical network, its signal quality degrades and its BER increases due to attenuation, noise, dispersion, crosstalk, jitter, and non-linear effects. This is because there is no signal regeneration on the path. Note that increasing BER of an optical signal decreases OSNR in optical receivers. This will finally result in optical packet drop in an egress switch or even in an intermediate OPS switch.

In the Quality of Transmission (QoT)-based scheduling mechanism, the network must switch network traffic in such a way that optical packets can be detectable at their egress switches. The QoT-based switching is different in OPS from OCS networks. Many studies on wavelength-routed networks have considered a non-ideal physical layer and provided complex models for evaluating different aspects of physical layer impairments for establishing a light-path since in these networks a light-path is first set up and then data are transmitted. However, using complex models for evaluating physical layer impairments for each optical packet cannot be practical in OPS networks because of its finest granularity. Therefore, evaluating complex models should be avoided in OPS. Instead, simple physical layer impairment models or indirect physical layer impairment models should be used in OPS. In the former, transmission impairments must be taken into account in switching of optical packets towards their destinations. To do this, OPS switches must be aware of the signal quality of an optical packet (i.e., QoT-aware OPS) and should not switch the optical packet toward its egress node if it does not have enough quality to traverse its path. In the latter, optical packets are routed through short-path or short-hop routes toward their destinations. This indirectly takes into account the impairments.

There are three kinds of Physical Layer Impairments (PLIs):

- Linear impairments (such as path loss, Chromatic Dispersion (CD) or Group Velocity Dispersion (GVD), Polarization Mode Dispersion (PMD), and insertion loss) are the impairments that do not depend on signal power.

- Nonlinear impairments (such as intra-channel Self-Phase Modulation (SPM), Stimulated Brillouin Scattering (SBS), Stimulated Raman Scattering (SRS), inter-channel crosstalk originated from the non-linear interaction within fiber spans of several signals co-propagating on different wavelengths such as inter-channel Cross-Phase Modulation (XPM) and inter-channel Four-Wave Mixing (FWM) crosstalk between channels) are signal power dependent and may have a time-varying property.

- Other impairments such as Amplified Spontaneous Emission (ASE) noise and intra-channel crosstalk (resulted from optical leaks in optical switches). Different physical impairments increase the BER of an optical packet.

For a detailed description of the impairments, an interested reader is referred to [6, 31, 32, 48, 136, 137].

1.4.6.3 QoS Support by GMPLS   In conventional IP networks, each switch/router routes every packet toward its destination. However, there are some applications such as video on demand, IP Television, and voice over IP that generate a flow of packets destined to the same destination. In this case, the same routing must be repeated for every packet at each router. By providing a connection-oriented communication between the source and destination of a real-time communication and using the MPLS protocol, one can reduce the complexity of routing in routers/switches and improve packet performance in the core of the networks. MPLS forwards data from one network node to the next node based on short and fixed-length labels rather than long-length IP addresses. MPLS operates at layer 2.5 that lies between layer 2 (data link layer) and layer 3 (network layer). Generalized MPLS (GMPLS) has been developed to generalize the MPLS concept in optical networks. Both MPLS and GMPLS can provide QoS support in optical networks as discussed in the following.

1.4.6.3.1 MPLS Operation   The MPLS maps IP routing into a type of link layer connection referred to as Label Switching Path (LSP), which is actually a unidirectional virtual circuit. In other words, an LSP is functionally an IP level route. However, the individual packet forwarding is optimized since most of the packets only follow the LSP at the link layer (i.e., switching) and do not need an IP level routing operation. MPLS-aware routers are in fact routers/switches, each called a Label Switching Router (LSR). A set of LSRs creates an MPLS domain [138].

MPLS can separately support various communication sessions between a given source–destination pair. For this purpose, MPLS introduces the concept of Forwarding Equivalent Class (FEC), where an FEC is the set of all packets sharing the same list of criteria (say belong to the same IP traffic class). By coupling FEC and LSP, a flexible way is offered to perform traffic engineering. Traffic engineering controls how traffic flows through a network, redirects the traffic flows in the core network to avoid congestion, provides QoS to end users, and optimizes the utilization of a network resources [138, 139].

An LSP is created by distributing labels to routers/switches on the path. Label distribution can be performed by extension of routing protocols (such as Open Shortest Path First (OSPF) and Border Gateway Protocol (BGP)) or by the Resource Reservation Protocol (RSVP). There is also a special protocol, called Label Distribution Protocol (LDP), designed for this purpose in order to distribute labels between Label Edge Routers (LERs) and LSRs [138].

An MPLS network includes a core with a number of LSRs and a number of Label Edge Routers (LERs) at the edge of the network. Each LER consists of two parts: ingress router (for managing traffic transmission) and egress router (for managing the received traffic). The functions of MPLS components are as follows [139]:

- **MPLS Ingress Router**: An ingress router can establish, modify, reroute, and terminate LSPs by using IP signaling and routing protocols such as BGP and RSVP. Note that an LSP is a kind of virtual circuit identified by its label that is formed by a sequence of LSRs.

There are two messages used by RSVP to establish an LSP. The RSVP PATH message is generated by the ingress router and is forwarded through the network along the path of a LSP. At each hop, the PATH message checks the availability of requested resources and stores this information. The RSVP RESERVATION message is created by the egress router in the MPLS domain and used to confirm the reservation request that was sent earlier with the PATH message.

After establishing an LSP, traffic transmission for that LSP can be performed. For an arriving IP packet, the ingress router determines its FEC, assigns it the label of its relevant LSP, encapsulates it in an MPLS header and creates a labeled IP packet (this process is called label pushing), and then sends the labeled IP packet to the MPLS core router.

- **MPLS Core Router**: An LSR does not examine IP header during forwarding. Instead, it forwards labeled IP packets according to the label swapping paradigm, where the LSR maps particular input label/port of an arriving labeled IP packet to a predetermined output label/port, which is provided during the LSP setup. In other words, packets forwarding decisions in LSRs are made solely based on their labels. An LSR may need to change the label of a labeled IP packet before forwarding it to the next LSR.

- **MPLS Egress Router**: An IP packet is restored from a labeled IP packet at the end of LSP by egress router. This process is called label popping.

The QoS tasks in an MPLS controller include the classification of client packets in ingress switches, the differentiated servicing of optical packets in OPS switches, and traffic engineering that performs operations such as LSP admission control and traffic load management. After creating an LSP, the forwarding hardware on the path is provisioned with the requested traffic engineering parameters to guarantee requested levels of QoS metrics such as traffic bandwidth, delay, jitter, and loss. When traffic of the LSP flows, network devices monitor and report the performance parameters and actual level of resource utilization at each core switch interface to the MPLS controller. The controller can use this information on setting up the future coming LSPs and improving the quality of the current LSP [6].

1.4.6.3.2 GMPLS Operation   GMPLS is one of the key techniques to integrate IP over WDM. It can manage different types of switchings other than packet switching. A GMPLS-capable LSR is capable of handling five switching mechanisms such as [138]:

- **Fiber-Switch-Capable (FSC) Switching**: In space domain switching, data flow of an entire input fiber is forwarded to another output fiber. For this, a label represents a single-fiber in a bundle.

- **Waveband-Switch-Capable (WSC) Switching**: In waveband domain switching, data are carried by an incoming waveband and are forwarded to an outgoing waveband. For this, a label represents a single waveband within a fiber.

- **Lambda-Switch-Capable (LSC) Switching**: In wavelength domain switching, data are carried by an incoming wavelength channel and is forwarded to an outgoing wavelength channel. For this, a label represents a single wavelength within a waveband or fiber.

- **Time-Division Multiplex (TDM) Capable Switching**: In time domain switching, data are carried and switched in time slots. For this switching, a label represents a set of time slots within a wavelength.

- **Packet-Switch-Capable (PSC) Switching**: This is the original MPLS switching format where packets are switched based on their labels.

Since MPLS deals only with PSCs, there may be a number of LSPs on the same physical link, and a fine bandwidth allocation can be performed on them. For non-PSC LSPs in GMPLS, bandwidth allocation can be performed in discrete units with far fewer choices and small number of LSPs. In optical networks, capacity of a wavelength is too much and there are a number of wavelength on each fiber. Therefore, handling of huge number of LSPs at the PSC (or even TDM) level in GMPLS and communicating labels for them are complicated issues. One approach to alleviate label communication is to use the LSP hierarchy function of MPLS in which a number of hierarchical labels are assigned to a packet. In this hierarchy-label architecture [138],

- At the top level, there are FSC-LSPs so that each one can accommodate several LSC-LSPs or WSC-LSPs.

- At the next level, there are WSC-LSPs so that each one can group several LSC-LSPs.

- At the next level, an LSC-LSP can aggregate several TDM-LSPs.

- At the lowest level, several PSC-LSPs can be grouped into a TDM-LSP.

In a GMPLS-enabled optical network, there are three types of service blocking [140, 141]:

- **Routing Blocking (RB)**: blocking in the route computation phase when the path cannot be found between a source–destination pair.

- **Forward Blocking (FB)**: blocking in the forward phase of RSVP (i.e., during sending RSVP PATH messages) due to resource unavailability (say wavelength unavailability) on the computed path links.

- **Backward Blocking (BB)**: blocking in the backward phase of RSVP due to outdated information or reservation collision, e.g., more than two RSVP RESERVATION messages concurrently attempt to reserve the same wavelength or part of spectral segments in their common links.

GMPLS can be used in connection-oriented OPS networks, where each optical packet is labeled with a connection ID. In this case, a virtual connection must be setup. Each OPS core switch must be able to switch the optical packets belonging to the same connection through the path the connection passes. In GMPLS-based WDM optical packet-switched networks, optical packets are forwarded and switched in the optical domain, and the LSPs they follow are controlled and established by GMPLS protocols. Different techniques proposed for label coding and detection in optical packet headers under GMPLS have been analyzed and studied in [142].

A QoS routing algorithm in GMPLS can select a near-optimal path to meet the desired QoS of a traffic request, considering the combined topology and resource usage information of both IP and WDM layers [114]. GMPLS can set up connections that satisfy the QoS requirements of end users in terms of quality of transport level, reliability, and other QoS-related constraints. In the following, there is a list of LSP setup request examples made by an application generating high-priority traffic that will create a high-priority FEC:

- The application may request from GMPLS to set up an LSP on which packets will experience low crosstalk. This is possible when the LSP passes through the shortest hop path.

- The application may request from GMPLS to set up an LSP on a reliable path. This is possible when the LSP passes through the links with more protection fibers.

- The application may request from GMPLS to set up an LSP in a balanced manner in the network to reduce traffic loss and network congestion. This is possible when the LSP passes through less-congested links.

- The application may request from GMPLS to set up an LSP on which packets will experience low delay. This is possible when the LSP passes through the shortest distance path.

- The application may request from GMPLS to set up an LSP on a secure path. The GMPLS network must monitor the network and find the paths that are more secure and route a high-priority FEC LSP through those secure paths. Clearly, a low-priority FEC LSP can be routed through a less-secure path.

1.5 Optical OFDM-Based Elastic Optical Networking (EON)

In Section 1.3, a number of optical networking mechanisms (such as OPS, OBS, OCS) have been studied. These mechanisms are called fixed-grid optical networks since they provide rigid nature and inflexibly assign resources to users' traffic demands. For example, in a fixed-grid OCS network, a demand must occupy the whole capacity of a wavelength, even if the demand is less than the full wavelength capacity. Hence, the wavelength capacity may be wasted. The OBS is more flexible than OCS, but it still may waste some bandwidth since traffic is usually dynamic and fluctuates over time. Note that it is common to over-provision network resources to protect network performance at high traffic loads, and this over-provisioning increases the network cost. In short, an optical network with fine-granularity can save network resources and use them better than a coarse-granularity optical network.

With the growth of users' traffic demands, the need for a network to flexibly assign its existing resources to the demands becomes important. Therefore, a new generation of optical networking called Elastic Optical Network (EON) has been introduced. An EON is a flex-grid optical network that can provide fine nature and flexibly assign the network resources to demands exactly according to the requested amount of arriving demands, thus avoiding wasting of the resources.

The Optical Orthogonal Frequency Division Multiplexing (OOFDM)-based EON is studied in the following. As a new and promising network architecture, there are a variety of issues that should be resolved in the OOFDM-based EON such as [143]:

- improving network planning, traffic engineering, control plane technologies

- enhancing current optical network standards

- developing novel Bandwidth-Variable (BV) transponders and BV Cross Connects (BV-WXCs)

- designing flexible spectrum allocation and routing algorithms

- designing traffic grooming approaches

- designing survivability mechanisms

- designing energy efficient mechanisms and components

1.5.1 OOFDM Modulation

The main difference between the fixed and flex grid optical networks lies in the modulation format. A flex-grid optical network can use the OOFDM modulation to provide superior advantages like high spectrum efficiency, superior tolerance to CD/PMD impairments, robustness against inter-carrier and inter-symbol interference, scalability to ever-increasing data rates based on its subcarrier multiplexing technology, and adaptability to channel conditions.

OOFDM is a method of encoding data on multiple carrier frequencies. OOFDM is a type of FDM used as a multi-carrier modulation method. A large number of closely spaced orthogonal subcarrier signals are used to carry traffic on several parallel channels. Each subcarrier can be modulated with a conventional modulation scheme at a low symbol rate while maintaining total data rate as the conventional single carrier modulation scheme in the same bandwidth.

Since different frequency components of an optical pulse travel with different speeds, the optical pulse is spread out after transmission (i.e., dispersion phenomenon). Therefore, an OOFDM symbol with a large delay spread may cross its symbol boundary after traversing a long-distance, thus leading to interference with its neighboring symbols (called Inter-Symbol Interference (ISI)). Moreover, OOFDM symbols of different subcarriers are not aligned due to the delay spread, and therefore the critical orthogonality condition for the subcarriers may be lost, resulting in an Inter-Carrier Interference (ICI) penalty [143].

Using OOFDM as a bandwidth-variable and highly spectrum-efficient modulation format can provide flexible and scalable sub-wavelength and super-wavelength granularity in contrast to the fixed-grid optical networks [144, 145]. Unlike in WDM networks where all wavelength channels must be separated by a guard band in the frequency domain in order to prevent from interference, the subcarriers in EON can partially overlap in the frequency because of their orthogonalities. A comprehensive survey of OFDM-based optical high-speed transmission and networking technologies have been presented in [143].

Using OOFDM, EON can make the optical network more efficient and flexible. Besides, EON can generate elastic optical paths (the paths with variable bit rates), use bandwidth variable features, and divide available spectrum flexibly according to traffic demands of clients. In EON, if an arriving demand size is more than a wavelength capacity, multiple channels can be grouped together to construct a super-channel. A super-channel includes multiple very closely spaced channels which can transmit the demand as a single entity. As a multi-carrier system, the OOFDM modulation technique used in optical networks distributes data on several low bit rate subcarriers. The spectrum of orthogonally modulated neighbor subcarriers can overlap, thus increasing transmission spectral efficiency. The OOFDM can serve connections with fine-granularity by the elastic allocation of low bit rate subcarriers according to the connection demands [143, 146, 147].

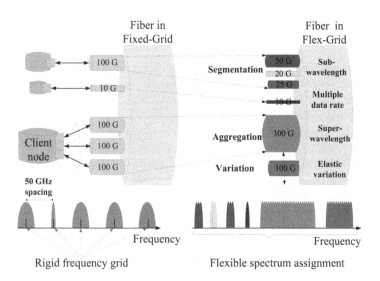

Figure 1.15: An example for spectrum assignment [148]

Figure 1.15 [148] illustrates an example for spectrum assignment in both fixed and flex-grid optical networks, where traffic from different clients with rates 10, 20, 25, 50, 100, and 300 Gbits/s arrive at an ingress switch. Assume data rate of wavelength channels to be 100 Gbits/s in a flex-grid optical network. In this case, the ingress switch must aggregate traffic 20, 25, 50 on one wavelength, use three wavelengths for transmitting 300 Gbits/s traffic, and use a wavelength for transmitting 10 Gbits/s traffic. Under flex-grid networks, these traffic can be separately assigned to different spectrum bands and transmitted (called sub-wavelength traffic accommodation). As it can be observed, the 300 Gbits/s traffic is sent by a number of subcarriers as a whole (called super-wavelength traffic accommodation), and the 10 Gbits/s traffic is directly accommodated in the spectrum using a smaller number of subcarriers, i.e., efficient accommodation of multiple data rates because of the flexible assignment of spectrum. Expansion and contraction of an elastic optical paths is one of the major unique features in EON and some subcarriers should be used for expansion purposes (i.e., variation bandwidth).

1.5.2 EON Components

In OOFDM networks, BV transponders and BV-WXCs are used to generate traffic in ingress switches and switch subcarriers in core nodes, respectively. To obtain high spectral flexibility and serve a client demand, a BV transponder generates an optical signal according to the traffic demand using just enough spectral resources in terms of subcarriers with an appropriate modulation level. Hence, it can provide efficient use of available spectrum resources. For instance, if a subcarrier can provide bandwidth of b bits/s (say 2 Gbits/s), and a client traffic demand needs bandwidth of r bits/s, then the BV transponder can only utilize $\lceil \frac{r}{b} \rceil$ contiguous subcarriers [144].

In traditional WDM networks, all light-paths are assigned the same spectrum width, regardless of the transmission distance of each light-path. This leads to an inefficient utilization of the spectrum. The distance-adaptive spectrum allocation concept can be used in OOFDM-based EON using adaptive modulation and bandwidth variable transponders to improve the spectrum efficiency and QoS for network traffic. There are different modulation formats to be used in OOFDM such as the M-PSK group [e.g., the BPSK (or 2-PSK) and the QPSK (or 4-PSK) groups] and the M-QAM group (e.g., 4-QAM, 8-QAM, 16-QAM, 64-QAM, and 256-QAM). A low-level modulation format with wide spectrum is adapted for long-distance paths, and a high-level modulation format with narrow spectrum can be used for short-distance paths. For this purpose, an adaptive modulation technology can be used to decide what modulation format to be used on which subcarrier under OOFDM according to channel conditions such as reach and OSNR. Here, a subcarrier with high OSNR should be loaded with a high-level modulation format (i.e., more bits loading per symbol), while a low-OSNR subcarrier should use a low-level modulation format (i.e., less bits loading per symbol). For example, let the capacity of a subcarrier using one-bit-per-symbol BPSK modulation be b bit/s. Then, QPSK with 2 bits per symbol provides a capacity of $2 \times b$ bit/s, and 64-QAM with 6 bits per symbol provides a capacity of $6 \times b$ bit/s [143, 147].

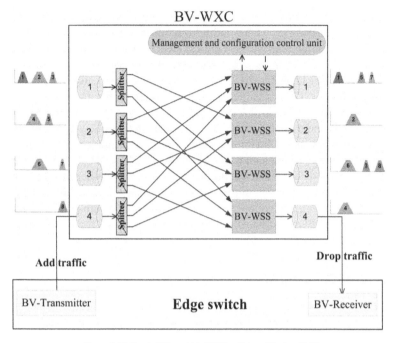

Figure 1.16: Bandwidth-variable WXC switch architecture [149]

A BV-WXC needs to flexibly configure its switching window according to the spectral width of the incoming optical signal. Figure 1.16 shows a $4 * 4$ bandwidth-variable WXC switch architecture that consists of splitters at its input side and Bandwidth Variable Wavelength Selective Switches (BV-WSS) at its output side. The main functions of an intermediate BV-WXC in EON are add-and-drop for local optical signals, and switching/routing of transit optical signals. Splitters split incoming spectrums and send them to appropriate BV-WSSs. A BV-WSS is based on a Reconfigurable Optical Add/Drop Multiplexer (ROADM) to support the flexibility feature of EON networks. The BV-WSSs can receive the spectrums and perform multiplexing/demultiplexing and switching functions using integrated spatial optics. Liquid crystal or MEMS-based BV-WSSs can be employed as switching elements to realize an optical cross-connect with flexible bandwidth and center frequency. The management and configuration control unit assigns optical resources to the setup paths, and it routes the incoming signals by configuring the BV-WSS modules according to the information obtained from a forwarding table that includes the information of the traversing paths. The information of the table is updated when either a connection is established or terminated [143, 149].

Figure 1.16 also depicts switching of traffic coming on different subcarriers. For example, the traffic coming on spectrum segment 4 (where a segment is the contiguous subcarriers allocated to a given connection demand) at input port 2 is delivered to the local edge switch through output port 4, and the traffic on spectrum segment

7 from input port 3 is switched to output port 1. It should be noted that switching in spectrum is possible if there is no overlapping (i.e., no collision) between the switched segments in spectrum at the same output port. In Fig. 1.16, there is no collision among the switched segments at output ports 1 and 3.

1.5.3 Routing and Spectrum Assignment in EON Networks

When a connection demand arrives between a given source–destination pair, the OOFDM network should establish the connection using the Routing and Spectrum Assignment/Allocation (RSA) technique, which can be divided into two subproblems as routing and Spectrum Assignment (SA). There are different routing algorithms and various SA mechanisms developed for RSA. The conventional routing schemes used in OCS networks (such as fixed shortest path routing, fixed alternate routing and K-shortest path routing) can be used for the OOFDM network as well. The aim of routing is to find the shortest feasible path, where the feasible path is determined by the availability of common idle subcarriers along the shortest path.

In SA, there are two constraints that must be satisfied when assigning subcarriers to a connection as [149]:

- **Continuous Constraint**: Under this constraint, the same subcarriers must be used in all links of a path.

- **Contiguous Constraint**: Under this constraint, the subcarriers must be contiguous in the spectrum.

In SA, two issues must be taken into consideration. The first is to avoid spectrum collision, say by randomly assignment of different spectrum segments to different demands. The other issue is to avoid spectrum fragmentation and retain contiguous spectrums as much as possible. There are some heuristic SA mechanisms used for selecting the spectrum resources (i.e., subcarriers) for a given connection demand as [150–152]:

- **First-Fit (FF)**: The simple and popular FF mechanism chooses the first idle segment in the spectrum and assigns the starting frequency of the connection to the starting point of the first idle segment. By this way, the FF selects frequency segments in the low-frequency domain and reduces the probability of spectrum fragmentation. However, collision is a significant problem in FF as different nodes may choose the same spectrum resource simultaneously, and this may lead to connection blocking.

- **Random Spectrum (RS)**: The RS mechanism chooses the spectrum resource and the left or right side of spectrum randomly. Compared with FF, the RS tends to choose different resources, hence it can reduce the probability of spectrum collision. Obviously, RS splits the spectrum into more fragments, thus increasing the routing blocking.

- **Minimum Residual Spectrum (MR)**: The MR mechanism selects the segment that has minimal, but sufficient bandwidth. Other segments with more contiguous subcarriers are kept for future coming connection demands.

- **Adaptive Frequency Assignment-Collision Avoidance (AFA-CA)**: This mechanism minimizes the allocated frequency spectrum width in the network. The main idea behind AFA-CA is to assign the contiguous subcarriers to requests without happening any overlapping in contiguous spectrums.

Bandwidth in OOFDM networks can be provided for a connection demand by using either offline (static) RSA or online (dynamic) RSA [153]:

- **Offline Bandwidth Provisioning**: A traffic matrix is given with the requested transmission rates of all connections, and the network should allocate paths and spectrum resources in a way to both serve the connections and minimize the utilized spectrum. However, no spectrum overlapping on any given link is allowed among these connections. An offline RSA may be formulated as an Integer Linear Programming (ILP) problem that returns the optimum solution through a combined routing and spectrum assignment. The ILP formulations can find optimum or near-optimum solutions for small EON networks. However, they are not scalable to large networks [143, 151, 154–156].

For example, Balanced Load Spectrum Allocation (BLSA) provides routing with balancing load within a network in order to minimize the maximum number of subcarriers used on each fiber. The BLSA supports static traffic and performs routing and spectrum assignment according to a given traffic matrix. The BLSA includes three steps as path generation that uses the k-shortest path algorithm to generate k paths, path selection that selects the path based on load balancing within all the fibers in the network, and spectrum allocation [156].

- **Online Bandwidth Provisioning**: OOFDM provides bandwidth dynamically for the demands that their traffic rates fluctuate over time. In other words, each BV transmitter can adaptively change its modulation/coding format to deal with either physical layer impairments or traffic flow variances. In this case, OOFDM should assign a smaller or a larger number of subcarriers to the setup connection that its traffic rate is changing over time. However, the assigned subcarriers must be adjacent. Performance evaluation results show that OOFDM networks can provide lower connection blocking than OCS networks [141, 144, 149, 150, 157].

QoS issues should also be considered in RSA. QoS, as stated in Section 1.4.6.2, can be provided in three ways. There are few works that have considered QoS in RSA. In the following, RSA with QoS improvement, RSA considering QoS differentiation, and RSA considering QoT-aware are reviewed:

- As stated in Section 1.4.6.3.2, backward blocking (collision) is an important reason for blocking in GMPLS-enabled optical networks. The backward collision would happen in EON more frequently than that in OCS networking. Therefore, a collision avoiding routing mechanism with maximum available spectrum allocation is proposed in [141], where the SA searches for the spectrum segment with minimum collision possibility in order to reduce backward blocking. This issue can improve the network performance parameters by distributing the traffic more evenly among the links (i.e., load-balancing), thus improving QoS.

- In allocating bandwidth, the network can pre-reserve r contiguous subcarriers for a connection demand to be established between a given source–destination pair. The number of reserved subcarriers depends on different parameters such as average traffic rate, maximum traffic rate, burstiness parameter, number of hops of the shortest-hop path, and quality of service type of the connection demand. For Best Effort (BE) connection demands, we usually have $r = 0$, while for guaranteed high-priority connection demands, we must have $r > 0$. Therefore, there are two types of subcarriers: pre-reserved (determined during connection setup) and shared (the subcarriers that can be allocated or deallocated to the connections according to their time-varying traffic rates). Note that if the traffic rate of a high-priority connection reduces and the number of required subcarriers m is less than r, the pre-reserved subcarriers are not released. However, when $m > r$, the network should assign $m - r$ additional contiguous subcarriers to the connection. However, considering both continuous and contiguous constraints, this extension may not be possible, and therefore the extension of bandwidth for the connection may be blocked. Now, assume that $m - r$ additional contiguous subcarriers can be allocated to the connection, and after a while traffic rate reduces. Then, the additional subcarriers can be released, but the network should always keep at least r subcarriers for the connection [144].

- The Quality of Transmission (QoT) can be taken into account in EON so that RSA tries to assign paths with the least amount of Optical Signal-to-Noise Ratio (OSNR), however without any traffic classification. In an EON, when a new connection demand arrives, the proposed dynamic QoT-aware RSA starts finding all the candidate paths and specifying the common unoccupied spectrum of each path with the appropriate modulation format. Then, it chooses a path that qualifies the required QoT constraint in terms of QoT metrics (such as OSNR) and spectrum efficiency. Finally, it assigns spectrum segment based on the allocated modulation level and requested bit rate using traffic balancing-spectrum assignment method, which uniformly distributes the traffic load over the optical spectrum. If there is no path available in the path computation phase, the demand is blocked. If no path can be chosen in the path selection phase, the blocking is because of the poor QoT conditions [149].

The impairment-aware routing and subcarrier allocation (IARSA) finds the shortest feasible path, where the availability of free subcarriers along a path determines the feasibility of the path by considering impairment levels and use of regenerators. The IARSA tries to balance traffic flows evenly across the network to reduce the blocking probability. It finds the path that does not violate the QoT threshold, where maximum distance is considered as the impairment constraint [152].

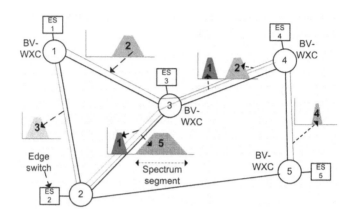

Figure 1.17: An example for connection establishment

Figure 1.17 displays an example for connection establishment under EON, where there are five setup connections: (1) one-hop connection from edge switch 1 (ES 1) to ES 2 on spectrum segment 3, (2) one-hop connection from ES 4 to ES 5 on spectrum segment 4, (3) one-hop connection from ES 2 to ES 3 on spectrum segment 5, (4) two-hop connection from ES 2 to ES 4 on spectrum segment 1, and (5) two-hop connection from ES 1 to ES 4 on spectrum segment 2. As can be observed, these connections are similar to the setup light-paths depicted in Fig. 1.6, where a whole wavelength channel is allocated to each light-path.

One important technique to have a successful RSA is fragmentation. It is possible that there are no sufficient resources to accommodate a connection demand in available paths. One way to solve this problem is to fragment the demand into multiple low-bit-rate demands and route them from separate paths [153].

1.5.4 IP over OOFDM

As studied in previous sections, OOFDM networks focus on connection-oriented networks and can be a good replacement for OCS networks. The OFDM-based EON offers finer bandwidth granularity than WDM networks, and a coarser granularity than OPS networks. Hence, it can be considered as a middle-term alternative to the OPS technology. However, few works have considered packet-based networking under OOFDM [157–161].

A programmable and adaptive IP over OOFDM network architecture and subcarrier resource assignment for IP/OOFDM network have been proposed in [157, 159, 160], where virtual links are set up between every ingress–egress switch pair to build a virtual network topology among them. The virtual links are isolated from each other using different OOFDM subcarriers on each physical link. According to traffic rate variations, the bandwidth of each virtual link can be changed adaptively by adjusting either the modulation format used on a subcarrier or the number of subcarriers. As stated in Section 1.5.2, an OOFDM network can use different modulation formats on each subcarrier; therefore, different bandwidths can be provided for the traffic on different subcarriers (e.g., using QPSK instead of BPSK doubles the subcarrier capacity). Backbone routers can periodically measure traffic for the virtual links. Then, when a significant change occurs in arriving traffic rate, the bandwidth adjustment decisions can be made. Performance evaluation results show that the adaptive IP/OOFDM network can save up to 30% of receivers and up to 10% total cost in comparison to an IP over TDM/WDM network [143, 159].

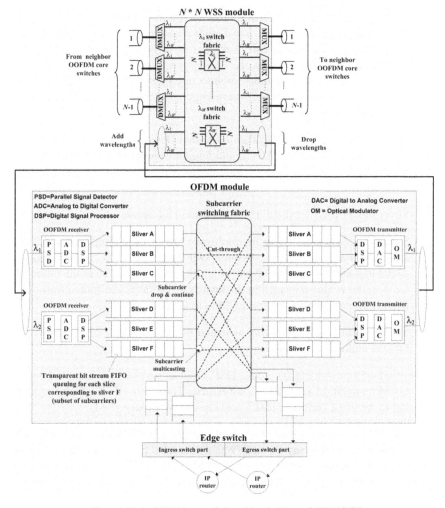

Figure 1.18: An OOFDM core switch model under IP over OOFDM [159]

To implement IP over OOFDM, a two-level hierarchical switching mechanism is used in OOFDM core switches (see Fig. 1.18). The top switching level (i.e.,

Wavelength-Selective Switch (WSS) module) switches wavelengths, whereas the bottom switching level (i.e., sub-wavelength OOFDM module) switches subcarriers. There are a number of dropped wavelengths from the WSS module to the OOFDM module and some add wavelengths from the OOFDM module to the WSS module. In addition, there are some add/drop traffic flow links from/to local ingress/egress switch. Each edge switch may be connected to a number of IP routers. Figure 1.18 shows the OOFDM module for the case that there are two wavelengths λ_1 and λ_2 in the network (i.e., $W = 2$).

Define sliver to be a set of subcarriers that can carry arbitrary format of user-defined data. Hence, each sliver may carry different types of packets, bursts, and even jumbo frames for various types of future Internet traffic. In Fig. 1.18, it is assumed that there are three slivers in each wavelength channel. For example, Slivers A, B, and C inside wavelength λ_1 may carry burst traffic, variable-length packet traffic, and fixed-length packet traffic, respectively.

Figure 1.18 presents the detailed architecture of the OOFDM module for two wavelengths and three slivers in each wavelength. The following components can be found in this architecture:

- OOFDM receivers including Parallel Signal Detector (PSD), high-speed Analog to Digital Converter (ADC), and Digital Signal Processing (DSP) for the functionalities such as QAM demodulation and Fast Fourier Transform (FFT) computation.

- Sliver input FIFOs for queuing of transparent bit stream.

- Subcarrier switching fabric for multi-rate sliver switch supporting functionalities such as subcarrier cut-through switching, subcarrier drop-and-continue switching, and subcarrier multicasting switching as depicted in the figure. The local traffic is also managed in the switching fabric as the source traffic flows from IP router and the sink traffic flows to IP router.

- Sliver output FIFOs for saving switched bit stream which is ready for optical OFDM multiplexing/modulation.

- OOFDM transmitters including DSP module for the functionalities such as QAM modulation and Inverse Fast Fourier Transform (IFFT), high-speed Digital to Analog Converter (DAC), and Optical Modulator (OM).

It should be noted that a time-division multiplexing/multiple access (TDM/TDMA)-based grooming switch (say SONET/WDM) can also be used as a subwavelength switching mechanism. However, the proposed OOFDM switch architecture provides unique flexibility with high resource efficiency when dealing with bandwidth resource sharing. It utilizes DSP for subcarrier multiplexing and grooming combined with adaptive transmission and parallel signal detection. In addition, each OOFDM transmitter can dynamically modify its modulation/coding methods in order to deal with packet traffic variance at the subwavelength level. Moreover, adaptive load balancing can be obtained by mapping virtual topology to variable packet flows in real time. Besides, the OOFDM switch utilizes a receiver structure (i.e., PSD) that can receive multiple non-overlapping baseband signals on the same wavelength simultaneously. Note that no more than one wavelength can enter a PSD. Finally, no global synchronization and complicated multiplexing hierarchy are needed in the proposed OOFDM switch [162].

1.6 Summary

Future Internet requires both bandwidth and QoS because of an ever-increasing number of Internet users and a growing number of real-time applications (such as video on demand, voice over IP, IPTV, video conferencing, etc.) that require different levels of QoS. This could be a problem when the network bandwidth is limited, when the network supports only the best effort traffic, or when the Internet traffic does not have a uniform characteristic.

Different optical switching mechanisms in two versions of asynchronous and slotted have been detailed in this chapter. Among the proposed switching schemes, the OCS network is very popular in simplifying traffic transmission. However, it is neither efficient on the bandwidth usage nor scalable. Contention-based schemes seems to be appropriate in quickly responding to traffic dynamics such as Internet traffic. The OPS as a contention-based switching scheme appears to provide the finest granularity compared with OCS and OBS. Combination of huge bandwidth offered by optical technology with the flexibility of packet switching in OPS can be a very good candidate for future optical networks, for both metro and backbone networks.

Different aspects of OPS including OPS switch architecture, edge switch architecture, signaling in OPS, contention problem in OPS, and QoS in OPS have been studied in this chapter. As mentioned, contention is the major problem in OPS networks. Unfortunately, many proposed architectures resolve the contention problem in OPS by using immature optical buffering technology (using optical fiber delay lines), which is expensive, complex, and bulky. In addition, optical buffering degrades the signal power of an optical packet, increases its delay, and makes its arrival at its destination out of order. To reduce these problems, OPS core switches should reduce using optical buffers for contention resolution. In addition, slotted networks can provide a much finer granularity in bandwidth sharing and can improve the bandwidth usage. As stated, next generation optical networks must support the applications that require different levels of QoS. Based on all the above discussions, synchronized buffer-less all-optical packet-switched networking under the DiffServ domain could be a very good candidate for efficient transmission of traffic in optical networks.

In short, since QoS support is a requirement for any next-generation network and OPS networks are the promising next generation optical networks, this book will provide a comprehensive review of handling QoS issues in OPS networks.

Finally, OOFDM networks have been reviewed that can flexibly provide bandwidth for connection demands, while reducing the network bandwidth wastage. Using the OOFDM technology, the novel elastic optical network architecture with great flexibility and scalability on allocating spectrum and accommodating data rate has been studied. Most of the current OOFDM network designs have focused on connection-oriented networks, and therefore they can be a good replacement for OCS networks. However, few works have considered packet-based networking under OOFDM.

REFERENCES

1. M. OMahony, D. Simeonidou, D. Hunter, and A. Tzanakaki. The application of optical packet switching in future communications networks. *IEEE Communications Magazine*, 39(3):128–135, Mar. 2001.

2. D. K. Hunter and I. Andonovic. Approaches to optical Internet packet switching. *IEEE Communications Magazine*, 38(9):116–122, Sept. 2000.

3. J. Ryan. Fiber considerations for metropolitan networks. *Alcatel Telecommunications Review*, pages 52–56, 2002.

4. Fiber types in gigabit optical communications. Cisco Public Information, 2008.

5. R. Ramaswami. Optical fiber communication: From transmission to networking. *IEEE Communications Magazine*, 40:138–147, 2002.

6. S. V. Kartalopoulos. *DWDM: Networks, Devices, and Technology*. Wiley-IEEE Press, 2003.

7. Y. Huang, J. P. Heritage, and B. Mukherjee. Connection provisioning with transmission impairment consideration in optical WDM networks with high-speed channels. *IEEE Journal of Lightwave Technology*, 23(3):982–993, 2005.

8. R. Ramaswami, K. Sivarajan, and G. Sasaki. *Optical Networks: A Practical Perspective*. 3rd edition, Morgan Kaufmann, 2009.

9. A. M. Odlyzko. Internet traffic growth: sources and implications. In *SPIE Optical Transmission Systems and Equipment for WDM Networking II*, volume 5247, pages 1–15, Orlando, FL, 2003.

10. X. Xiao and L. M. Ni. Internet QoS: a big picture. *IEEE Network*, 13(2):8–18, 1999.

11. F. Callegati, G. Corazza, and C. Raffaelli. Exploitation of DWDM for optical packet switching with quality of service guarantees. *IEEE Journal on Selected Areas in Communications*, 20:190–201, 2002.

12. G. I. Papadimitriou, M. S. Obaidat, and A. S. Pomportsis. Advances in optical networking. *International Journal of Communication Systems*, 15(2-3):101–113, 2002.

13. A. K. Somani, M. Mina, and L. Li. On trading wavelengths with fibers: a cost performance based study. *IEEE/ACM Transactions on Networking*, 12:944–951, 2004.

14. O. Gerstel and H. Raza. Merits of low-density WDM line systems for long-haul networks. *IEEE/OSA Journal of Lightwave Technology*, 21(11):2470–2475, Nov. 2003.

15. A. G. Rahbar and O. Yang. Fiber-channel tradeoff for reducing collisions in slotted single-hop optical packet-switched (OPS) networks. *Journal of Optical Networking*, 6:897–912, 2007.

16. G. Muretto and C. Raffaelli. Combining contention resolution schemes in WDM optical packet switches with multifiber interfaces. *Journal of Optical Networking*, 6:74–89, 2007.

17. M. Dhodhi, S. Tariq, and K. Saleh. Bottlenecks in next generation DWDM-based optical networks. *Computer Communications*, 24(17):1726–1733, 2001.

18. T. E. Stern, G. Ellinas, and K. Bala. *Multiwavelength Optical Networks: Architectures, Design, and Control*. Second edition, Cambridge University Press, 2008.

19. G. Shen and R. S. Tucke. Translucent optical networks: the way forward. *IEEE Communications Magazine*, 45(2):48–54, Feb. 2007.

20. M. J. Potasek. All-optical switching for high bandwidth optical networks. *Optical Networks Magazine*, 3(6):30–43, 2002.

21. C. Papazoglou, G. Papadimitriou, and A. Pomportsis. Design alternatives for optical-packet-interconnection network architectures. *OSA Journal of Optical Networking*, 3(11):810–825, 2004.

22. M. J. OMahony, C. Politi, D. Klonidis, R. Nejabati, and D. Simeonidou. Future optical networks. *IEEE Journal of Lightwave Technology*, 24(12):4684–4696, Dec. 2006.

23. M. Klinkowski, D. Careglio, and J. Sol-Pareta. Wavelength vs. burst vs. packet switching: comparison of optical network models. In *The e-Photon/ONe Winter School workshop*, Aveiro, Portugal, 2005.

24. G. N. Rouskas and H. G. Perros. A tutorial on optical networks. *Lecture Notes in Computer Science*, 2497:155–193, 2002.

25. H. Zang, J. P. Jue, and B. Mukherjee. A review of routing and wavelength assignment approaches for wavelength routed optical WDM networks. *Optical Networks Magazine*, 1(1):47–60, Jan. 2000.

26. T. E. Stern and K. Bala. *Multiwavelength Optical Networks: A Layered Approach*. Prentice Hall, 2000.

27. A. G. Rahbar. A review of dynamic impairment-aware routing and wavelength assignment techniques in all-optical wavelength-routed networks. *IEEE Communications Surveys and Tutorials*, 14(2):1065–1089, 2012.

28. S. Barkabi and A. G. Rahbar. MC-BRM: A distributed RWA algorithm with minimum wavelength conversion. *Elsevier Optical Fiber Technology*, 18(4):230–241, 2012.

29. S. Barkabi and A. G. Rahbar. LC-DLA: A fair congestion-aware distributed lightpath allocation mechanism. *Elsevier OPTIK–International Journal for Light and Electron Optics*, 123(15):1370–1377, Aug. 2012.

30. A. G. Rahbar. Dynamic impairment-aware RWA in multi-fiber wavelength-routed all-optical networks supporting class-based traffic. *IEEE Journal of Optical Communications and Networking*, 2(11):915–927, Nov. 2010.

31. I. Tomkos, D. Vogiatzis, C. Mas, I. Zacharopoulos, A. Tzanakaki, and E. Varvarigos. Performance engineering of metropolitan area optical networks through impairment constraint routing. *IEEE Communications Magazine*, 42(8):S40–S47, Aug. 2004.

32. C. Saradhi and S. Subramaniam. Physical layer impairment aware routing (PLIAR) in WDM optical networks: issues and challenges. *IEEE Communications Surveys & Tutorials*, 11(4):109–130, Dec. 2009.

33. A. Mokhtar and M. Azizoglu. Adaptive wavelength routing in all-optical networks. *IEEE/ACM Transactions on Networking*, 6(2):197–206, Apr. 1998.

34. H. Zang, J. P. Jue, L. Sahasrabuddhe, R. Ramamurthy, and B. Mukherjee. Dynamic light path establishment in wavelength-routed WDM networks. *IEEE Communications Magazine*, pages 100–117, 2001.

35. A. Askarian, Y. Zhai, S. Subramaniam, Y. Pointurier, and M. Brandt-Pearce. QoT-aware RWA algorithms for fast failure recovery in all-optical networks. In *IEEE/OSA OFC'08*, San Diego, CA, 2008.

36. N. A. Shalmany and A. G. Rahbar. On the choice of all-optical switches for optical networking. In *IEEE International Symposium on High Capacity Optical Networks and Enabling Technologies (HONET07)*, UAE, Dubai, 2007.

37. C. Clos. A study of non-blocking switching networks. *Bell System Technical Journal*, 32:406–424, 1953.

38. X. Chu, H. Yin, and X. Y. Li. Lightpath rerouting in wavelength routed WDM networks. *Journal of Optical Networking*, 7(8):721–735, 2008.

39. G. Mohan and C. S. R. Murthy. A time optimal wavelength rerouting algorithm for dynamic traffic in WDM networks. *IEEE Journal of Lightwave Technology*, 17(3):406–417, Mar. 1999.

40. K. Zhu and B. Mukherjee. A review of traffic grooming in WDM optical networks: architectures and challenges. *Optical Networks Magazine*, 4(2):55–64, 2003.

41. J. Liu, G. Xiao, and W. Wang. On the performance of distributed light-path provisioning with dynamic routing and wavelength assignment. *Photonic Networks Communications*, 17:191–201, 2009.

42. G. S. Pavani, L. G. Zuliani, H. Waldman, and M. Magalhes. Distributed approaches for impairment-aware routing and wavelength assignment algorithms in GMPLS networks. *Computer Networks*, 52:1905–1915, 2008.

43. P. Singh, A. K. Sharma, and S. Rani. Minimum connection count wavelength assignment strategy for WDM optical networks. *Optical Fiber Technology*, 14:154–159, 2008.

44. C. S. R. Murty and M. Gurusamy. *WDM Optical Networks: Concepts, Design, and Algorithms.* Prentice-Hall India, 2002.

45. A. Szymanski, A. Lason, J. Rzasa, and A. Jaiszczyk. Grade-of-service-based routing in optical networks. *IEEE Communications Magazine*, 45(2):82–87, Feb. 2007.

46. M. Gagnaire and S. Zahr. Impairment-aware routing and wavelength assignment in translucent networks: State of the art. *IEEE Communications Magazine*, 47(5):55–61, May 2009.

47. E. Salvadori, Y. Ye, C. Saradhi, A. Zanardi, H. Woesner, M. Carcagni, G. Galimberti, G. Martinelli, A. Tanzi, and D. La Fauci. Distributed optical control plane architectures for handling transmission impairments in transparent optical networks. *IEEE Journal of Lightwave Technology*, 27(13):2224–2239, Jul. 2009.

48. S. Azodolmolky, M. Klinkowski, E. Marin, D. Careglio, J. Pareta, and I. Tomkos. A survey on physical layer impairments aware routing and wavelength assignment algorithms in optical networks. *Computer Networks*, 53(7):926–944, May 2009.

49. S. Azodolmolky, Y. Pointurier, M. Klinkowski, E. Marin, D. Careglio, J. Sol-Pareta, M. Angelou, and I. Tomkos. On the offline physical layer impairment aware RWA algorithms in transparent optical networks: state-of-the-art and beyond. In *Optical Network Design and Modeling (ONDM2009)*, Braunschweig, Germany, 2009.

50. E. V. Breusegem, J. Cheyns, D. D. Winter, D. Colle, M. Pickavet, P. Demeester, and J. Moreau. A broad view on overspill routing in optical networks: a real synthesis of packet and circuit switching? *Elsevier Optical Switching and Networking*, 1(1):51–64, 2004.

51. L. Xu, H. Perros, and G. Rouskas. Techniques for optical packet switching and optical burst switching. *IEEE Communications Magazine*, 39(1):136–142, Jan. 2001.

52. M. Nord, S. Bjrnstad, and C. M. Gauger. OPS or OBS in the core network? In *IFIP Working Conference on Optical Network Design and Modeling*, Budapest, Feb. 2003.

53. S. Yao, B. Mukherjee, and S. Dixit. Advances in photonic packet switching: an overview. *IEEE Communications Magazine*, 38(2):84–94, Feb. 2000.

54. M. Yoo and C. Qiao. Optical burst switching (OBS): a new paradigm for an optical Internet. *Journal of High Speed Networks*, 8(1):69–84, 1999.

55. K. C. Chua, M. Gurusamy, Y. Liu, and M. H. Phung. *Quality of Service in Optical Burst Switched Networks.* Springer, 2007.

56. J. Liu and N. Ansari. The impact of the burst assembly interval on the OBS ingress traffic characteristics and system performance. In *IEEE International Conference on Communications (ICC)*, Paris, France, June 2004.

57. S. J. B. Yoo, F. Xue, Y. Bansal, J. Taylor, P. Zhong, C. Jing, J.Minyong, T.Nady, G.Goncher, K.Boyer, K.Okamoto, S. Kamei, and V. Akella. High-performance optical-label switching packet routers and smart edge routers for the next generation Internet. *IEEE Journal on Selected Areas in Communications*, 21(7):1041–1051, Sept. 2003.

58. C. Raffaelli and P. Zaffoni. TCP performance in optical packet-switched networks. *Photonic Network Communications*, 11(3):243–252, May 2006.

59. K. Long, R. S. Tucker, and C. Wang. A new framework and burst assembly for IP diffserv over optical burst switching networks. In *IEEE Globecom*, San Francisco, CA, 2003.

60. V. Vokkarane and J. P. Jue. Prioritized burst segmentation and composite burst assembly techniques for QoS support in optical burst switched networks. *IEEE Journal on Selected Areas in Commun*, 21(7):1198–1209, Sept. 2003.

61. S. Seo, K. Bergman, and P. R. Prucnal. Transparent optical networks with time-division multiplexing. *IEEE Journal on Selected Areas in Communications*, 14(5):1039–1051, June 1996.

62. N. F. Huang, G. H. Liaw, and C. P. Wang. A novel all optical transport network with time-shared wavelength channels. *IEEE Journal on Selected Areas in Communications*, 18(10):1863–1875, Oct. 2000.

63. A. Chen, A. K. Wong, and C. T. Lea. Routing and time-slot assignment in optical TDM networks. *IEEE Journal on Selected Areas in Communications*, 22(9):1648–1657, Nov. 2004.

64. B. Wen and K. M. Sivalingam. Routing, wavelength and time-slot assignment in time division multiplexed wavelength-routed optical WDM networks. In *IEEE Infocom*, New York, 2002.

65. T. S. El-Bawab and J. Shin. Optical packet switching in core networks: between vision and reality. *IEEE Communications Magazine*, 40(9):60–65, Sept. 2002.

66. Caida. Packet size reports. Technical report, http://www.caida.org/research/traffic-analysis/fix-west-1998/packetsizes/, 1998.

67. J. Ramamirtham and J. Turner. Time sliced optical burst switching. In *IEEE Infocom*, pages 2030–2038, San Francisco, Mar. 2003.

68. I. Chlamtac, V. Elek, A. Fumagalli, and C. Szabo. Scalable WDM network architecture based on photonic slot routing and switched delay lines. In *IEEE Infocom*, Kobe, Japan, Apr. 1997.

69. I. Chlamtac, A. Fumagalli, and G. Wedzinga. Slot routing as a solution for optically transparent scalable WDM wide area networks. *Photonic Network Communications*, 1(9):9–21, 1999.

70. M. Maier. *Optical switching networks*. Cambridge University Press, 2008.

71. J. Fang, W. He, and A. K. Somani. Optimal light trail design in WDM optical networks. In *IEEE ICC*, pages 1699–1703, Paris, June 2004.

72. M. Maier and M. Reisslein. AWG-based metro WDM networking. *IEEE Communications Magazine*, 42(11):S19–S26, Nov. 2004.

73. S. Bjornstad, M. Nord, D. R. Hjelme, and N. Stol. Optical burst and packet switching: node and network design, contention resolution and quality of service. In *IEEE International Conference on Telecommunications (ConTEL2003)*, Zagreb, Croatia, June 2003.

74. H. S. Yang, M. Herzog, M. Maier, and M. Reisslein. Metro WDM networks: performance comparison of slotted ring and AWG star networks. *IEEE Journal on Selected Areas in Communications*, 22(8):1460–1473, Oct. 2004.

75. A. Aggarwal, A. Bar-Noy, D. Coppersmith, R. Ramaswami, B. Schieber, and M. Sudan. Efficient routing in optical networks. *Journal of the ACM*, 43(6):973–1001, 1996.

76. H. Y. Tyan, J. C. Hou, B. Wang, and C. C. Han. On supporting temporal quality of service in WDMA-based star-coupled optical networks. *IEEE Transactions on Computers*, 50(3):197–214, 2001.

77. P. Sarigiannidis, G. Papadimitriou, and A. Pomportsis. CS-POSA: a high performance scheduling algorithm for WDM star networks. *Photonic Network Communications*, 11(2):211–227, 2006.

78. C. Fan, S. Adams, and M. Reisslein. The FT-FR AWG network: a practical single-hop metro WDM network for efficient uni- and multicasting. *IEEE Journal of Lightwave Technology*, 23(3):937–954, 2005.

79. M. Jin and O. Yang. APOSN: operation, modeling and performance evaluation. *Computer Networks*, 51(6):1643–1659, 2007.

80. C. Peng, G. v. Bochmann, and T. J. Hall. Quick Birkhoff–von Neumann decomposition algorithm for agile all-photonic network cores. In *IEEE ICC*, Istanbul, Turkey, 2006.

81. A. G. Rahbar and O. Yang. An integrated TDM architecture for AAPN networks. In *SPIE Photonics North*, volume 5970, Toronto, Canada, Sept. 2005.

82. A. G. Rahbar. EBvN: efficient BvN in multi-fiber/multi-wavelength overlaid-star optical networks. *Springer Journal of Annals of Telecommunications*, 67(11-12):575–588, Nov. 2012.

83. A. G. Rahbar and O. Yang. Agile bandwidth management techniques in slotted all-optical packet interconnection networks. *Elsevier Computer Networks*, 54(3):387–403, Feb. 2010.

84. N. Saberi and M. Coates. Scheduling in overlaid star all-photonic networks with large propagation delays. *Photonic Network Communications*, 17(2):157–169, 2009.

85. M. Jin and O. Yang. A TDM solution for all-photonic overlaid-star networks. In *CISS2006*, pages 1691–1695, Princeton, N. J., 2006.

86. L. Mason, A. Vinokurov, N. Zhao, and D. Plant. Topological design and dimensioning of agile all-photonic networks. *Computer Networks*, 50(2):268–287, Feb. 2006.

87. S. A. Paredes, G. Bochmann, and T. J. Hall. Deploying agile photonic networks over reconfigurable optical networks. In *IEEE Symposium on Computers and Communications (ISCC)*, pages 182–187, Sousse, Tunisia, 2009.

88. F. J. Blouin, A. W. Lee, A. J. M. Lee, and M. Beshai. Comparison of two optical-core networks. *Journal of Optical Networking*, 1(1):56–65, Jan. 2002.

89. A. Carena, V. D. Feo, J. M. Finochietto, R. Gaudino, F. Neri, C. Piglione, and P. Poggiolini. Ringo: an experimental WDM optical packet network for metro applications. *IEEE Journal on Selected Areas in Communications*, 22(8):1561 – 1571, Oct. 2004.

90. I. M. White, M. S. Rogge, K. Shrikhande, and L. Kazovsky. A summary of the HORNET project: A next-generation metropolitan area network. *IEEE Journal on Selected Areas in Communications*, 21(9):1478–1494, Nov. 2003.

91. L. F. K. Lui, L. Xu, P. K. A. Wai, L. Y. Chan, C. C. Lee, H. Y. Tam, and M. S. Demokan. 1*4 all-optical packet switch with all-optical header processing. In *IEEE Optical Fiber Communication Conference*, California, Mar. 2005.

92. N. Calabretta, M. Presi, G. Contestabile, and E. Ciaramell. Compact all-optical header processing for DPSK packets. In *European Conference on Optical Communications*, 2006.

93. P. K. AWai, L. Y. Chan, L. F. K. Lui, L. Xu, H. Y. Tam, and M. S. Demokan. All-optical header processing using control signals generated by direct modulation of a DFB laser. *Optics Communications*, 242(1-3):155–161, Nov. 2004.

94. J. P. Wang, B. S. Robinson, S. A. Hamilton, and E. P. Ippen. 40-gbit/s all-optical header processing for packet routing. In *Optical Fiber Communication Conference*, 2006.

95. A. G. Rahbar. Impairment-aware merit-based scheduling in QoS-capable multi-fiber OPS networks. *Journal of Optical Communications*, 33(2):103–121, June 2012.

96. A. Pattavina. Architectures and performance of optical packet switching nodes for IP networks. *IEEE/OSA Journal of Lightwave Technology*, 23(3):1023–1032, Mar. 2005.

97. B. Bostica, M. Burzio, P. Gambini, and L. Zucchelli. *Synchronization issues in optical packet switched networks*, pages 362–376. Springer-Verlag, 1997.

98. J. R. Feehrer and L. H. Ramfelt. Packet synchronization for synchronous optical deflection-routed interconnection networks. *IEEE Transactions on Parallel and Distributed Systems*, 7(6):605–611, 1996.

99. N. Brownlee and K. C. Claffy. Understanding Internet traffic streams: dragonflies and tortoises. *IEEE Communications Magazine*, 40(10):110–117, 2002.

100. V. Jacobson, K. Nichols, and K. Poduri. An expedited forwarding PHB. Technical report, RFC 2598, 1999.

101. J. Heinanen, F. Baker, W. Weiss, and J. Wroclawski. Assured forwarding PHB group. Technical report, RFC 2597, 1999.

102. H. Elbiaze, T. Chahed, T. Atmaca, and G. Hhuterne. Shaping self-similar traffic at access of optical network. *Performance Evaluation*, 53(3-4):187–208, 2003.

103. V. Sivaraman, D. Moreland, and D. Ostry. A novel delay-bounded traffic conditioner for optical edge switches. In *IEEE Workshop on High Performance Switching and Routing (HPSR)*, Hong Kong, May 2005.

104. A. G. Rahbar and O. Yang. Distribution-based bandwidth access scheme in slotted all-optical packet-switched networks. *Elsevier Computer Networks*, 53(5):744–758, Apr. 2009.

105. A. G. Rahbar. Improving throughput of long-hop TCP connections in IP over OPS networks. *Springer Photonic Network Communications*, 17(3):226–237, June 2009.

106. A. Shacham and K. Bergman. An experimental validation of a wavelength-striped, packet switched, optical interconnection network. *Journal of Lightwave Technology*, 27(7):841–850, Apr. 2009.

107. C. P. Lai and K. Bergman. Network architecture and test-bed demonstration of wavelength-striped packet multicasting. In *OFC*, Mar. 2010.

108. Z. Lu, D. K. Hunter, and I. D. Henning. Contention resolution scheme for slotted optical packet switched networks. In *International conference on Optical Network Design and Modelling (ONDM)*, Milan, Italy, Feb. 2005.

109. M. R. Salvador. *MAC protocols for optical packet-switched WDM rings*. PhD thesis, Centre of Telematics and Information Technology, University of Twente, 2003.

110. S. Yao, B. Mukherjee, S. J. B. Yoo, and S. Dixit. A unified study of contention-resolution schemes in optical packet-switched networks. *IEEE Journal of Lightwave Technology*, 21(3):672, 2003.

111. H. Øverby. How the packet length distribution influences the packet loss rate in an optical packet switch. In *Advanced International Conference on Telecommunications and International Conference on Internet and Web Applications and Services (AICT/ICIW 2006)*, 2006.

112. What is quality of service (QoS). http://www.igi-global.com/dictionary/quality-of-service-qos/24296. Accessed: 2014-04-30.

113. A. Kavitha. Performance of optical networks: a short survey. *International Journal of Engineering Science and Technology (IJEST)*, 4(2):600–605, Feb. 2012.

114. W. Wei, Q. Zeng, Y. Ouyang, and D. Lomone. Differentiated integrated QoS control in the optical Internet. *IEEE Communications Magazine*, 42(11):S27 – S34, Nov. 2004.

115. N. Golmie, T. D. Ndousse, and D. H. Sue. A differentiated optical services model for WDM networks. *IEEE Communications Magazine*, 38(2):68–73, Feb. 2000.

116. H. Øverby, N. Stol, and M. Nord. Evaluation of QoS differentiation mechanisms in asynchronous bufferless optical packet-switched networks. *IEEE Communications Magazine*, 44(8):52–57, 2006.

117. P. Ellerman. Calculating reliability using FIT & MTTF: Arrhenius HTOL model. Technical report, Microsemi, 2012.

118. R. Jain, D. M. Chiu, and W. Hawe. "a quantitative measure of fairness and discrimination for resource allocation in shared computer systems". DEC research report tr-301. Technical report, Digital Equipment Corporation, 1984.

119. H. Dahmouni, A. Girard, and B. Sans. An analytical model for jitter in IP networks. *Annals of Telecommunications*, 67:81–90, 2012.

120. N. Olifer and V. Olifer. *Computer Networks: Principles, Technologies and Protocols for Network Design*. John Wiley & Sons, 2006.

121. L. F. Mollenauer and J. P. Gordon. *Solitons in Optical Fibers: Fundamentals and Applications*. Academic Press, 2006.

122. A. Kaheel, T. Khattab, A. Mohamed, and H. Alnuweiri. Quality-of-service mechanisms in IP-over WDM networks. *IEEE Communications Magazine*, 40(12):38–43, Dec. 2002.

123. H. Øverby and N. Stol. Quality of service in asynchronous buffer-less optical packet switched networks. *Telecommunication Systems*, 27(2-4):151–179, Oct.–Dec. 2004.

124. D. Careglio, J. Sol-Pareta, and S. Spadaro. Novel contention resolution technique for QoS support in connection-oriented optical packet switching. In *IEEE ICC*, Seoul, Korea, May 2005.

125. R. Braden, D. Clark, and S. Shenker. Integrated services in the Internet architecture: an overview. Technical report, Internet informational RFC 1633, 1994.

126. S. Blake, D. Black, M. Carlson, E. Davies, Z. Wang, and W. Weiss. An architecture for differentiated services. Technical report, RFC 2475, 1998.

127. A. Striegel and G. Manimaran. Packet scheduling with delay and loss differentiation. *Computer Communications*, 25(1):21–31, Jan. 2002.

128. P. Siripongwutikorn, S. Banerjee, and D. Tipper. A survey of adaptive bandwidth control algorithms. *IEEE Communications Surveys and*, 5(1):2–14., 2003.

129. A. Habib, S. Fahmy, and B. Bhargava. Monitoring and controlling QoS network domains. *ACM/Wiley International Journal of Network Management*, 15(1):11–29, Jan. 2005.

130. W. Zhao, D. Olshefski, and H. Schulzrinne. Internet quality of service: An overview. Technical report, Columbia University, 2001.

131. Q. Zhang, V. M. Vokkarane, B. Chen, and J. P. Jue. Early drop scheme for providing absolute QoS differentiation in optical burst-switched networks. In *High Performance Switching and Routing (HPSR)*, pages 153– 157, Torino, Italy, 2003.

132. Y. Chen, C. Qiao, M. Hamdi, and D. H. K. Tsang. Proportional differentiation: a scalable QoS approach. *IEEE Communications Magazine*, 41:52–58, 2003.

133. J. Heinanen, F. Baker, W. Weiss, and J. Wroclawski. AF-forwarding. http://tools.ietf.org/html/rfc2597, Jun 1999. RFC2597.

134. D. K. Hunter, M. C. Chia, and I. Andonovic. Buffering in optical packet switching. *IEEE/OSA Journal of Lightwave Technology*, 16(10):2081–2094, Dec. 1998.

135. F. Callegati, W. Cerroni, G. Corazza, and C. Raffaelli. Optical packet switching: a network perspective. In *IEEE Conference on Optical Fiber communication/National Fiber Optic Engineers Conference*, San Diego, CA, Feb. 2008.

136. R. Martinez, C. Pinart, F. Cugini, N. Andriolli, L. Vakarenghi, P. Castoldi, L. Wosinska, J. Comellas, and G. Junyent. Challenges and requirements for introducing impairment-awareness into the management and control planes of ASON/GMPLS WDM networks. *IEEE Communications Magazine*, 44(12):76–85, Dec. 2006.

137. G. Agrawal. *Fiber-Optic Communication Systems*. 3rd edition, John Wiley & Sons, 2002.

138. K. Chen and W. Fawaz. *Optical Networks: New Challenges and Paradigms for Quality of Service*, chapter 5, pages 115–134. John Wiley & Sons, 2009.

139. X. Xiao, A. Hannan, B. Bailey, and L. M. Ni. Traffic engineering with MPLS in the Internet. *IEEE Network*, 14(2):28–33, Aug. 2000.

140. A. Giorgetti, N. Sambo, I. Cerutti, N. Andriolli, and P. Castoldi. Suggested vector scheme with crankback mechanism in GMPLS-controlled optical networks. In *Optical Network Design Modeling (ONDM)*, Kyoto, Japan, 2010.

141. B. Guo, J. Li, Y. Wang, S. Huang, Z. Chen, and Y. He. Collision-aware routing and spectrum assignment in GMPLS-enabled flexible-bandwidth optical network. *IEEE Journal of Optical Communications and Networking*, 5(6):658–666, 2013.

142. J. V. Gontn, P. P. Mario, J. V. Alonso, and J. G. Haro. A feasibility study of GMPLS extensions for synchronous slotted optical packet switching networks. In *The V Workshop in G/MPLS Networks*, pages 89–100, Girona, Spain, 2006.

143. G. Zhang, M. D. Leenheer, A. Morea, and B. Mukherjee. A survey on OFDM-based elastic core optical networking. *IEEE Communications Surveys and Tutorials Journal*, 15(1):65–87, FIRST QUARTER 2013.

144. K. Christodoulopoulos, I. Tomkos, and E. Varvarigos. Dynamic bandwidth allocation in flexible OFDM-based networks. In *IEEE Optical Fiber Communication Conference*, Los Angeles, California, Mar. 2011.

145. J. Armstrong. OFDM for optical communications. *IEEE Journal of Lightwave Technology*, 27(3):189–204, Feb. 2009.

146. O. Gerstel, M. Jinno, A. Lord, and S. J. Ben Yoo. Elastic optical networking: a new dawn for the optical layer. *IEEE Communications Magazine*, 50(2):s12–s20, Feb. 2012.

147. M. Jinno, B. Kozicki, H. Takara, A. Watanabe, W. Imajuku, Y. Sone, T. Tanaka, and A. Hirano. Distance-adaptive spectrum resource allocation in spectrum-sliced elastic optical path network. *IEEE Communications Magazine*, 48(8):138–145, Aug. 2010.

148. M. Jinno, H. Takara, B. Kozicki, Y. Tsukishima, Y. Sone, and S. Matsuoka. Spectrum-efficient and scalable elastic optical path network: architecture, benefits, and enabling technologies. *IEEE Communications Magazine*, 47(11):66–73, Nov. 2009.

149. H. Beyranvand and J. A. Salehi. A quality-of-transmission aware dynamic routing and spectrum assignment scheme for future elastic optical networks. *Journal of Lightwave Technology*, 31(18):3043–3054, Sept. 2013.

150. X. Wan, N. Hua, H. Zhang, and X. Zheng. Study on dynamic routing and spectrum assignment in bitrate flexible optical networks. *Photonic Network Communications*, 24:219–227, 2012.

151. M. Klinkowski and K. Walkowiak. Routing and spectrum assignment in spectrum sliced elastic optical path network. *IEEE Communications Letters*, 15:884–886, 2011.

152. S. Yang and F. Kuipers. Impairment-aware routing in translucent spectrum-sliced elastic optical path networks. In *17th European Conference on Networks and Optical Communication (NOC)*, June 2012.

153. S. Talebi, F. Alam, I. Katib, M. Khamis, R. Salama, and G. N. Rouskas. Spectrum management techniques for elastic optical networks: a survey. *Optical Switching and Networking*, 13:34–48, Feb. 2014.

154. K. Christodoulopoulos, I. Tomkos, and E. Varvarigos. Spectrally/bitrate flexible optical network planning. In *36th European Conference and Exhibition on Optical Communication (ECOC)*, 2010.

155. K. Christodoulopoulos, I. Tomkos, and E. Varvarigos. Elastic bandwidth allocation in flexible OFDM-based optical networks. *IEEE Journal of Lightwave Technology*, 29:1354–1366, 2011.

156. Y. Wang, X. Cao, Q. Hu, and Y. Pan. Towards elastic and fine-granular bandwidth allocation in spectrum-sliced optical networks. *IEEE Optical Communications and Networking*, 4(11):906–917, Nov. 2012.

157. W. Wei, C. Wang, and X. Liu. Adaptive IP/Optical OFDM networking design. In *IEEE Optical Fiber Communication (OFC)*, San Diego, CA, Mar. 2010.

158. Y. Yoshida, T. Kodama, S. Shinada, N. Wada, and K. Kitayama. Fixed-length elastic-capacity OFDM payload packet: Concept and demonstration. In *IEEE OFC/NFOEC*, 2013.

159. W. Wei, J. Hu, D. Qian, P. N. Ji, T. Wang, X. Liu, and C. Qiao. PONIARD: A programmable optical networking infrastructure for advanced research and development of future Internet. *IEEE Journal of Lightwave Technology*, 27(3):233–242, Feb. 2009.

160. W. Wei, C. Wang, J. Yu, N. Cvijetic, and T. Wang. Optical orthogonal frequency division multiple access networking for the future Internet. *IEEE Journal of Optical Communications and Networking*, 1:236–246, 2009.

161. W. Wei, J. Hu, C. Wang, T. Wang, and C. Qiao. A programmable router interface supporting link virtualization with adaptive Optical OFDMA transmission. In *IEEE OFC/NFOEC*, 2009.

162. W. Wei, C. Wang, and J. Yu. Cognitive optical networks: Key drivers, enabling techniques, and adaptive bandwidth services. *IEEE Communications Magazine*, 50:106–113, 2012.

CHAPTER 2

CONTENTION AVOIDANCE IN OPS NETWORKS

As discussed in Chapter 1, optical packet contention is the major problem for an OPS network. Contention resolution, contention avoidance, and their combinations are the schemes to manage the contention problem. Recall that a resolution scheme (as a reactive scheme used in an OPS switch) resolves the collisions that occurred in OPS switches, while an avoidance scheme (as a proactive scheme used in ingress switches) tries to send optical packets to the OPS network in a way to reduce the number of potential optical packet collisions in OPS switches. In this chapter, contention avoidance schemes proposed for both asynchronous OPS and synchronous (slotted) OPS networks are studied. A number of hardware-based and software-based schemes have been proposed to reduce the number of collision events in an OPS network. The objective of this chapter is to study the contention avoidance schemes that would minimize optical Packet Loss Rate (PLR), thus improving the throughput of OPS networks. Note that contention avoidance is a technique for providing QoS indirectly since it reduces traffic loss rate and improves the performance of the network (see 1.4.6.2.1). The contention avoidance schemes proposed for supporting QoS differentiation in QoS-capable OPS networks are also described with the objective of lowering the PLR of HP optical packets.

Quality of Service in Optical Packet Switched Networks, First Edition.
By Akbar Ghaffarpour Rahbar Copyright © 2015 IEEE. Published by John Wiley & Sons, Inc.

In practice, both contention resolution and avoidance schemes should be used in an OPS network. Due to the fact that optical buffering is bulky and wavelength conversion is expensive, some contention avoidance mechanisms with upper layer management must be utilized in the OPS network in order to decrease the burden of contention resolution resources required in OPS switches. In this case, network throughput can be improved efficiently.

It should be noted that there is a tradeoff between lowering PLR and network complexity or network performance parameters. For instance, reducing PLR may increase the complexity of edge switches, and therefore the cost of OPS network may increase. Other performance metrics may also be degraded such as increasing client packets delay, jitter, and loss in ingress switches.

2.1 Software-Based Contention Avoidance Schemes

In this section, software-based contention avoidance schemes are reviewed in an OPS network. These avoidance schemes have a number of features:

- They are implemented in ingress switches. However, in some cases, the OPS network may provide feedback on some performance parameters to them for a proper operation.

- Since these schemes are algorithmic-based schemes, they may only increase the processing burden of ingress switches.

- They impose no additional hardware cost on the network implementation cost. In other words, using software-based contention avoidance schemes, the probability of contention decreases without requirement for employing additional hardware in the network.

In short, the software-based contention avoidance schemes are some kinds of bandwidth access schemes that involve how to transmit optical packets in an ingress switch to an OPS network in order to reduce the amount of contention in OPS switches.

2.1.1 Single-Class Traffic

In this section, software-based contention avoidance schemes proposed for asynchronous OPS and slotted-OPS networks with single-class traffic are studied. The edge switch architecture depicted in Section 1.4.3 is used for single-class traffic.

2.1.1.1 Packet Transmission Based on the Scheduling of Empty Time Slots
This technique [1–3] is proposed for single-hop slotted-OPS networks with an OPS switch in center (see Section 1.4.1). In this technique, ingress switches connected to a given OPS switch are weakly coordinated. Each ingress switch sends its optical packets on a given wavelength channel in such a way that the variance of the number

of optical packets arriving at the OPS switch within any time slot is minimized. For example, consider that 10 edge switches are connected to the OPS switch and each one has an average traffic load of 0.65. Each ingress switch must send its optical packets to the OPS switch in a way to have either six or seven optical packets at each time slot arriving at the input ports of the OPS switch. Since the variance of optical packets arrivals is reduced, this scheme can lead to a lower PLR than the case that the variance of optical packets arrival to the OPS switch is high. This scheme is much more effective at low traffic loads, and the effectiveness of this technique is reduced by increasing network traffic load. This technique is easy to implement in ingress switches with electronic buffers, thus suitable for a single-hop multi-fiber OPS metro network. This technique will be detailed in Chapter 6.

To implement this technique in a multi-hop OPS network, ingress switches connected to a given OPS core switch must use optical buffers in order to save optical packets and then forward them to the given OPS core switch at scheduled time slots. The requirement to use optical buffers in multi-hop networks is a disadvantage for this technique.

2.1.1.2 Symmetric Traffic Transmission   In a synchronized OPS, it is proven that the smallest PLR in a given OPS switch s is achieved when the distribution of optical packets traversing OPS switch s is symmetric to all output links of OPS switch s [3, 4]. In other words, if OPS switch s has N output links, the probability of switching an optical packet to every output link of s is $\frac{1}{N}$. Otherwise, when traffic distribution to the output links of s is asymmetric, PLR becomes high.

Consider that an edge switch (as an ingress switch) is connected to OPS switch s. Under the symmetric traffic transmission technique, traffic arrival at different output links of s should be in a symmetric manner; i.e., with the same probability. The approach for symmetric optical packet transmission can be achieved by sending almost equal numbers of optical packets to each output link of OPS switch s within a specific time interval with size T seconds. In other words, within interval T, the number of optical packets switched to each output link of OPS switch s should almost be equal.

The operation of symmetric traffic transmission is the same for both single-fiber and multi-fiber OPS networks since what matters is the probability of $\frac{1}{N}$ in which N is the number of output links. In addition, the proposed scheme can be used for both slotted OPS and asynchronous OPS networks.

In a single-hop OPS network (see Section 1.4.1), symmetric traffic transmission can be easily provided by each ingress switch that uses electronic buffers, where the ingress switch sends symmetric traffic to the single OPS switch in the OPS network using the mentioned approach. Recall that in a single-hop OPS network, there is one OPS switch in which each one of input/output ports is connected to an edge switch. Each ingress switch maintains only $N-1$ electronic buffers for saving the traffic destined to $N-1$ egress switches, excluding itself. An ingress switch connected to OPS switch s sends its optical packets in a round-robin fashion from its $N-1$ electronic buffers on the available wavelength channels toward different egress switches.

Now, let us see what happens if ingress switches send their optical packets to egress switches of a single-hop slotted OPS network with one OPS switch in an asymmetric manner. Consider fixed-length client packets with Poisson inter-arrival times arrive at ingress switches, and each optical packet carries only one client packet. Consider three scenarios for core switch size at $L = 1.0$:

- $N = 10$: Let distribution probabilities of optical packets to egress switches be (0.43, 0.01, 0.04, 0.09, 0.01, 0.15, 0.15, 0.01, 0.08, 0.03). In this case, we find PLR of 0.533614.

- $N = 20$: Assume distribution probabilities of optical packets to egress switches to be (0.01, 0.03, 0.07, 0.01, 0.01, 0.12, 0.01, 0.02, 0.05, 0.01, 0.14, 0.09, 0.01, 0.12, 0.02, 0.01, 0.15, 0.01, 0.08, 0.03). In this scenario, network experiences PLR of 0.507164.

- $N = 30$: Consider distribution probabilities of optical packets to egress switches to be (0.01, 0.01, 0.01, 0.01, 0.07, 0.01, 0.01, 0.01, 0.10, 0.01, 0.01, 0.02, 0.05, 0.01, 0.11, 0.01, 0.01, 0.01, 0.09, 0.01, 0.03, 0.04, 0.04, 0.01, 0.02, 0.01, 0.15, 0.01, 0.08, 0.03). Here, we find PLR of 0.520263.

It is clear to see that the asymmetry of distribution of optical packets to different egress switches is very high and this is why high PLRs are observed. One can observe that the asymmetry rate is higher at $N = 10$ than at $N = 20$ and this is why for $N = 20$ we have a lower PLR than for $N = 10$. Note that the maximum PLR for symmetric traffic distribution at $L = 1.0$ is only 36.7% [3, 5].

On the other hand, in a multi-hop network, the ingress switch connected to OPS switch s has N electronic buffers for saving the traffic destined to N output links of OPS switch s. Here, the ingress switch must monitor transit optical packets traversing through s and adjust its transmission based on different optical packets switched to different output links of s. For example, if the number of transit optical packets destined to output link p of OPS switch s is too much within interval T, the ingress switch should reduce transmission of its optical packets to output link p. This can provide symmetric traffic transmission to some extent in OPS switches, but not as good as the single-hop OPS networks.

2.1.1.3 Load-Balanced Traffic Transmission Balanced traffic transmission in OPS networks can be grouped as balancing on wavelength channels and balancing on routing operations as follows.

2.1.1.3.1 Balancing on Wavelength Channels Here [3, 4], traffic on each connection link between either an edge switch to an OPS switch or an OPS switch to another OPS switch is transmitted in a balanced manner so that no wavelength is overwhelmed. In this technique, the controller of an ingress switch (both edge switch and OPS switch) connected to a given OPS switch s balances its traffic load by distributing optical packets uniformly among the wavelengths/fibers on its output link that connects the ingress switch to OPS switch s. Since the traffic is balanced on all wavelength channels of a fiber, it is expected to have the same PLR on all wavelengths. Balanced optical packet transmission is so simple; it just needs to uniformly select the wavelengths on a fiber. Note that in a multi-fiber OPS network, besides uniformly selection of wavelengths for transmission of optical packets, optical fibers must also be chosen uniformly. Note that uniform selection from a list with m items means selecting an item randomly with probability $\frac{1}{m}$. This technique is true for both slotted OPS and asynchronous OPS.

The PLR for a balanced traffic load distribution is less than an unbalanced traffic load distribution [3]. Consider two scenarios for optical packet transmission in a single-hop slotted OPS network (see Section 1.4.1) with one OPS switch with size 50 * 50 at $L = 0.7$ and $W = 4$. Fixed-length client packets with Poisson inter-arrival times arrive at ingress switches, and each optical packet carries only one client packet. Under unbalanced optical packet transmission, each ingress switch sends its optical packets with unequal loads of 1.0, 0.4, 0.6 and 0.8, respectively, on λ_0, λ_1, λ_2, and λ_3. For example, here the ingress switch sends its optical packets with probability of 0.8 on time slots of wavelength λ_3. The simulation results show that we have PLR of 36.42%, 17.31%, 24.48%, and 30.8% on these four wavelengths, respectively, and the average PLR in OPS switch is 29.53%. On the other hand, under the scenario of balanced optical packet transmission on wavelength channels, each ingress switch distributes its traffic on the four wavelengths in a balanced manner (i.e., on each wavelength with load 0.7). In this case, the average PLR on all four wavelength channels and the OPS switch is 27.73%. This means that the balanced optical packet transmission leads to a lower PLR than does the unbalanced optical packet transmission.

2.1.1.3.2 Balancing on Routing Operation In [6], adaptive routing schemes are studied in asynchronous connection-oriented OPS networks under MPLS traffic control (see 1.4.6.3.1). Under MPLS, client packets are classified into a number of Forwarding Equivalent Classes (FECs) based on identification address and quality of service requirements in MPLS provider ingress routers where this role is played with ingress switches in OPS networks. An FEC is identified by a label added to the client packets. Edge switches set up unidirectional connections, called Label-Switched Paths (LSPs), throughout the OPS network. The client packets belonging to the same FEC are transferred from a source to a destination along the LSP that corresponds to their labels. Each OPS switch (as an MPLS core router) performs a simple label-matching operation on a pre-computed LSP forwarding table, thus speeding up the forwarding operation. Since the MPLS operation returns back to both edge and core switches, this technique has been studied in this section.

The dynamic path selection balances the traffic load inside the OPS network, which in turn reduces PLR and improves network performance characteristics compared with the shortest path routing scheme. Consider that optical buffers (as queues) are used at every output port of any OPS switch. Assume that each ingress switch knows the congestion status of every OPS switch in the network at every time. Three routing techniques are proposed for establishing an LSP between a source–destination pair among k available paths:

- **Path Excluding**: Under path excluding, each node always picks the less congested output port among all the ports included in the set of available paths between the source–destination pair. This choice determines the next hop and excludes from the set of available paths all those paths that do not include this hop in their routes. In other words, from k original paths, each node removes some paths to obtain only one path. For example, consider that there are three paths A-D-E, A-B-C-E, and A-D-C-E between source node A and destination node E. If the less congested queue of A is on the output port toward node B, then node A definitely selects A-B-C-E, excluding the other paths. Then, the rest of the nodes in this chosen path cannot take other routing decisions.

- **Multiple Choice**: Here, for every optical packet, each intermediate OPS switch chooses the best path between k available paths, with the objective of avoiding loops. This not only can decrease global network load, but also can reduce end-to-end delay of optical packets. For example, assume there are three paths A-D-E, A-B-C-E, and A-D-C-E between source node A and destination node E. Consider node A picks node D because the output port to D has the less congested queue. In node D, there are three paths: D-E, D-C-E, and D-F-E. Now, node D selects the best path among them with the least congested queue (where there is one queue for each output port).

- **By-Pass (BP)**: Under BP, an ingress switch selects a single path for each optical packet among k available paths. The path can only be changed when the traveling optical packet experiences a congested link on its way in an OPS switch. In this case, the OPS switch attempts to by-pass the optical packet using an intermediate OPS switch to reach the next hop. Assume that node A sends an optical packet to destination node E through intermediate node D (i.e., the path is A-D-E). Assume that when the optical packet arrives at node D, no resources are available to reach node E. Then, node D finds two by-pass paths in its forwarding table, say D-C-E and D-F-E. Node D selects the one with the less congested queue.

The mentioned routing algorithms not only can solve the contention in space domain, but also can balance the traffic load inside the OPS network. Performance evaluation results show that BP is the best choice for reducing PLR. In addition, after the shortest path routing algorithm, the BP is the best algorithm for finding the routes with small number of hops.

2.1.1.4 *Reservation-Based OPS*   Here, the reservation-based contention avoidance schemes that try to provide loss-less optical packet switching are introduced.

2.1.1.4.1 *Reservation in Single-Hop OPS*   To remove contention completely, this technique relies on bandwidth reservation in OPS networks. This reservation-based scheme is implemented in a single-hop slotted OPS network (see Section 1.4.1) with n edge switches connected to a core switch. This scheme is executed periodically at the OPS switch of the star network to recompute a loss-free schedule in

response to the traffic demands of edges switches. Then, ingress switches send their traffic following the predetermined schedule [7–9].

This reservation scheme for a single-hop OPS network can use the idea of switching in Input-Queued (IQ) switches [10–13]. In this network, the input buffers are shifted to ingress switches to benefit from electronic buffers. Thus, the central scheduler in the OPS switch schedules a delayed version of the traffic for the ingress switches.

The scheduling in input-queued switches can be mapped to the bipartite matching problem. Building a perfect (maximum size) matching [10, 14] costs a complexity of $O(n^{2.5})$ [15]. To reduce this complexity, a number of heuristic (maximal size) matching algorithms [11, 16] have been proposed, but at the expense of bandwidth wastage because they do not obtain a perfect matching.

The scheduling in input-queued switches can be either slot-by-slot or frame-based. In the former, the scheduling is performed in a time slot level. The slot-by-slot scheduling is used in [7, 8] to perform the scheduling in a single-hop all-optical OPS network, where the time slots at the output ports of the OPS switch are reserved based on the signaling requests from the ingress switches. These algorithms account for the propagation delay because the buffers are remotely located at the ingress switches. The slot-by-slot design is not suitable for a WAN topology because of the performance degradation due to long distance between an edge switch and an OPS switch [7]. The work in [8] uses a multi-processor architecture to overcome the complexity of the scheduling algorithm, but a multi-processor system is expensive.

In the frame-based scheduling, the scheduling algorithm is run once in a time frame (a frame is a number of time slots) to obtain the OPS switch configuration in each time slot inside the frame. A frame-based scheduling can decrease the frequency of matching computation compared with the slot-by-slot algorithms [17, 18]. In addition, the frame-based scheduling can reduce the communication overhead during scheduling [19] and can accommodate more scheduled time slots within a frame period of F time slots than F times of slot-by-slot scheduling [20]. Furthermore, the frame length can be tailored by the network manager based on the running time of the scheduling algorithm [21]. Finally, it decouples the time constraints of the control plane and the data plane [21]. However, a frame-based scheduling may suffer from a larger delay and a larger jitter [17, 21]. In the frame-based scheduling based on the Birkhoff and von Neumann (BvN) decomposition algorithm [10, 14, 22], a traffic matrix sent from ingress switches to the OPS switch can be scheduled in a frame by using the matrix decomposition algorithm. Since it is a maximum matching algorithm, it guarantees to schedule the demanded traffic. However, it has a computation complexity of $O(n^{4.5})$ and the memory complexity of $O(n^3 \times \log(n))$.

Borrowing from the permutation-based scheduling algorithms [10, 14], a matrix decomposition algorithm can be applied to the traffic demand matrix sent from the ingress switches to the OPS switch to form a frame-based scheduling. After collecting traffic demands from all ingress switches at the OPS switch, the time-slot assignment process is simultaneously determined for all ingress switches. To perform the scheduling, the traffic matrix must first be converted to a doubly stochastic matrix because the traffic matrix is decomposable if and only if it is doubly stochastic

[12]. The lowest complexity for this conversion is $O(n^2)$ [12] where n is the number of edge switches. By using the Birkhoff and von Neumann (BvN) theorems [14], a doubly stochastic matrix is converted to the permutation matrices with $O(n^2)$ iterations. For each iteration, a perfect matching with the complexity of $O(n^{2.5})$ using the maximum flow algorithm [15] is carried out, leading to total complexity of $O(n^{4.5})$.

After determining the scheduling, the information of the new traffic scheduling is sent back to ingress switches and they arrive at ingress switches after two times the propagation delay plus some scheduling time. After the new schedule arrives at an ingress switch, the ingress switch must wait for some synchronization period with the furthest edge switch from the OPS switch. This may drastically increase the queuing delay in the edge switches, which is not appropriate for real-time applications. The reservation-based OPS technique will be fully detailed in Chapter 6.

2.1.1.4.2 Network Global Control (NGC)

This technique [23] has been proposed for contention avoidance in which traffic of all edge switches is monitored in a central core switch (called central node), and therefore a better contention avoidance can be achieved. In this technique, each edge switch sends its traffic information through a (low bandwidth) control channel to a central node located close to the topological and geographical center of the network. After collecting traffic information of all edge switches, the central node schedules traffic transmission of each ingress switch in a way to reduce collision of optical packets in the OPS network. Then, the scheduling information is returned back to each ingress switch through the control channel. Each ingress switch should obey its scheduling to transmit its optical packets. Note that the central node is equipped with some powerful processors to calculate the required scheduling. Since the failure of the central node is an important drawback of these schemes, backup central nodes should be provisioned in order to increase the reliability of NGC.

In NGC, a service traffic between any two given edge switches is called a torrent. The NGC makes traffic periodic. There is a Global Period (GP) time T, where T is the average transmission time between two client packets of a torrent among every ingress–egress switches pair in the network. Suppose that each torrent has at most K client packets to send with rate B packets/second. Then, $U = B \times T$ shows the number of client packets to be transmitted within a GP. Let V denote the number of time slots per GP, where each time slot can carry only one client packet as an optical packet.

Each ingress switch monitors its traffic within 10 GPs and collects the information of each torrent such as the number of packets a torrent has to send. Every 10 GPs, each ingress switch sends its control information (including the volume of traffic to be sent to each specific egress switch) to the central node using the control channel. Then, the central node runs a time-Slot Assignment Algorithm (SAA) to schedule time slots to each ingress switch traffic in a way that no collision happens in the OPS network. Note that SAA provides the scheduling only for the first GP, and the schedulings of all the following GPs are the same as the first GP. For a periodic traffic under NGC, if contention is solved within the first GP, it will be automatically solved in the following GPs. After receiving the scheduling information of time slots by an ingress switch, they cannot be used immediately. To avoid the contention between the optical packets with the new time-slot assignment and the optical packets already using the previous time-slot assignment, another Transient time-slot Assignment Algorithm (TAA) is used, where an additional GP is required for transition from the old scheduling to the new one. As an example for SAA, if the first and the third time slots of the first GP are assigned to transmit two client packets of a given torrent between a given ingress–egress switch pair, then in all the following nine GPs, the first and the third time slots are used to transmit the client packets of the torrent. In general, we have the following cases for transmission of a torrent traffic in an ingress switch:

- $K \leq U$: In this case, K time slots S_1, S_2, ..., S_K are chosen to send the traffic within the first GP, where $1 \leq S_1 < S_2 < ... < S_K \leq V$.

- $K = k \times U + U_r$, where $k > 0$ and $0 \leq U_r < U$: In this case, U time slots S_1, S_2, ..., S_U are chosen in the first k GPs, and U_r time slots are chosen in the last GP to send the traffic of the torrent, where $1 \leq S_1 < S_2 < ... < S_U \leq V$.

- $K \gg U$: In this case, the distribution of packets in GPs will have a very strong periodicity.

Implementing NGC keeps PLR at zero because no collision happens in any OPS switch. However, it is not scalable and a network using NGC is not robust because of a single point of failure (i.e., the central node). In addition, for some traffic cases of an ingress switch, the loss-less time-slot assignment could be impossible. In this situation, traffic information of the ingress switch is modified accordingly; i.e., some traffic of the ingress switch cannot be sent and must be buffered in the ingress switch. Moreover, a symmetric network topology (in which the number of input/output ports in all OPS switches are equal, all edge switches have the same distance from each other, and bandwidths of all channels are equal) should be used under NGC. Furthermore, the computational complexities of SAA and TAA are high (with the upper limit of $O(n \times V^2)$, where n is the number of edge switches). On the other hand, all the computations must be performed within a GP. Finally, each central node should use a multiprocessing system (including one main processor and n subprocessors for n edge switches in the network).

Performance evaluation results show that OPS under NGC has a nearly ideal throughput compared with the conventional slotted OPS even equipped with Fiber Delay Line (FDL) buffering. This is because all potential optical packet contentions can be eliminated by SAA. In addition, it is shown that OPS under NGC has the best average end-to-end delay and jitter when traffic load is above $L = 0.7$ compared with the conventional slotted OPS even equipped with FDL buffering. However, below load $L = 0.7$, OPS under NGC provides higher end-to-end delay and jitter than conventional slotted OPS.

2.1.1.5 Renegotiation-Based Transmission of Optical Packets   This technique [24–26] is proposed for slotted OPS. Edge switches learn from the status of contending optical packets and change the position of optical packets transmitted within the frame period in order to reduce loss in future frames. In this technique, each ingress switch keeps one electronic buffer for saving traffic of each egress switch. The ingress switch sends its traffic information to other edge switches in the OPS network so that the edge switches can provide a scheduling with less traffic contention in the core network. Here, traffic is smoothed at each ingress switch using a renegotiated heuristic service model to allow a regulated traffic to arrive in an OPS network. This model considers both the user level and the network level conditions and prevents too much traffic from flowing into the network under a critical network condition.

In this technique, traffic rate prediction and a new renegotiated service model are used for traffic shaping and smoothing traffic at the edge of the network. Here, a stream implies a circuit switch connection that is set up in either direction between each pair of edge switches. Each stream bandwidth is updated at regular intervals (which are not synchronized between different streams) in order to consider both user requirements and network conditions. Renegotiation is a connection-based service. The scheduling and topology information are distributed throughout the network using the Open Shortest Path First (OSPF) protocol at the beginning renegotiation intervals. In addition, it is necessary for each edge switch to be informed of transmission-rate changes in other edge switches. Based upon this information and the bandwidth requested by a user application, each ingress switch can periodically adjust its sending rate. In each renegotiation interval T, an ingress switch computes its transmission rate for its streams according to user traffic arrival and network conditions and distributes it to other edge switches if necessary. During a negotiation interval, the transmission rate remains fixed. For this, a time unit $\tau \in [5, 100]$ms is considered in which traffic arrival from users in the relevant buffer is measured. The renegotiation interval includes $m = \lfloor \frac{T}{\tau} \rfloor$ time units. Define B to be the buffer occupancy, $S_{available}$ to be the available network bandwidth, $S_{threshold}$ to be the threshold rate for network bandwidth below which the critical state happens in which the network resources are scarce, U_i to be the measured arrival rate during time unit i, parameter R_i to be the smoothed arrival rate in the buffer during time unit i, the value of $\frac{B}{\tau}$ to be the additional capacity needed to empty the buffer in the next time unit, v to be the factor that shows to what extent previous experience and to what extent the current measured rate must affect the new service rate, and α to be the factor that shows to what extent the algorithm responds to use traffic requirements. Let S be the current service time of a stream buffer in an ingress switch at current renegotiation interval, where one packet is transmitted in $\gamma \propto \frac{1}{S}$ time slots. Then, the following equations can be developed in a heuristic way:

$$R_{i+1} = v \times R_i + (1 - v) \times \left(U_i + \frac{B}{\tau} \right), \qquad (2.1)$$

with

$$v = \begin{cases} \min\left(\frac{R_i - S}{S_{available}}, 1\right) & \text{if} \quad S_{available} < S_{threshold} \text{ and } R_i > S, \\ \alpha & \text{otherwise.} \end{cases} \quad (2.2)$$

For v, the first condition occurs when the network condition is in the critical state, and therefore the network conditions must be taken into account mostly compared with the user requirements. On the other hand, the second condition for v shows that the network is in the non-critical state, and therefore the user requirements can be taken into account more than the network conditions.

Define S_f to be the next service time of the stream at the next renegotiation interval averaged among m time units, where $S_f = \frac{\sum_{j=1}^{m} R_j}{m}$. If $|S_f - S| > \delta$; then renegotiation is performed for S_f, and the edge switch starts transmitting with the new service rate S_f to other edge switches for necessary rescheduling purposes. Otherwise, the stream keeps S as its service rate.

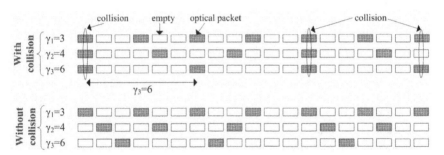

Figure 2.1: An example for smoothing and shifting in renegotiation-based transmission of optical packets

As stated, one optical packet is transmitted in γ time slots for the traffic stream between each ingress–egress switch pair. For example in Fig. 2.1, consider that three streams are multiplexed with $\gamma_1 = 3$, $\gamma_2 = 4$, and $\gamma_3 = 6$. For the first stream, one optical packet is scheduled to be sent at the first time slot of the frame with size three time slots. Similarly, one optical packet is scheduled to be transmitted at the first time slot of the frame with size four time slots for the second stream. In the upper image, as depicted, there are some collisions among these three streams at some time slots in a given link of an intermediate OPS core switch. The intermediate core switch should announce this collision information (through signaling) back to the relevant ingress switches. By rescheduling of traffic transmission in their relevant ingress switches, the lower image shows a collision-less scheduling. Since the first stream in this example has the highest rate, it does not need any change in traffic transmission. In the rescheduling, the scheduling of the second stream is updated (by right shifting from the collision point) to avoid collision with the first stream (as the highest rate stream). In other words, each ingress switch should update its scheduling according to the higher rate streams. Similarly, the scheduling of the third stream (as the lowest rate stream) is updated (by right shifting from the collision point) to avoid collision with the first and the second streams. The rescheduling should occur when a new stream is generated or when the rate of a stream changes significantly. Performance evaluation results show that the proposed technique can reduce PLR in slotted OPS networks, and therefore the volume of required optical buffering, wavelength conversion, and deflection routing can be reduced. The proposed scheme could be suitable for networks with small network diameter; therefore, small number of links between an ingress–egress switch pair. Performance results show that the longer the negotiation interval, the more delay introduced in client traffic, but smaller PLR. Under fixed traffic load, PLR dramatically decreases as the negotiation interval increases, and then the PLR curve flattens out for larger negotiation intervals. On the other hand, the client packets buffering delay performance increases with negotiation interval size. Hence, the negotiation interval size should be chosen in a way to obtain an appropriate tradeoff between delay and PLR.

On the other hand, propagation delay is high in large networks; therefore, out-of-date scheduling information may arrive at ingress switches. This problem can be resolved by taking high values for T such as 0.2 s and higher. However, this may reduce the effectiveness of the proposed technique. Another solution is to partition the large network into some areas, where each area can implement packet scheduling independently using optical buffering used in the OPS switches located in the over-lapping areas. However, this makes a complicated system. Finally, it should also be mentioned that resolving contention on one link may not resolve the contention on other links. Note that the rescheduling may even cause additional contention on other links for streams in the network. In addition, this protocol is proposed for an OPS network using wavelength-insensitive OPS switches, and may be suitable for semi-static traffic where traffic arrival status changes at long intervals.

As stated, this technique requires special signaling among edge switches and core switches to distribute scheduling information in any renegotiation interval and synchronization of scheduled data to local clock in ingress switches. However, it is not always possible to find a free time slot sequence through the path of a stream and have complete knowledge of all network elements with right timing. Therefore, FDLs must be used in core switches to apply delays for contending optical packet sequences, and then send them at appropriate times when wavelength channel becomes available. Since optical packets are buffered in FDLs, they are sent in the FIFO manner that leads to a regulated flow to become bursty and make unevenly spaced stream after passing through multiple core switches.

2.1.1.6 Monitoring of Core Switch Traffic
Consider that an edge switch (with two ingress and egress parts) is directly connected to OPS switch s, where the edge switch is called local edge switch. Define a local optical packet to be the packet either added to OPS switch s by its local ingress switch (as an add packet) or dropped by the OPS switch to its local egress switch (as a drop packet). On the other hand, a transit optical packet is the packet that passes through OPS switch s toward another OPS switch.

Under monitoring of core switch traffic [27], the ingress switch part of the edge switch stores arriving client packets in its electronic transmission buffers. On the other hand, a scheduler located in OPS switch s monitors the status of all wavelengths at all output ports of the OPS switch. When there is no transit optical packet occupying wavelength channel λ_i in output port p, the scheduler immediately informs the ingress switch to transmit an optical packet, or an aggregated optical packet (see Section 2.1.1.8) to OPS switch s on wavelength λ_i that must be switched toward output port p. The same function is performed by all ingress switches in the OPS network.

By this technique, the chance of contention of local optical packets with transit optical packets are either reduced or eliminated in OPS switch s. In this case, the contention resolution resources (such as wavelength converters and optical buffers) used in OPS switch s can be used solely for transit optical packets. Note that in asynchronous OPS, transit optical packets may arrive at any time, and therefore the contention of local optical packets with transit optical packets cannot be fully eliminated. However, in slotted OPS, this contention can be fully eliminated. In slotted OPS, however, monitoring, informing the ingress switch, and making ready the optical packets by the ingress switch must all be carried out within a small fraction of time-slot period in order to have a proper switching.

By this monitoring of optical packets and not sending those optical packets that are likely to cause collision in OPS switch s, the ingress switch reduces the PLR at switch s, thus reducing PLR in the whole network. This is unlike the other schemes that send their optical packets without considering the status of the core switch traffic. Note that in this technique, the ingress switch must be very close to OPS switch s for a proper operation. In addition, each OPS switch needs a complex controller for monitoring traffic information.

2.1.1.7 Virtual Optical Bus (VOB) This technique [28] has been proposed for contention avoidance in OBS networks, but it can be used in OPS networks as well. Here, a node denotes a core switch and its relevant edge switch together. The Virtual unidirectional Optical Bus (VOB) technique forms a cluster out of the nodes that potentially share the same set of connected links in sequence on their path, and therefore their local optical packets may collide with each other. A VOB can accommodate a number of origin–destination (O-D) flows, where an O-D flow is a flow of packets originating at an ingress switch and destined to a specific egress switch. An O-D flow must be associated with only one VOB. In short, a VOB is a directed simple path with a single origin and destination. A VOB can use any wavelength channel on any link. However, the flows associated with a VOB must use only one wavelength channel at any given time.

An insertion buffer is used in each intermediate node on a VOB to actively delay transit optical packets flowing on the VOB in order to prevent them from colliding with the optical packets in transmission by the nodes along the VOB path. By this way, a transit optical packet is held until the transmission of an ongoing local optical packet is completed. This technique needs an insertion buffer per VOB per any intermediate node being part of that VOB. In other words, in a network with n nodes, each node would require at most $n - 1$ insertion buffers since its relevant edge switch can generate traffic destined to at most $n - 1$ other edge switches in the network.

The operation of the VOB MAC protocol in a node is as follows. At ingress switch i of node i being part of a VOB, there is a token bucket shaper filled with new tokens with rate t_i bytes/s. The bucket has the maximum size of $T_{max,i}$ bytes. Define T_i (in bytes) to be the available tokens in the bucket at any given time. Ingress switch i can inject a packet, located at the head of its local transmit buffer, into the VOB as soon as all the following conditions are met:

- The corresponding wavelength channel is idle.

- The packet size is smaller than the available token T_i in the corresponding token bucket.

- The insertion FDL is either idle or the gap between the time that the packet inside the FDL starts leaving the buffer and the current time is larger than the time needed to transmit the local packet.

After packet transmission, T_i is reduced by the packet size. If a transit optical packet arrives at an intermediate node on the VOB while a local packet is being transmitted, it will be delayed by the insertion FDL. This operation guarantees a collision-free optical packet transmission among the flows within a VOB.

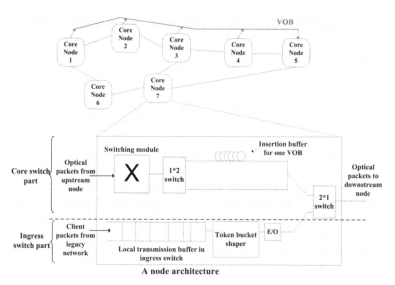

Figure 2.2: An example for VOB and its node architecture

Figure 2.2 illustrates an example for VOB. Consider that there are four O-D flows from nodes $1 \rightarrow 5$, $2 \rightarrow 5$, $3 \rightarrow 5$, and $4 \rightarrow 5$. These four flows can make a VOB (as depicted in the figure) on the same wavelength channel if no wavelength converter is used in the nodes. This figure also shows the node architecture that includes two parts as ingress switch part and core switch part (egress switch part is not shown). The insertion buffer in the core switch part is only displayed for one VOB. Clearly, for every VOB, there is a separate insertion buffer. As can be observed, on an output wavelength, either an optical packet right from the ingress switch, or a transit optical packet from the VOB (without arriving at the insertion buffer), or a delayed transit optical packet from the VOB (after delaying in the insertion buffer) is transmitted toward the downstream node.

2.1.1.8 Packet Aggregation (PA) The nature of IP traffic is bursty. Similar to an electronic network, the performance of an OPS network is degraded dramatically under bursty traffic [29, 30]. On the other hand, statistical traffic analysis shows that almost half of the IP packets are 40–52 bytes. The lengths of other IP packets are 576 and 1500 bytes. Due to the bursty traffic pattern and the irregular packet-size distribution, common contention resolution schemes alone may not be sufficiently maintain reasonable network performance, especially under high traffic loads. Using traffic shaping, one can decrease the burstiness of traffic and make a more uniform traffic pattern and decrease probability of contention in OPS networks [31].

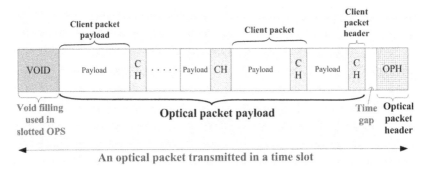

Figure 2.3: An aggregated optical packet

One approach to reduce the burstiness of optical packet arrival in an OPS network could be Packet Aggregation (PA) [27, 31, 32] as a traffic shaping scheme. Under PA, instead of immediately sending a client packet in an optical packet, a number of client packets are aggregated in a large optical packet and then sent to the OPS network. Figure 2.3 illustrates an aggregated optical packet that includes an integer number of client packets, where the client packets may have different payload and header sizes. An Optical Packet Header (OPH) is transmitted at the head of the optical packet that includes different information for the client packets. Since an optical packet should be sent within a time slot in slotted OPS networks, the optical packet size should be equal to time-slot size in bits (for a network with time-slot size S_{ts} in seconds and bandwidth of B_C in bps for a wavelength channel, time-slot size in bits equals $B_C \times S_{ts}$). However, the client packets sizes may not suffice such a constraint. Therefore, a void section must be used at the end of the optical packet filled with dummy bits. However, in asynchronous OPS networks, there is no need for the void section since the optical packet could have any size.

In PA, when client packets arrive at an ingress switch, they are collected in electronic buffers. The buffers may be either dedicated to each egress switch or dedicated to each flow. In the former (mixed-flow aggregation), client packets of any flow of the same egress switch are saved in the buffer dedicated to the egress switch. In the latter (per-flow aggregation), client packets of the same flow are saved in a dedicated buffer. Clearly, the latter aggregation is not scalable because of huge number of flows available in networks. This is why the mixed-flow aggregation is simpler and more common than the per-flow aggregation [27, 33]. Mixed flow has better packet aggregation efficiency because of high probability of filling optical packets before reaching the aggregation time out [33].

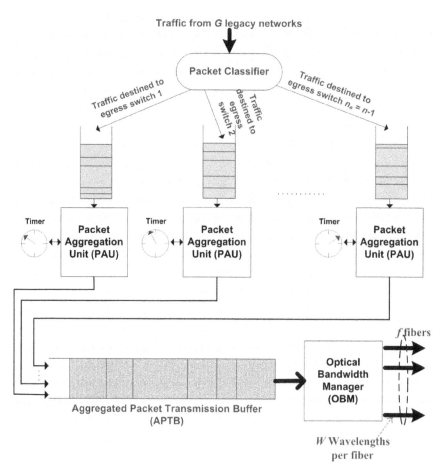

Figure 2.4: Ingress switch architecture for single-class OPS network

Figure 2.4 displays the common ingress switch architecture used for the PA technique under single-class traffic. This architecture has the following features:

- There are G legacy networks connected to the ingress switch.

Algorithm: Packet Aggregation in Single-Class OPS

Definitions:
Q_d :   Queue dedicated to egress switch d
L_d :   Length of queue Q_d in bits
S_{th} :   Maximum size of an optical packet in bits
$AgPk$:   A data structure to contain a number of clients packets up to S_{th}
$APTB$:   Aggregated packet transmission buffer

PacketArrival (client packet p, egress switch d):
1: **if** $L_d = 0$ **then**
2:      SetTimer (d) {set a new timer for buffer d if it is empty}
3: **end if**
4: Append p to Q_d {append client packet p to the end of Q_d buffer}
5: Update L_d { update the size of buffer d}
6: **if** $L_d = S_{th}$ **then**
7:      GenOpticalPacket (d, 0) {aggregate clients packets from buffer d}
8: **else**
9:      **if** $L_d > S_{th}$ **then**
10:         GenOpticalPacket (d, 1) {aggregate clients packets from buffer d, except packet p}
11:     **end if**
12: **end if**
13: END

Timeout (egress switch d):
1: GenOpticalPacket (d, 0) {aggregate all clients packets from buffer d}
2: END

GenOpticalPacket (egress switch d, mode m):
1: ResetTimer (d) {clear timer of buffer d}
2: $AgPk$= aggregated packets from Q_d up to S_{th} {create aggregated packet ($AgPk$)}
3: **if** $m = 1$ **then**
4:      SetTimer (d) {set a new timer for buffer d if it is empty}
5: **end if**
6: Update L_d {update the size of buffer d}
7: Append $AgPk$ to APTB buffer {append aggregated packet $AgPk$ to the APTB buffer}
8: END

Figure 2.5: Pseudo code of packet aggregation in single-class OPS networks

- The packet classifier unit in the ingress switch determines the egress switch of each client packet and then forwards it to the relevant First-Come-First-Served (FCFS) buffer.

- For n edge switches in the OPS network, each ingress switch maintains $n_e = n - 1$ FCFS buffers.

- The Packet Aggregation Unit (PAU) j is responsible for aggregation of client packets from Buffer j.

- A dedicated timer is assigned to each buffer. A threshold value is also specified on the optical packet size.

- The aggregated client packets are sent to a FCFS Aggregated Packet Transmission Buffer (APTB), as an FCFS buffer.

- The Optical Bandwidth Management unit (OBM) creates optical packets from the aggregated packets and then sends them to the OPS network on the available wavelength channels and fibers at a suitable time (say, at time slot boundary in synchronized OPS). The wavelength channels/fibers are selected randomly.

Using PA for single-class traffic, an integer number of client packets in an ingress switch are aggregated to form an optical packet. There are three common aggregation methods in order to form an aggregated packet $AgPk$:

- **Threshold-Based**: A threshold value, S_{th}, is specified on the optical packet size. The threshold S_{th} could be based on either traffic volume (say 1 Mbits) or a number of client packets (say 10 client packets). The traffic of an optical packet is collected for egress switch d from the client packets waiting in queue d whenever there is a sufficient number of client packets to make traffic of an optical packet considering the threshold value S_{th}. Here, when a new client packet arrives at the ingress switch, it is added to the queue and then the volume of traffic in the queue is updated. When this volume is higher than S_{th}, the aggregated packet is generated, but without the new client packet.

  In slotted OPS, the value of S_{th} equals the time-slot size in bits (where in a network with time-slot size S_{ts} in seconds and bandwidth of B_C in bps for a wavelength channel, time-slot size in bits equals $B_C \times S_{ts}$). However, in asynchronous OPS, there is no specific rule for determining S_{th}. Choosing a large S_{th} makes coarse switching granularity that increases client packet aggregation delay and jitter. On the other hand, selecting a small S_{th} leads to high overhead of optical packet headers and high PLR. Therefore, selecting a right S_{th} is a tradeoff between fine switching granularity and reduced header overhead.

- **Timer-Based**: A timer is assigned to queue d which is dedicated to egress switch d in order to limit the waiting time of the client packets destined to the egress switch. The timer of queue d is started whenever a client packet destined to egress switch d arrives at this queue, and queue d is empty. The traffic of an optical packet is collected for egress switch d from all the client packets waiting in queue d whenever the timeout event for queue d has happened.

- **Combined**: Both threshold value and timer are used for each queue under the combined packet aggregation. The traffic for an optical packet is collected for egress switch d from the client packets waiting in queue d whenever there are enough number of client packets to make traffic of an optical packet with the known threshold S_{th}, or whenever the timeout event for queue d has happened. The combined aggregation is more common than both the threshold-based and timer-based packet aggregation schemes.

Figure 2.5 shows the pseudo code of the aforementioned combined aggregation. After creating an aggregated packet ($AgPk$), it is sent to a APTB unit (see Fig. 2.4) even if there are not enough number of client packets in queue i to make a full traffic for an optical packet [27, 31, 32, 34, 35]. Without using this FCFS transmission buffer, the aggregated client packets may be discarded at ingress switches [36]. When a wavelength/fiber becomes available, the Optical Bandwidth Manager

(OBM) unit (see Fig. 2.4) removes an aggregated traffic unit from the FCFS transmission buffer, converts it to an optical packet, and then transmits the optical packet on the wavelength/fiber. As a slight modification to the pseudo code in Fig. 2.5, the work in [37] triggers the aggregation process even if the size of client packets waiting in a buffer becomes larger than $S_{th} - 44$ bytes. This is because no client packet usually has a size smaller than almost 44 bytes in practice.

Table 2.1: Performance of packet aggregation at load $L = 0.6$ [31]

	Inter-Arrival Time			Optical Packet Length (Bytes)		
	No Shaping	$S_{th} = 3000B$	$S_{th} = 9000B$	(No Shaping)	$S_{th} = 3000B$	$S_{th} = 9000B$
min	0.13	2.56	17.70	40	277	4185
max	51.10	93.50	239.93	1500	3000	9000
mean	2.12	13.13	45.18	405.69	2462.2	8470.4
STD	2.04	4.82	14.17	512.50	455.01	428.69
CoV	0.962	0.367	0.314	1.26	0.19	0.051
ACF(1)	0.170	0.0406	-0.0997	—	—	—

Packet aggregation has a number of advantages:

- PA reduces the coefficient of variation and the first-lag autocorrelation function for both inter-arrival time and lengths of optical packets sent to the OPS network (see Table 2.1) [31]. As the optical packet size goes up, the arrivals of optical packets at the network tends to be smoother. Reducing traffic burstiness leads to the reduction of PLR; i.e., contention avoidance. It has been shown that for any given S_{th}, the efficiency of aggregation tends to decrease with the increase of traffic load. This is because the traffic load becomes less bursty at high traffic load, and therefore the effect of traffic aggregation is less prominent. In addition, the effect of different aggregation sizes has been studied and shown that PLR reduces by increasing S_{th} [37].

- PA allows network designers to use relatively large optical packet sizes, and large time-gaps between time slots in slotted OPS networks.

- PA alleviates the requirement to use a very fast optical switch in an OPS switch.

- PA reduces the complexity of an OPS switch because the number of entities per unit time to be processed by the OPS switch is decreased [38].

- Consider that an optical packet carrying a number of client packets has successfully arrived at the destination TCP. When the acknowledge of these multiple client packets arrive at the sender TCP at the same time, it causes rapid expansion of TCP congestion window at the sender TCP, and therefore the TCP source transmit rate is increased. This feature is called correlation benefit. The correlation benefit can lead to rapid growth of the congestion window after an optical packet loss event. The effect of correlation benefit is greater for fast TCP senders since they usually need to aggregate more client packets in a single optical packet than slow TCP senders [33].

The PA technique appears to be quite popular [27, 31] because of its advantages. However, a number of disadvantages can be listed for the PA technique:

- PA does not reduce burstiness in general [39].

- In an ingress switch, the management of load balancing on the transmission channels is difficult because the OBM unit would not know the number of optical packets to be generated from the fluctuating traffic arrival in future. Therefore, an obvious approach is to aggressively transmit optical packets. Consequently, transmission channels may be fully utilized at some times and they may be almost empty at some other times.

- Although aggregation decreases PLR and improves throughput, it has negative effect on TCP performance because of increasing RTT of client packets resulted from queuing delay in ingress switches. By increasing the threshold size S_{th}, it takes more time to aggregate client packets, and therefore the TCP transmission rate decreases due to late arrival of acknowledge packets [33]. There is a trade-off between S_{th} and TCP transmission time. The time needed for successfully transmission of a 1.6 MB file using FTP with TCP window size of 8 KB has been studied under no aggregation and aggregation with different threshold values of $S_{th} = 10$ packets, $S_{th} = 30$ packets, and $S_{th} = 100$ packets. It has been shown that 10 packets aggregation has the best performance in TCP with 8 KB window, followed by no aggregation, 30 packets aggregation, and 100 packets aggregation. In other words, increasing threshold size S_{th} does not decrease the file transfer time [27].

- When an optical packet is lost, a number of client packets of the same flow carried in the optical packet are lost. This results in traffic loss indication for the relevant sender TCP, and therefore the sender TCP shrinks TCP congestion window size to one segment [33].

2.1.1.9 Offset-Based Traffic Transmission To avoid contention, a mandatory fixed amount waiting time (as an offset time) is applied to every optical packet in each ingress switch; from the time the optical packet is ready until its transmission to an asynchronous OPS network [40]. This is similar to the OBS offset time. During the waiting phase and using the link state dynamic routing protocol, edge switches communicate with each other and report their traffic information to each other. In other words, each ingress switch will send and receive information updates from other ingress switches in the network. These updates include information about the traffic load and the routes of the traffic that will be injected into the network by the adjacent edge switches in the near future. Note that the ingress switch should have enough time to estimate the traffic that will be entered on different links of the OPS network in the near future.

When it is time to transmit buffered client packets, the ingress switch knows the traffic state of network links, and therefore it can dynamically route its optical packets through the best possible routes (with less optical packet contention), thus increasing the throughput of the OPS network and reducing PLR. Using this mechanism, each ingress switch regulates its optical packet flow into the OPS network accordingly. In short, this technique helps the ingress switches to time-shift the possible contention of optical packets before they flow in the network.

By increasing the offset time in all ingress switches, every client packet will experience a mandatory delay which may even be unnecessary. However, this allows the ingress switches to find better routes, thus reducing PLR. In addition, by increasing the information exchange rate (i.e., by decreasing the information exchange interval), better routing decisions can be made. This is because more and more information about the traffic that will be sent on the network links can be exchanged; i.e., more accurate information can be achieved. However, this increases the overhead in the OPS network. Note that within the information exchange interval, ingress switches exchange routing tables as well.

It should be stated that the complexity of this approach is high since the process of obtaining information from adjacent edge switches is performed for each optical packet. In addition, the delay of all optical packets increases by the offset time. However, it is claimed that due to the constant delay applied to all optical packets, no additional jitter is introduced; therefore, throughput would not be degraded. The last disadvantage is the additional overhead of routing packets due to the communication between edge switches.

Network PLR under different waiting times and packet sizes are evaluated in [40], and the simulation results show that by increasing waiting time, the PLR decreases due to increasing regulating effect of the algorithm and decreasing contention probability in the network. In addition, it is shown that that PLR decreases with increasing packet size until a certain size. Obviously, throughput increases by decreasing PLR.

2.1.1.10 Network Layer Packet Redundancy   The objective of this technique [41] is to duplicate transmission of the same traffic in the network in order to increase the chance of successful traffic delivery to egress switches. In this technique, a set of m_s subsequent client packets going to the same egress switch is called a packet set. In an ingress switch, upon arrival of a client packet belonging to a given packet set, it is converted to an optical packet (called data optical packet) and then transmitted to the OPS network, while its copy is temporarily saved in the ingress switch buffer. When all client packets in a given packet set have arrived, r_s redundancy optical packets are created from the copies of m_s client packets stored in the ingress switch. Then, the redundancy optical packets are transmitted to the same egress switch as well. In the egress switch, at most r_s lost data optical packets can be reconstructed from successfully received optical packets, but under certain conditions.

Redundancy optical packets are created based on the Reed–Solomon (RS) error correcting codes, which have been extensively used in communication systems to correct bit errors at the data link layer. Assume that of m_s data optical packets and r_s redundancy optical packets transmitted for a packet set, only m_r and r_r optical packets have arrived successfully at an egress switch, respectively. This means that $m_s - m_r$ data optical packets and $r_s - r_r$ redundancy optical packets have been lost in the network. All the $m_s - m_r$ lost data optical packets can be reconstructed from successfully arrived data and redundancy optical packets in the egress switch provided that $m_r + r_r \geq m_s$. Otherwise, reconstruction is not possible and the number of lost data optical packets is $m_s - m_r$. For instance, at $m_s = 4$, $r_s = 2$, $m_r = 2$, $r_r = 1$, we have $m_r + r_r = 3 < m_s = 4$, and therefore the two lost data optical packets cannot be reconstructed.

Although this technique provides possible reconstruction of lost packets, it increases the network load and traffic burstiness. However, this technique reduces loss rate more than the case that no redundancy is employed in the network. Finally, the packet redundancy scheme requires additional processing power in the network edge switches compared with the other techniques.

Let r and m be, respectively, the number of redundancy and data optical packets per packet set. Here, fraction $\frac{r}{m}$ shows the percentage of used redundancy optical packets with respect to data optical packets. Clearly, $\frac{r}{m} = 0$ indicates the original OPS without using the packet redundancy mechanism, where PLR is equal to 1.8×10^{-2} under the simulated scenario in a ring topology in [41]. Performance of PLR for data optical packets (denoted by DPLR) shows that increasing $\frac{r}{m}$ leads to a reduced DPLR. However, this reduction continues until the DPLR reaches its minimum at point $\frac{r}{m} = 1.2$, where DPLR reduces to $DPLR = 4.2 \times 10^{-4}$ at $m = 5$, $DPLR = 3.4 \times 10^{-5}$ at $m = 10$, and $DPLR = 8.7 \times 10^{-6}$ at $m = 15$. This also shows that increasing m at a fixed value of $\frac{r}{m}$ can reduce DPLR. After this minimum point, DPLR goes up by increasing $\frac{r}{m}$ because the network traffic load increases due to the additional generated redundancy optical packets. It is also shown that by increasing the number of ring nodes, DPLR goes up.

2.1.1.11 Even-Spaced Optical Packet Transmission
This technique has been proposed for synchronized OPS. It is shown that the overall loss in the network can be minimized if traffic transmission on every link of an ingress switch and OPS switch is evenly spaced [42]. In this technique, edge switches use their electronic buffers to condition the traffic and evenly space the transmission of optical packets into the OPS network. It is proved that PLR is minimized, but at the expense of increasing end-to-end delay.

Consider a window of N time slots on a wavelength channel and traffic load L in an ingress switch. The ingress switch must send either $k = \lceil N \times L \rceil$ or $k = \lfloor N \times L \rfloor$ optical packets in each window. These k optical packets can be sent in three different distributions:

- **Block Transmission**: In this case, all k optical packets are transmitted in k contiguous time slots within the window.

- **Random Transmission**: In this case, an optical packet is transmitted in a time slot with probability L, and therefore some random time slots will be allocated.

- **Even Transmission**: In this case, all k optical packets are transmitted with even spacing within the window. The spacing should be $\lfloor \frac{1}{L} \rfloor$ with which the smallest PLR can be achieved. For a stable system, the spacing can get any value from 1 to $\lfloor \frac{1}{L} \rfloor$.

It is shown that even transmission can lead to smaller PLR than the random transmission, and the random transmission can achieve smaller PLR than the block transmission. The even transmission technique is easy to implement in ingress switches with electronic buffers, thus suitable for a single-hop OPS metro network (see Section 1.4.1). In a multi-hop network, however, each OPS switch must use optical buffers and maintain this spacing when multiplexing traffic streams. The requirement to use optical buffers in a multi-hop network could be the disadvantages of this technique.

2.1.1.12 Active Congestion Control   In [43], a TCP-aware active congestion control mechanism has been proposed for asynchronous OPS networks that presents the coordination among the core switches, edge switches, and the NC&M (network control and management) system. The key idea is to detect congestion and convey congestion notifications early enough back to TCP sources so that they can reduce their transmission rates before increasing PLR.

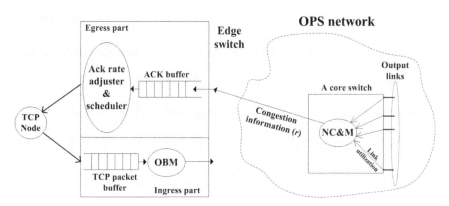

Figure 2.6: Active congestion control mechanism for TCP traffic management in asynchronous OPS networks

An NC&M agent embedded in each OPS core switch continuously monitors the traffic rate and infers the network congestion status directly from the estimated link utilization (see Fig. 2.6). For each output fiber link, average traffic rate r is estimated by using an exponential averaging given by Eq. (2.3):

$$r_{new} = \left(1 - e^{-\frac{T}{K}}\right) \times \frac{s}{T} + \left(e^{-\frac{T}{K}}\right) \times r_{old} , \qquad (2.3)$$

where T is the inter-arrival time between the current and the previous optical packet, s is the size of the arriving optical packet, and K is a constant. Assuming that OPS core switches balance traffic load among multiple wavelengths on a fiber, the link utilization could be approximated only on one wavelength. Define MAX_{th} to be maximum utilization threshold. Then, the NC&M can detect a link to be congested if we have $r \geq B_C \times MAX_{th}$ at all times during an interval of length τ, where B_C is the wavelength bandwidth. Once the NC&M detects the incipient congestion, it sends out a congestion notification message back to the relevant edge switch x, which is responsible for providing congestion control feedback to the TCP sources.

Note that in TCP congestion control, ACK messages operate as regulator to clock out the transmission of new data from TCP sources. In other words, TCP data transmission in the forward path greatly depends on the transmission rate of ACK segments over the reverse path. Under the active congestion control technique (see Fig. 2.6), an ACK transmission rate adjustor is employed at the egress part of edge switch x for saving the ACK segments over the reverse path. The egress part of edge switch x can provide fast feedback to the relevant TCP sources by dynamically adjusting the ACK transmission rate according to the current status of the OPS network.

Define r_a to be the ACK transmission rate, r_n to be the ACK transmission rate for normal situation, α to be a decrease factor, β to be an increase factor, and MIN_{th} to be the minimum utilization threshold which indicates the link under-load status. This threshold can trigger the egress switch to increase the ACK transmission rate r_a. There are three cases for adjusting ACK transmission rate:

- When a congestion notification arrives from the NC&M system (i.e., when $r \geq B_C \times MAX_{th}$ for a time interval of length τ), the egress part of edge switch x initiates congestion control by decreasing the ACK transfer rate to $r_a = \frac{r_n}{\alpha}$.

- When we have $r \leq B_C \times MIN_{th}$ for a time interval of length τ, the egress part of edge switch x increases the ACK transmission rate by $r_a = r_n \times \beta$.

- Otherwise, when we have $B_C \times MIN_{th} < r < B_C \times MAX_{th}$ for a time interval of length τ, the egress part of edge switch x sets $r_a = r_n$.

Performance evaluation results show that by dynamically adjusting the ACK transmission rate based on the network status, the proposed scheme can provide a smooth traffic load and can prevent the adverse load oscillations observed in the OPS network without active congestion control. In addition, active congestion control can provide a noticeable lower PLR, higher fairness, and higher good-put compared with the case that no congestion control is used in the OPS network.

2.1.1.13 Delay-Controlled Traffic Shaping   As is common, traffic shaping introduces additional delay on client packets. To overcome this drawback, the traffic shaping technique proposed in [44–46] implements a traffic conditioner in which the packet transmission rate to OPS network is variable, but the packet delay is bounded. In other words, this technique produces the smoothest output traffic that releases optical packets within their time constraints. The traffic conditioner in each ingress switch consists of a FIFO queue and a rate controller (see the top image in Fig. 2.7). In this system, a packet arrival curve A and a packet deadline curve D are utilized where the deadline curve D is the right-shifted version of the arrival curve A with a horizontal distance equal to the delay bound d (see the bottom image in Fig. 2.7). Here, the deadline time of a client packet arriving at time t_i is $\tau_i = t_i + d$. The rate controller generates service rate for the scheduler by which the ingress switch system can maintain the least bursty exit curve for the client packets currently in the queue by considering their deadline times, but without accounting for future coming client packets arrivals (see the dotted line in below image of Fig. 2.7).

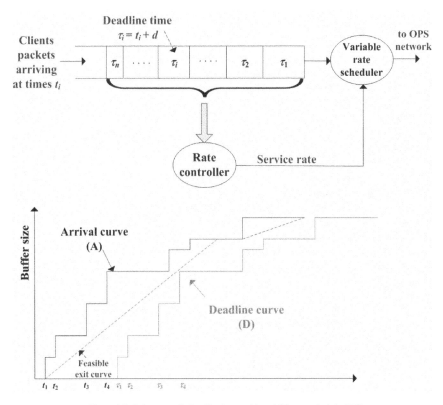

Figure 2.7: Delay controlled traffic shaper with variable rate scheduler [46]

Algorithm in Fig. 2.8 shows the pseudo code of computing the deadline curve in the delay-controlled traffic shaping scheme. A linked list *Hull* is maintained to keep the information relevant to different piecewise-linear hulls of the shortest-path exit curve (including *startTime*, *endTime*, and *slope*). At time t, when client packet p arrives at an ingress switch, list item h is created (as a line segment) with an appropriate slope from the packet p information, and then saved at the end of the linked list *Hull*. The newly appended line segment may have a slope larger than its preceding line pieces. Therefore, the backward linear segments are scanned in order to restore the convexity of the deadline curve. If segment line h has a slope larger than its preceding slope *Prev*, both line segments h and *Prev* are combined into a single line segment in which their end points are joined and a new slope is computed. This process repeatedly fuses line segments until the slope of the last segment is smaller than its preceding segment.

The scheduler releases client packets according to the computed exit curve (see Fig. 2.9 for slotted OPS). The scheduler first determines the current time slot and

`Algorithm: Delay Controlled Traffic Shaping`
`Input:`
p: `client packet`
t: `time of arrival`

`Description:`
1: $L = Length(p)$
2: $\tau = t + d$ {create deadline for packet p}
3: $h = newlistItem$ {create a new item}
4: **if** $Hull$ is empty **then**
5:    $h.startTime = t$ {as the first item}
6: **else**
7:    $h.startTime = Hull.tail().endTime$
8: **end if**
9: $h.endTime = \tau$
10: $h.slope = \frac{L}{(\tau - h.startTime)}$
11: Append item h to list $Hull$ {add item h to the end of linked list $Hull$}
12: $h = Hull.tail()$ {search backward for Hull convexity}
13: **while** $((Prev = h.prev) \neq null) \wedge (Prev.slope \leq h.slope)$ **do**
14:    $a = \max(t, Prev.startTime)$
15:    $h.slop = \frac{(h.slope \times (h.endtime - h.startTime) + Prev.slope \times (Prev.endTime - a))}{h.endTime - a}$
16:    Delete item $Prev$ from linked list $Hull$
17: **end while**
18: END

Figure 2.8: Pseudo code of computing deadline curve in delay-controlled traffic shaping [46]

`Algorithm: Packet Scheduling in a Time Slot in Delay`
`Controlled Traffic Shaping`

`Description:`
1: p = packet at the head of queue
2: $L = Length(p)$ {length of packet p}
3: $T = currenttime$ {obtain current slot time }
4: $credit = credit + Hull.head().slope$ {create a new item}
5: **if** $credit \geq L$ **then**
6:    dequeue p,
7:    make optical packet from p
8:    send optical packet to network in the time slot
9:    $credit = credit - L$
10: **end if**
11: **if** $Hull.head().endTime \leq T$ **then**
12:    Delete the item from head of $Hull$
13: **end if**
14: END

Figure 2.9: Pseudo code for scheduling of client packets in each time slot

the length of packet p at the head of the queue. Then, the credit value based on the service rate in the hull is computed for packet p. If there are sufficient credits, an optical packet is created from p, and the optical packet is sent to the OPS network within the time slot. Finally, the hull piece is removed from list $Hull$ if the piece has been expired.

Performance evaluation data in a slotted OPS show that the delay-controlled traffic shaping can reduce burstiness of optical packet arrival to the network, especially under high values of d, thus reducing PLR in the OPS network. Although this technique can reduce PLR taking into account the timing restrictions of client packets, it increases the computational complexity of ingress switches. As mentioned, this technique can produce a smooth traffic when packet deadline is relatively high. However, high value of d is not suitable for HP traffic. Therefore, for HP traffic with small d, the output traffic would not be so smooth.

2.1.1.14 Adaptive IP/OOFDM The OOFDM is a facilitator for providing contention avoidance through subcarrier allocation. As an application of the OOFDM switch designed for IP networks in Section 1.5.4, a novel method has been introduced in [47, 48] to adaptively reconfigure the logical link bandwidth between ingress–egress switch pairs. Recall that in the proposed switch architecture, each OOFDM transmitter can dynamically modify its modulation/coding methods in order to deal with packet traffic variance or even physical layer impairments at the subwavelength level. This mechanism is clearly suitable for the bursty Internet traffic. Here, assume that there is one edge switch connected to every OOFDM core switch, where a number of IP routers may be connected to the edge switch (see Fig. 1.18). The proposed IP/OOFDM architecture shares the bandwidth in a dedicated frequency-domain subcarrier-based fashion. When combined with adaptive virtual link data rate transmission, it can maintain well the fidelity of the slivers in terms of jitter, delay, and loss QoS performance parameters. This technique provides a loss-free OPS network since subcarriers are reserved for traffic flows, and this reservation is updated when traffic flow fluctuates.

The ingress switch design using OOFDM-based link virtualization technique has been presented in [48], where each ingress switch contains multiple virtual interfaces that correspond to virtual OOFDM links. The virtual interfaces maintain separate FIFO electronic buffers for each virtual link to save its client packets.

Unlike the inefficient and expensive traditional approaches that over-provide adequate capacity on each network link to meet the maximum traffic demand over all time, in the current proposal a virtual link (called pipe) is set up between every ingress and egress switch pair with the help of OOFDM switches in order to build an all-to-all full mesh logical network topology (see Fig. 2.10) among all the edge switches in the network [47]. The virtual links are isolated from each other using different OOFDM subcarriers on each physical link. Since the traffic rate between each ingress–egress switch pair can change over time quickly, it is desirable to update the virtual topology accordingly.

Each virtual pipe in the logical topology corresponds to a dedicated set of subcarriers that provide 100% guaranteed bandwidth for the relevant traffic flow. As stated,

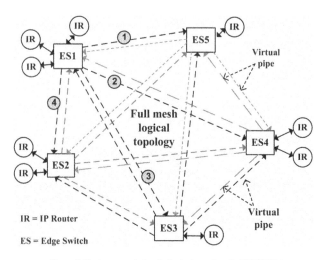

Figure 2.10: An example for the logical topology in IP/OOFDM

this bandwidth may adaptively change with variations of traffic flow (by adjusting either the modulation format or the number of subcarriers). Given a time period, each ingress switch should measure the traffic rate for each virtual pipe, and based on the measurements, it predicts the required size of its virtual pipes. It then initiates the reservation for each of the virtual pipes on the OOFDM switches along the physical path.

Consider the five-node network topology depicted in Fig. 1.17, where an edge switch is connected to each OOFDM core switch. In Fig. 2.10, Edge Switch 1 (ES1) has established four outgoing virtual pipes 1, 2, 3, and 4 to four peers displayed with dotted lines. As can be observed, in the full mesh logical topology, every edge switch has a setup connection with other edge switches. Note that in a full-mesh logical topology, all virtual connections are displayed as one-hop connections, but each one of them may have used several physical links. For instance, virtual pipe 2 established from ES1 to ES4 passes through two physical links in the network under discussion. No matter how the traffic from the IP routers (IRs) connected to ES1 fluctuates, the network can resize the virtual pipes bandwidth (with no conflicts in the physical links) in a given short time period to adaptively track the fluctuations of the traffic flows to ES2, ES3, ES4, and ES5. This can be performed, say, by either assigning more subcarriers or changing the modulation format of the virtual pipe to ES2 when the traffic volume to ES2 increases, and decreasing the number of subcarriers of the virtual pipe to ES3 when its traffic volume decreases. Consequently, the network can achieve adaptive load balancing by adapting the virtual topology to the real traffic demand in a real-time manner.

The IP/OOFDM networking method has several merits:

- Due to using virtual pipes, it is a single-hop scheme with less routing processing involved, thus resulting in high QoS performance for optical packets (e.g., less delay, less jitter, and less packet loss ratio).

- It theoretically avoids from traffic congestion in the core network regardless of the traffic distribution pattern, and therefore it is a contention avoidance scheme.

- Each core node has a low cost since only one transponder and router port are needed for the establishment of a full mesh virtual topology.

However, there are two major issues in the proposed IP/OOFDM architecture:

- When the network needs to dynamically adjust multiple virtual pipes bandwidth (by either changing the modulation format or adjusting the number of subcarriers), contention may happen among virtual pipes since some virtual pipes may be mapped onto the same physical link on which no subcarrier extension is possible.

- It requires accurate traffic measurement mechanisms for dynamically reserving resources.

Algorithm: Adaptive OOFDM Subcarrier Allocation and Assignment

1: **while** TRUE **do**
2:     **while** TRUE **do**
3:         Measure real arrival data rate
4:         Monitor each queue length variance to decide the next scheduling interval
5:         **if** (queue length variance exceeds threshold) or (time expires the configured maximum scheduling interval) **then**
6:             Exit loop
7:         **end if**
8:     **end while**
9:     Initialize each subcarrier allocation variable for each queue
10:     **while** there is a subcarrier not assigned **do**
11:         Choose queue k that leads subcarrier j' to have maximal utility by
$$k = arg \ max \left(\frac{\lambda_i(\vec{T})}{\mu_i(\vec{T})} \times e^{\frac{b_i(\vec{T})}{B_i}} \right)$$
12:         Assign subcarrier j' to queue k
13:         Update assignment variable for each queue
14:     **end while**
15: **end while**
16: END

Figure 2.11: Adaptive OOFDM subcarrier allocation and assignment algorithm [49]

The adaptive algorithm for dynamic OOFDM subcarrier allocation when traffic fluctuates has been proposed in [49], which combines the advantage of statistical multiplexing in packet-switched networks with bandwidth guarantees of circuit switched networks (see Fig. 2.11). Assume that the packet flows arrive randomly at a given ingress switch and are then buffered in several output queues, where there is one queue per egress switch flow. The algorithm considers both the traffic characteristics and buffer occupancy/variance when determining the dynamic subcarrier allocation in such a way that the available subcarriers are always allocated to the nodes that mostly require them (this is achieved using the equation used in the pseudo code), thereby maximizing the utilization of each subcarrier in every short time period.

In the pseudo code, there are some variables defined as follows. Define $\Lambda_i(\overrightarrow{t})$ to be the measured arrival data rate for queue i during the previous scheduling interval $\overrightarrow{t}$, and $x_{i,j} = 1$ if subcarrier j is assigned to queue i; and 0 otherwise. Let $d_{i,j}$ denote the corresponding data rate of each subcarrier when the adaptive modulation format based on the transmission quality (such as OSNR) is used. Then, the service rate for queue i can be computed by $\mu_i(\overrightarrow{t}) = \sum_j x_{i,j} \times d_{i,j}$. Note that a number of subcarriers can be assigned to each queue traffic. Define the buffer usage at queue i as $e^{\frac{b_i(\overrightarrow{t})}{B_i}}$, where $b_i(\overrightarrow{t})$ is the measured average queuing length during the last scheduling interval $\overrightarrow{t}$, and B_i is the buffer size of queue i.

Performance evaluation results show that the proposed IP/OOFDM can provide less end-to-end packet delay compared with the Valiant Load Balancing (VLB) over SONET technique. In addition, the IP/OOFDM solution can save up to 48% capacity resources compared with VLB/SONET. Note that VLB/SONET also provides logical topology and uses a fixed two-hop logical routing scheme. It is quite flexible since it can accommodate arbitrary traffic patterns between all edge switch pairs.

2.1.1.15 Minimum Interference in GMPLS-Enabled OPS

In a GMPLS-enabled connection-oriented OPS network, an LSP (requiring a fraction of a wavelength channel bandwidth) is assigned a route and a wavelength. Each intermediate core switch routes the optical packets belonging to the same LSP (based on its forwarding table specified during the LSP setup) to the desired output port, identified by a fiber–wavelength pair. Similar to connectionless OPS, contention in a core switch may arise among the optical packets belonging to different LSPs coming from different input ports and directed to the same output port of the core switch. The core switch should either use optical buffering or wavelength conversion to resolve the contention. However, optical buffering offers very limited and discrete buffering capability. On the other hand, changing the wavelength used by an LSP (called LSP-to-wavelength relocation) has several drawbacks: (1) increasing the switch control burden due to forwarding table updates, (2) causing out-of-sequence optical packet arrivals, and (3) making contention even on the new wavelength, thus impairing previously unaffected LSPs. Therefore, contention avoidance finds importance in connection-oriented OPS. As a contention avoidance mechanism, contention of the traffic of different LSPs can be reduced by the network-wide end-to-end LSP admission control and efficient LSP-to-wavelength assignment during LSP setup [50].

Assume that a connection request arrives at an ingress switch to an egress switch that needs bandwidth b. The GMPLS signaling finds the shortest route that can guarantee to provide the requested bandwidth (admission control); i.e., there is at least one wavelength with capacity greater than b on each link of the selected route. Then, the LSP setup procedure between the ingress–egress switch pair is started on the route using RSVP. Number the core switches from 1 to k from the ingress switch to the egress switch.

The contention avoidance can be achieved by minimizing the sharing of the same wavelength among the LSPs coming from different input ports of a core switch. The Minimum Interference (MI) contention avoidance mechanism adds a suggested vector and a label set to the PATH message under RSVP (see Section 1.4.6.3.1).

During propagation of the PATH message from core switch 1 to core switch k, the suggested vector stores the preference level for each wavelength channel present in the label set, which corresponds to the bandwidth not interfering with already-established LSPs. Note that a label set keeps the reference of the wavelengths that can provide the required bandwidth b. Both suggested vector and label set are arrays with length W.

At the egress switch, the GMPLS backward reservation process starts, where the wavelength assignment is made. The egress switch chooses the wavelength with the highest preference level (i.e., the wavelength with the greatest value in the received suggested vector) and sends a RESERVATION message hop-by-hop back to the ingress switch. During this process, two different mechanisms can be used:

- **MI-Hop (MI-H)**: Core switch i selects its most preferable wavelength. When the chosen wavelength is different from the preferred wavelength in downstream core switch $i + 1$, wavelength conversion should be used in core switch i for the LSP. This scheme can minimize the interference with other LSPs on each hop of the path, but at the expense of more frequent usage of WCs.

- **MI-Keep (MI-K)**: Core switch i keeps the wavelength chosen by the downstream core switch $i + 1$ as long as it is possible. For example, the wavelength preferred by core switch $i + 1$ may not be able to support required bandwidth b in core switch i. In this case, core switch i must choose another preferred wavelength; therefore, wavelength conversion should happen in core switch i. The MI-K tries to reduce the number of required WCs in the network.

Performance evaluation results show that MI-H can provide smaller PLR than MI-K, where wavelength conversion probability is higher in MI-H than MI-K; i.e., wavelength conversion cost is high in MI-H.

2.1.1.16 Comparison of Software-Based Contention Avoidance Schemes for Single-Class Traffic   Now, a comparison is made among the software-based contention avoidance schemes proposed for single-class OPS networks. A discussion is also provided on the combination of these schemes with each other. By all accounts, the contention avoidance schemes appear to be much cheaper than the resolution schemes since most of the avoidance schemes are software level tools. However, the software-based schemes cannot reduce PLR in an OPS network to very small values. In addition, they may increase the waiting delay of client packets in ingress switches buffers.

The operations of studied contention avoidance schemes can be grouped into three categories:

- **No Coordination (NC)**: Ingress switches in some techniques (like PA, delay-controlled traffic shaping, symmetric traffic transmission, balancing on wavelength channels, even-spaced optical packet transmission, and network layer packet redundancy) send their optical packets without any coordination with OPS core network. All the techniques in NC can be easily combined with the balancing on wavelength channels technique. They can also be combined with the network layer packet redundancy technique. For example, PA can be combined with both the balancing on wavelength channels technique and the network layer packet redundancy technique. The combination of other techniques is a challenging issue. For example, PA reduces PLR by reducing the burstiness of traffic arrival to the OPS network; however, it may slightly increase the delay of client packets in an ingress switch. Although delay-controlled traffic shaping controls the delay of client packets by a threshold value in ingress switches, its effectiveness in reducing PLR may decrease when the threshold delay is small. Combining delay-controlled traffic shaping with PA could be a challenging issue because of existing of different timing required in these techniques.

• **Loose Coordination (LS):** Ingress switches in some other techniques (like monitoring core switch traffic, balancing on routing operation, renegotiation-based transmission of optical packets, active congestion control, and packet transmission based on the scheduling of empty time slots) are loosely coordinated with the OPS network in order to have a controlled optical packet transmission. The schemes in LS cannot be combined with each other since these techniques send the optical packets at appropriate times or through appropriate paths so that PLR can be reduced. All the LS techniques can be combined with the network layer packet redundancy technique.

• **Strict Coordination (SC):** Ingress switches in techniques like offset-based traffic transmission, reservation in single-hop OPS, and NGC are strictly coordinated with the OPS core network in order to achieve an almost loss-free OPS network. For these techniques, there is a strict time for sending optical packets, and therefore they cannot be combined with any other technique in NC, LS, and SC categories. In addition, these techniques are usually suitable for single-hop OPS networks.

Table 2.2 compares single-class software-based contention avoidance schemes in terms of suitability for slotted or asynchronous OPS, suitability for single-hop or multi-hop OPS, suitability for single-fiber or multi-fiber OPS, and operation complexity.

2.1.2 QoS Differentiation

In this section, software-based contention avoidance schemes proposed for differentiating QoS suitable for both asynchronous OPS and slotted OPS networks are studied. The schemes proposed for relative differentiation and absolute differentiation are also discussed in this section. Reducing the PLR of HP optical packets is an important issue in OPS core networks in order to provide a desired QoS [51]. In the following, the software-based contention avoidance schemes that can support QoS in OPS networks are studied, where there are M classes of service for client packets.

Table 2.2: Comparison of single-class software-based contention avoidance schemes

Category	Algorithm	Slotted OPS/ Asynchronous OPS	Single-Hop/ Multi-Hop	Single-Fiber/ Multi-Fiber	Complexity
NC	Symmetric Traffic Transmission	both	both	both	low
	Balancing on Wavelength Channels	both	both	both	low
	Packet Aggregation (PA)	both	both	single-fiber	moderate
	Delay-Controlled Traffic Shaping	asynchronous OPS	both	single-fiber	moderate
	Network Layer Packet Redundancy	asynchronous OPS	both	single-fiber	moderate
LC	Even-Spaced Optical Packet Transmission	slotted OPS	suitable for single-hop	single-fiber	low
	Balancing on Routing Operation	connection-oriented asynchronous OPS	multi-hop	single-fiber	moderate
	Packet Transmission Based on the Scheduling of Empty Time Slots	slotted OPS	single-hop	both	moderate
	Renegotiation-Based Transmission of Optical Packets	slotted OPS	both	single-fiber	moderate
	Active Congestion Control	asynchronous OPS	both	single-fiber	moderate
	Monitoring Core Switch Traffic	both	muti-hop	single-fiber	moderate
SC	Offset-based Traffic Transmission	asynchronous OPS	both	single-fiber	high
	Reservation in Single-Hop OPS	slotted OPS	single-hop	both	high
	Network Global Control (NGC)	slotted OPS	both	single-fiber	high

Before detailing the software-based contention avoidance schemes, two ingress switch architectures proposed to support QoS in OPS networks are studied in the following.

- **Ingress Switch Architecture 1 (ISA1)**: Figure 2.12 shows the OPS ingress switch architecture for QoS-capable OPS network (referred to as ISA1) in which all the same-class client packets from G different legacy networks (each as a source) going to the same egress switch are saved in a shared FCFS buffer. Figure 2.12 illustrates the ISA1 under multi-class traffic [33, 34, 52–54]. This architecture has the following features:

  - There are G legacy networks connected to an ingress switch. Each legacy network sends its client packets to the ingress switch in M classes of service.

  - The packet classifier unit in the ingress switch determines the egress switch and class of each coming client packet and then sends it to the relevant FCFS Buffer Pool (BP). Therefore, if the class of a client packet is c and its destination egress switch is found to be j, it is saved in the class c buffer of BP j.

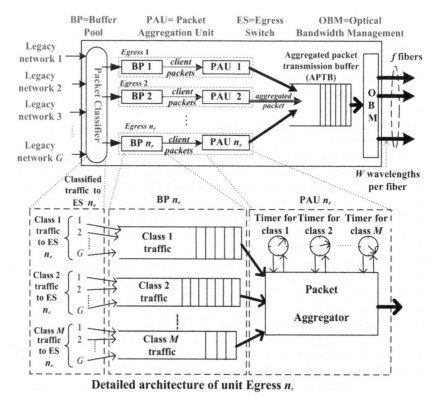

Detailed architecture of unit Egress n_e

Figure 2.12: Ingress switch architecture 1 (ISA1) for QoS-capable OPS network with M classes of traffic

– For an OPS network with n edge switches and M classes of service, each ingress switch maintains $M \times n_e$ FCFS electronic buffers for $n_e = n - 1$ egress switches.

– The Packet Aggregation Unit (PAU) j is responsible for aggregation of client packets from buffer pool j.

– A dedicated timer is assigned to each class (see the lower picture in Fig. 2.12). A threshold value is also specified on the optical packet size.

– The assembled client packets are sent to a APTB.

– The Optical Bandwidth Manager (OBM) unit displayed in this figure transmits the optical packets to the network on the available wavelengths/fibers, where the choice of both fiber number and wavelength number in the fiber follows the uniform distribution.

■ **Ingress Switch Architecture 2 (ISA2)**: Figure 2.13 is the second ingress switch architecture [55] (referred to as ISA2), proposed for a slotted OPS network. Each edge switch is physically connected to G immediate routers (i.e., legacy

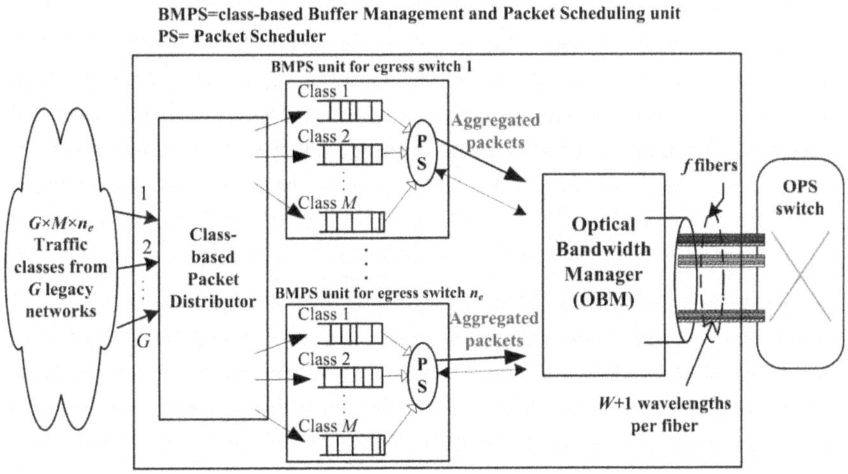

BMPS=class-based Buffer Management and Packet Scheduling unit
PS= Packet Scheduler

Figure 2.13: Ingress Switch Architecture 2 (ISA2) for QoS-capable OPS network with M classes of traffic. Reprinted from *Computer Networks*, Vol. 53, Akbar Ghaffarpour Rahbar and Oliver Yang, Distribution-based bandwidth access scheme in slotted all-optical packet-switched networks, 744–758, copyright 2009, with permission from Elsevier.

networks), where each router could be either an edge router or a core router in the DiffServ domain. An edge switch either aggregates traffic from G upstream immediate routers (hereafter referred to as a source router) connected to the edge switch or delivers received traffic to them. Totally, $G \times M \times n_e$ class-based traffic streams arrive at the ingress switch which are destined to n_e egress switches. There are n_e class-based Buffer Management and Packet Scheduling (BMPS) units in each ingress switch. After arriving a client packet from an immediate source router at the packet distributor unit, it is directed to BMPS j if it is going to egress switch j (where $j=1, 2, ..., n_e$). To fairly process all client packets belonging to a given class, a dedicated buffer is allocated to the class in BMPS j. Hence, total number of buffers used in an ingress switch is $G \times M \times n_e$. There is a class-based Packet Scheduler (PS) inside each BMPS unit for maintaining the transmission order of client packets from each class to the relevant egress switch when required. The difficulty in controlling the service order of client packets stated for ISA1 (see Section 2.1.2.1.3) can be resolved in ISA2 because of using a dedicated buffer to the traffic of each class from any source router and also using packet schedulers in an ingress switch to provide packet differentiation.

2.1.2.1 Packet Aggregation (PA)   Following the discussion on packet aggregation for single-class networks in Section 2.1.1.8, this technique is now discussed for multi-class traffic under ISA1 [56]. Here, a dedicated timer is assigned to each class. Traffic arrived from legacy networks is classified (in terms of class of service and egress switch) and then saved in appropriate buffers as described in the ISA1 architecture.

The timeout mechanism is used to provide packet differentiation in which small timeouts are assigned to HP traffic and large timeouts are assigned to LP traffic. However, the choice of timeout value is a challenging problem. To choose a timeout, a network designer must consider different network parameters such as the network traffic load, the number of QoS classes, the time-slot size in synchronized OPS, the wavelength channel rate, and the number of egress switches [34, 57, 58]. Since the volume of HP traffic is usually smaller than LP traffic, the average aggregation time for the HP traffic is longer than the LP traffic [53]. This is the reason that smaller timeouts must be chosen for the HP traffic than the LP traffic. However, this may increase the load on the OPS network and increase the PLR as well. Moreover, small timeouts or even large timeouts at high traffic load may result in an unstable ingress switch operation in which the optical packet generation rate to the OPS network is higher than the network capacity. Therefore, the waiting time in the Aggregated Packet Transmission Buffer (APTB) may increase the client packets waiting time in an ingress switch unbounded.

For differentiated traffic, two types of aggregation techniques have been proposed:

2.1.2.1.1 Composite Packet Aggregation (CPA) The CPA technique [59, 60] allows client packets from different classes of service to be aggregated in the same optical packet (see the CPA algorithm in Fig. 2.14). In this pseudo code, $Q_{d,c}$ denotes the buffer allocated to class c that saves clients traffic of class c destined to egress switch d, parameter S_{th} is the maximum traffic size from different classes (in bits) that can be carried in an optical packet, $L_{d,c}$ is the length of traffic (in bits) buffered in $Q_{d,c}$, and $AgPk$ is an aggregated packet to contain a number of client packets up to size S_{th}. Note that in synchronized OPS, we have $S_{th} = B_C \times S_T$, where B_C is the channel bandwidth in bits/s and S_T is the time-slot size (in seconds) that can carry traffic. For example, for $B_C = 40$ Gbits/s and $S_T = 100 \times 10^{-6}$s, we have $S_{th} = 4$ Mbits. There is one timer dedicated to each class in every buffer pool. The timer for class c is either set or reset when a user packet of this class arrives or departs.

The main objective of CPA is to avoid unused bandwidth in an optical packet. In other words, if there is available room in an optical packet created mainly for a desired class, CPA allows the optical packet to carry traffic from other classes, but starting from HP class. Under CPA, class-based traffic boundaries are moveable within an optical packet.

Under CPA, an aggregated packet is generated to egress switch d in one of the following cases:

- The volume of traffic in all class-based buffers in buffer pool d is equal to S_{th}: in this case, all user packets in the class-based buffers of egress switch d are aggregated in the same optical packet.

- The volume of traffic in all class-based buffers in buffer pool d is larger than S_{th}: here, all client packets in the class-based buffers of egress switch d, except the newly arrived client packet, are aggregated in the same optical packet.

Algorithm: Composite Packet Aggregation in Multi-Class OPS

Definitions:
$Q_{d,c}$:   Queue dedicated to class c of egress switch d
$L_{d,c}$:   Length of queue $Q_{d,c}$ in bits
S_{th}:   Maximum size of an optical packet in bits
$AgPk$:   A data structure to contain a number of client packets up to S_{th}
APTB: Aggregated packet transmission buffer

PacketArrival (packet p, class c, egress switch d):
1: **if** $L_{d,c} = 0$ **then**
2:     SetTimer (d,c) {set a new timer for class c if its relevant buffer is empty}
3: **end if**
4: Append p to $Q_{d,c}$ {append client packet p to the end of $Q_{d,c}$ buffer}
5: Update $L_{d,c}$ {update the size of class c buffer in buffer pool d }
6: $L_q = L_{d,1} + L_{d,2} + ... + L_{d,M}$ {combined traffic size in all classes in buffer pool d }
7: **if** $L_q = S_{th}$ **then**
8:     GenOpticalPacket $(d, 0)$ {aggregate clients packets from buffer d}
9: **else**
10:     **if** $L_q > S_{th}$ **then**
11:        GenOpticalPacket (d,c) {aggregate clients packets from buffer d, except packet p}
12:     **end if**
13: **end if**
14: END

Timeout (egress switch d):
1: GenOpticalPacket $(d, 0)$ {for any timeout event happened from any class 1 to M, aggregate all client packets from M buffers of buffer pool d}
2: END

GenOpticalPacket (egress switch d, mode m):
1: ResetTimer (d) {clear timer of all buffers in buffer pool d}
2: $AgPk=$" {initialize aggregated packet $(AgPk)$}
3: **for** each class c **do**
4:     $AgPk= AgPk$ + aggregated packets from $Q_{d,c}$ up to S_{th} {build aggregated packet $(AgPk)$}
5:     **if** $m = c$ **then**
6:        SetTimer (d,c) {set a new timer for class c since it has a client packet}
7:     **end if**
8:     Update $L_{d,c}$ {update the size of class c buffer in buffer pool d}
9: **end for**
10: Append $AgPk$ to APTB buffer {append aggregated packet $AgPk$ to the APTB buffer}
11: END

Figure 2.14: Pseudo code of composite optical packet aggregation (CPA) in multi-class OPS networks. Reproduced from [59] with permission from De Gruyter.

- A timeout happened for class c traffic in buffer pool d: here, the whole traffic from class c is first collected. If the volume of the collected traffic is still less than S_{th}, then the available traffic in other classes in buffer pool d, starting from HP class, are collected until the aggregation size reaches S_{th}. For example, assume that there are three classes in a network and timeout has happened for mid-priority class in buffer pool d. Consider the aggregated traffic from the mid-priority class in AgPk is still less than S_{th}. Then, the client packets from HP class are also aggregated in AgPk until the size of aggregated packets in AgPk reaches S_{th}, but not more than S_{th}. If there is still room and no client traffic available in the HP class, then the traffic from the LP class is also aggregated in AgPk until size S_{th}, but not more than S_{th}. It should be mentioned that the CPA algorithm ensures that the aggregation of a client packet in AgPk would never result in exceeding the size of AgPk from S_{th}. This feature can improve bandwidth usage because of reducing unused bandwidth in an optical packet.

Aggregated packet *AgPk* is then saved in APTB. At an appropriate time and at time slot boundary, the OBM unit converts each aggregated packet waiting in APTB to an optical packet, and sends to the OPS network on one of available wavelengths. Each ingress switch can transmit up to $f \times W$ simultaneous optical packets at the same time, called an Optical Packet set (OP set). Note that the fiber number (in a multi-fiber OPS network) and the wavelength number in the picked fiber are uniformly selected.

Algorithm: Non-Composite Packet Aggregation in Multi-Class OPS

Definitions:
$Q_{d,c}$:   Queue dedicated to class c of egress switch d
$L_{d,c}$:   Length of queue $Q_{d,c}$ in bits
S_{th} :   Maximum size of an optical packet in bits
$AgPk$:   A data structure to contain a number of clients packets up to S_{th}
APTB :   Aggregated packet transmission buffer

PacketArrival (packet p, class c, egress switch d):
1: **if** $L_{d,c} = 0$ **then**
2:    SetTimer (d,c) {set a new timer for class c if its relevant buffer is empty}
3: **end if**
4: Append p to $Q_{d,c}$ {append client packet p to the end of $Q_{d,c}$ buffer}
5: Update $L_{d,c}$ {update the size of class c buffer in buffer pool d}
6: **if** $L_{d,c} = S_{th}$ **then**
7:    GenOpticalPacket $(c,d, 0)$ {aggregate client packets from class c buffer of buffer pool d}
8: **else**
9:    **if** $L_{d,c} > S_{th}$ **then**
10:       GenOpticalPacket $(c,d, 1)$ {aggregate client packets from class c buffer of buffer pool d, except the new packet p}
11:    **end if**
12: **end if**
13: END

Timeout (class c, egress switch d
1: GenOpticalPacket $(c,d, 0)$ {for timeout event from class c, aggregate all client packets from class c buffer of buffer pool d}
2: END

GenOpticalPacket (class c, egress switch d, mode m):

1: ResetTimer (c,d) {clear timer of class c in buffer pool d}
2: $AgPk$= aggregated packets from $Q_{d,c}$ up to S_{th} {create aggregated packet $(AgPk)$}
3: **if** $m = 1$ **then**
4:    SetTimer (d,c) {set a new timer for class c since it has a client packet}
5: **end if**
6: Update $L_{d,c}$ {update the size of class c buffer in buffer pool d}
7: Append $AgPk$ to APTB buffer {append aggregated packet $AgPk$ to the APTB buffer}
8: END

Figure 2.15: Pseudo code of non-composite optical packet aggregation (CPA) in multi-class OPS networks

2.1.2.1.2 *Non-Composite Packet Aggregation (NCPA)* Figure 2.15 depicts the pseudo code of the NCPA mechanism with many similarities to CPA. Unlike the CPA technique, this technique only aggregates client packets of the same class in one optical packet [31, 33, 34]. There is one timer dedicated to class c traffic of each

buffer pool. For egress switch d traffic, the timer of class c is started whenever a client packet arrives at class c buffer and this buffer is empty. An aggregation for class c traffic related to egress switch d occurs when:

- Timeout occurs for class c: Here, all client packets in class c buffer are aggregated, even if there is not enough traffic to completely fill an optical packet, thus wasting network bandwidth. Then, the relevant timer is stopped and reactivated later on when a new client packet arrives at class c buffer; and

- A new client packet arrives at class c buffer and the size of traffic waiting in class c buffer becomes equal or larger than S_{th}: In this case, client packets up to size S_{th} are aggregated. Then, if the new client packet is still in class c buffer, the timer for class c is adjusted according to the arrival time of the new packet.

Then, optical packets can be made and then sent to the OPS network similar to CPA.

As a slight modification to the pseudo code in Fig. 2.15, the work in [54] triggers the aggregation process even if the size of client packets waiting in a buffer becomes larger than $S_{th} - 40$ bytes. This is because no client packet usually has a size smaller than 40 bytes in practice. The performance evaluation results in [54] show the effectiveness of packet aggregation in reducing hurst parameter (i.e., burstiness) of optical packets arrived at the OPS network, especially at high traffic load. In other words, at high traffic load, transmission of optical packets to the OPS network becomes smoother (i.e., PLR reduces), but the average waiting time of client packets waiting in ingress switches increases.

2.1.2.1.3 Comparison Between CPA and NCPA Since traffic from different classes can be carried in an optical packet payload, the CPA mechanism can reduce traffic load sent to an OPS network; i.e., a smaller number of optical packets per unit time arrive at OPS network compared with NCPA. Thus, CPA experiences a lower PLR than NCPA. In addition, the queuing delays and end-to-end delay for all classes can be reduced under CPA. An undesirable effect in CPA is that LP traffic may even experience lower queuing delay than higher-priority traffic; i.e., as a disadvantage for CPA.

We have mentioned a number of benefits and disadvantages for PA under single-class traffic in Section 2.1.1.8. Under differentiated traffic, PA may have additional disadvantages under ISA1:

- PA may not provide enough delay differentiation among traffic classes whenever a timeout event is deactivated due to a higher traffic arrival rate. This is because under high traffic arrival rate, before the timeout event, optical packets may have been generated.

- The choice of timeout is a challenging problem. Since in practice the volume of HP traffic is usually less than LP traffic, the average aggregation time for the HP traffic is higher than the LP traffic [61]. Thus, a smaller timeout must be chosen for the HP class to provide an appropriate differentiation compared with the LP class. By reducing timeout, however, many optical packets would be generated (with a small number of client packets inside), thus increasing optical packet load in the network. Hence, the number of collisions goes up. This in turn increases PLR. Furthermore, small timeouts or even large timeouts at high traffic loads may lead to the case in which the optical packet generation rate becomes higher than the optical packet transmission rate to the OPS network. Therefore, the waiting time in the transmission buffer may increase client packets waiting time in an ingress switch.

- By increasing timeout values, client packets experience high delay in ingress switches, and therefore generated optical packets can include more client packets inside than the previous case. Therefore, the number of optical packets arrived at the network goes down. This in turn reduces collision in the network, thus increasing the network throughput.

- There may be an unfair bandwidth allocation for well-behaved traffic sources. For example, consider a burst of client packets arrives at an ingress switch going to a given egress switch. If the threshold-based aggregation nature of PA generates a train of (malicious) optical packets going to a given egress switch, these malicious optical packets can block and delay other well-behaved optical packets going to other egress switches when accessing to the optical network. While these malicious optical packets are being served, well-behaved optical packets going to other egress switches would have to wait in APTB. This issue becomes even worse when the malicious optical packets carry LP traffic so that the LP traffic would experience even a lower queuing delay than HP traffic.

- It is difficult to control the service order of client packets coming from different legacy networks in a shared buffer because a malicious legacy network in a class may cause a high delay and even loss for well-behaved sources within that class.

- The management of load balancing on the transmission channels is usually difficult because an OBM unit would not know the number of optical packets that will be generated from the fluctuating arrival rate of client packets. Therefore, an obvious approach is that the OBM unit should aggressively transmit optical packets. Consequently, transmission channels may be fully utilized at some periods and they may be almost empty at some other periods.

2.1.2.2 Distribution-Based Bandwidth Access (DA) Smooth transmission of optical packets to an OPS network has been proposed in order to reduce PLR in [55, 62], called Distribution-based bandwidth Access (DA), which is suitable for slotted OPS under the ISA2 architecture (see Section 2.1.2). In DA, an ingress switch allocates some bandwidth to egress switch i in a number of time slots within a frame period, where the frame is a collection of time slots from all wavelengths on all fibers over a fixed duration of F time slots. Then, the allocated number of time slots to the desired egress switch are scheduled within the frame and among the output wavelengths/fibers of the ingress switch. When transmitting traffic, the ingress switch knows from the scheduling in which time slot, on which fiber, on which wavelength channel, and to which egress switch it should send the relevant traffic.

Note that DA does not require any timeout mechanism for packet differentiation required for QoS-based traffic. Instead, packet differentiation is implemented by packet scheduling algorithms used at the ingress switch. DA uses the Output Controlled Grant-based Round Robin (OCGRR) [63] scheduler as its class-based packet schedulers. In DA, an integer number of client packets from different classes can be carried in an optical packet. Since DA transmits traffic in a smooth way, not as greedy as the PA techniques that use timeout for releasing traffic to OPS network, the traffic load on the network is reduced. As a result, PLR can be decreased through the whole network.

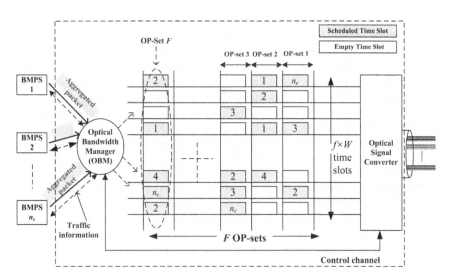

Figure 2.16: The OBM operation in the DA technique. Reprinted from *Computer Networks*, Vol. 53, Akbar Ghaffarpour Rahbar and Oliver Yang, Distribution-based bandwidth access scheme in slotted all-optical packet-switched networks, 744–758, copyright 2009, with permission from Elsevier.

Figure 2.16 shows $f \times W$ time slots available from the fibers and wavelengths for each OP set. There are all together $F \times f \times W$ time slots within a frame when the frame width is F OP sets. Note that an OP set is a set of optical packets transmitted at the same time slot over the available fibers/wavelengths in an ingress switch. A gray time slot in Fig. 2.16 shows a scheduled time slot within the frame, where the number inside the time slot represents an egress switch (alternatively a torrent) address. On the other hand, a white time slot shows a non-scheduled time slot within the frame in which no optical packet will be transmitted.

The OBM unit in an ingress switch (as shown in Fig. 2.16) has two main functionalities:

- It provides the required bandwidth in time slots for each torrent at frame boundaries. The OBM unit computes the required number of time slots to transmit its traffic to each egress switch at each frame boundary. The criterion to allocate time slots to egress switch i is based on the current traffic volume in the buffers dedicated to egress switch i plus the traffic arriving in these buffers within the previous frame period.

- To achieve smooth traffic transmission, the OBM unit schedules the optical packets to be transmitted in the time slots that have the same distance from each other within the frame. The OBM unit also distributes them evenly among all wavelengths/fibers of the ingress switch so that the traffic is load-balanced at each time-slot period.

2.1.2.2.1 Bandwidth Provisioning in DA In DA, bandwidth is provided at a given ingress switch for n_e torrents at frame boundaries by the OBM unit. A traffic estimator algorithm is used to compute the number of time slots required for Torrent-i in the ingress switch. Define B_C to be a wavelength channel bandwidth in bps, and S_T be the time duration of a time slot that carries traffic in seconds. Let L_a be average client packet length in bits, Q_i be the current queue size of traffic for Torrent-i in bits, $N_{e,i}$ be the average number of client packet arrivals for Torrent-i, $N_{c,i}$ be the number of client packet arrivals for Torrent-i during the last frame period, and $0 \leq \sigma \leq 1$ be a smoothing factor. The simple running average mechanism is used to compute $N_{e,i} = \sigma \times N_{e,i} + (1 - \sigma) \times N_{c,i}$. Since the average number of client packets that can be carried within a time slot as an optical packet is $N_S = \left\lfloor \frac{B_C \times S_T}{L_a} \right\rfloor$, the average number of time slots required to carry Torrent-i traffic is given by

$$\theta_i = \min \left(\left\lceil \frac{N_{e,i} + Q_i}{N_S} \right\rceil, \theta_{max} \right), \tag{2.4}$$

where θ_{max} is set by the network manager to limit the number of time slots that can be allocated to Torrent-i in order to avoid over-provisioning of bandwidth for malicious sources. It is suggested to have $\frac{F \times f \times W}{n_e} \leq \theta_{max} \leq \frac{F \times f \times W}{2}$ so that at most half of a bandwidth can be used by a torrent.

In Eq. (2.4), θ_i includes enough number of time slots to transmit the traffic in Torrent-i buffers as well as the amount of the traffic that will arrive during the next frame period for Torrent-i. Define $\theta = \sum_{i=0}^{n_e-1} \theta_i$ to be the total number of requested time slots by all torrents in the ingress switch. Since θ may exceed $F \times f \times W$ (i.e., the total number of available time slots within a frame), the number of requested time slots must be adjusted accordingly. For adjusting, a time slot scaling coefficient is defined by $\gamma = \min \left(1, \frac{F \times f \times W}{\theta} \right)$. Then, the actual number of time slots that can be assigned to Torrent-i is given by $\Lambda_i = \lfloor \gamma \times \theta_i \rfloor$. Clearly, if $\theta \leq F \times f \times W$, we have $\gamma = 1$, and the requested time slots to Torrent-i can be completely allocated.

The frame length F in DA must be selected appropriately. A large frame length can reduce the processing work of ingress switches. However, it may result in an inaccurate bandwidth provisioning, especially when traffic arrival is bursty. On the other hand, a small frame length can lead to frequent bandwidth provisioning and time-slot assignment process, which increases the processing load of an ingress switch. In addition, it may still result in an inaccurate bandwidth provisioning due to the lack of enough traffic arrival. Better results may be obtained whenever we have $40 \leq F \leq 100$ for a time slot of 10 μs, where bandwidth provisioning can be provided in 0.4 msec to 1.0 msec intervals.

2.1.2.2.2 Time-Slot Scheduling in DA   Figure 2.17 shows the pseudo code of the time-slot scheduling algorithm used by an OBM unit to evenly distribute Λ_i time slots for Torrent-i within a frame boundary and among the fibers and wavelengths. All of the wavelength channels of the fibers are numbered serially from 1 to $f \times W$. After initializing each item of time-slot scheduling (SS) matrix to an empty time slot, a capacity of $f \times W$ time slots is assigned to each OP set within the frame.

A Linked List (LL) is used to keep the reference of the torrents with $\Lambda_i > 0$, where $0 \leq i \leq (n_e - 1)$. Define a distribution round to be a the cycle that would include all the torrents with $\Lambda_i > 0$. In each distribution round, $\min(F, \Lambda_i)$ time slots of Torrent-i can only be distributed within the frame. Then, if $\Lambda_i > F$, the reference of Torrent-i is added to the end of the next distribution round so that the left-over time slots of Torrent-i are distributed in the next distribution round. This process continues until list LL becomes empty.

Within the outer while-loop in the pseudo code, the reference of Torrent-i is first obtained and then removed from LL. To distribute the time slots assigned to Torrent-i, a starting OP set p is first randomly selected, from which the algorithm tries to maintain equal distance (referred to as optimum distance) between any two consecutive time slots assigned to Torrent-i. The optimum distance η between two time slots of Torrent-i is ideally equal to $\frac{F}{\Lambda_i}$ time slots. In practice, however, the positions of Λ_i time slots need to be adjusted to satisfy the criterion of optimum-distance distribution; i.e., the distance between any two consecutive time slots (belonging to the same torrent) is equal. This is because the desired optimum-distance of $\frac{F}{\Lambda_i}$ time slots may not always be an integer value, and therefore the assigned time slots to Torrent-i cannot be always equally spaced. Hence, parameter δ is introduced, where $0 \leq \delta < 1$, for providing an optimum-distance distribution. Note that when F is divisible by Λ_i, we always have $\delta = 0$. Let the current occupied OP set be p. Then, the next OP set for occupation should be $p + z$, where z is given by $z = \left\lfloor \delta + \frac{F}{\Lambda_i} \right\rfloor$. Parameter δ is then updated by $\delta = \delta + \frac{F}{\Lambda_i} - z$. If the desired OP set is fully occupied, the next empty OP set is linearly searched and allocated within the distance of z.

After finding an empty OP set p for a time slot of Torrent-i, a non-occupied wavelength w is randomly chosen using function $findChannel()$. Note that this wavelength is actually mapped to wavelength channel $w \bmod f$ (i.e., remainder of w divided by f) on fiber $\left\lfloor \frac{w}{f} \right\rfloor$. In a multi-fiber OPS network, DA tries to avoid collision at OPS core switches by not distributing the time slots of the same torrent on the same wavelength

Algorithm: Time-Slot Scheduling

1: Call $EmptySS()$ {initialize matrix SS elements with EMPTY}
2: **for** $p = 1$ to F **do**
3:    $Capacity[p] = f \times W$ {initialize capacity of each column of frame (each OP set)}
4: **end for**
5: Call $Initialization()$ {required initializations}
6: **while** TRUE **do**
7:    $i = getNextRef()$ {get the next torrent in linked list LL}
8:    **if** $i = NULL$ **then**
9:       Return {finish time-slot scheduling if all time slots of all torrents have been scheduled}
10:    **end if**
11:    $p = selectRandomOPSet()$ {choose a random OP set in the frame}
12:    $\eta = \max(1.0, \frac{F}{\Lambda_i})$ {desired distance between OP sets}
13:    $k = 0$
14:    $\delta = 0$
15:    **while** $\Lambda_i > 0$ **do**
16:       $\delta = \delta + \eta$
17:       $z = \lfloor \delta \rfloor$ {compute the real distance between OP sets}
18:       $\delta = \delta - z$
19:       $p = getEmptyOPSet(p, z)$ {get the reference of the next empty OP set within the frame}
20:       $w = findChannel(p)$ {randomly choose non-occupied wavelength w in OP set p}
21:       $SS[w][p] = i$ {schedule one time slot for Torrent-i at column p and wavelength w}
22:       $Capacity[p] = Capacity[p] - 1$ {update capacity of each column (each OP set)}
23:       $k = k + 1$ {update the number of time slots scheduled for Torrent-i}
24:       $\Lambda_i = \Lambda_i - 1$ {update the number of time slots required for Torrent-i}
25:       **if** $k = F$ **then**
26:          **if** $\Lambda_i > 0$ **then**
27:             $appendList(i)$ {append Torrent-i to list LL if there are still non-scheduled time slots for Torrent-i}
28:          **end if**
29:          Exit loop {exit internal while loop}
30:       **end if**
31:    **end while**
32: **end while**
33: END

Figure 2.17: Time-slot scheduling (assignment) under the DA technique. Reproduced from *Computer Networks*, Vol. 53, Akbar Ghaffarpour Rahbar and Oliver Yang, Distribution-based bandwidth access scheme in slotted all-optical packet-switched networks, 744–758, copyright 2009, with permission from Elsevier.

of different fibers. After saving the reference of Torrent-i on row w and column p in matrix SS, the capacity of OP set p is decremented by 1, Λ_i is decremented by 1, while counter k is incremented by 1. If F time slots are already distributed for

Algorithm: **Function** $getEmptyOPSet(p,z)$

1: $q = (p - z) \bmod F$ {q is the reference of the next desired OP set in frame}
2: **If** $Capacity[q] > 0$ **then**
3:    Return q {since q is not a full OP set}
4: **end if**
5: $\omega = q \bmod \phi$ {the status of OP set q is kept by bit ω}
6: $u = h = \lfloor \frac{q}{\phi} \rfloor$ {obtain relevant E_u for OP set q}
7: $\omega_2 = 1$
8: $\phi_2 = 1$
9: Shift left ω_2 for ω times
10: Shift left ϕ_2 for ϕ times
11: $x = E_u \wedge (\omega_2 - 1)$ {reset all bits in E_u from ω to $\phi - 1$ to 0 and save the result in x. When $x \neq 0$, there are some bits equal to 1 in E_u}
12: **while** $x = 0$ **do**
13:    $u = (u - 1) \bmod m$ {point to the next word}
14:    $x = E_u$
15:    **if** $u = h$ **then**
16:      $x = x \wedge (\phi_2 - \omega_2)$ {reset all bits from 0 to $\omega - 1$ in E_u to zero}
17:    **end if**
18: **end while**
19: $y = findPosition(x)$ {find the position of the most significant set bit in x}
20: return $y + u \times \phi$ {return the desired OP set reference}
21: END

Figure 2.18: The function $getEmptyOPSet$ used in Fig.2.17. Reproduced from *Computer Networks*, Vol. 53, Akbar Ghaffarpour Rahbar and Oliver Yang, Distribution-based bandwidth access scheme in slotted all-optical packet-switched networks, 744–758, copyright 2009, with permission from Elsevier.

Torrent-i and we still have $\Lambda_i > F$, the reference of Torrent-i is appended to end of LL to be processed in the next distribution round. After distributing F time slots for Torrent-i, the next torrent is processed. When there is no torrent left in LL, the distribution process is over.

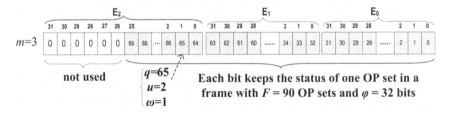

Figure 2.19: An example for bit memory management at $F = 90$ OP sets and $\phi = 32$ bits. Reprinted from *Computer Networks*, Vol. 53, Akbar Ghaffarpour Rahbar and Oliver Yang, Distribution-based bandwidth access scheme in slotted all-optical packet-switched networks, 744–758, copyright 2009, with permission from Elsevier.

To implement the function *getEmptyOPSet()* in Fig. 2.17 that searches for an empty time slot within a frame, the bit-pattern memory management technique is used (see Fig. 2.18). Using this technique, the status of OP sets (in terms of empty or non-empty) within a frame can be represented by an array of bits, where only one bit per OP set is used to indicate the occupancy status of the OP set. A "1" in an OP set means that there is at least one time slot empty in the OP set. On the other hand, "0" means that all the time slots in the OP set are already occupied. Here, Boolean operator "and" (denoted by $\land$) is applied on a word of size ϕ bits, where ϕ is multiples of 16 bits depending on the processor used in an edge switch. Hence, an array of F bits must be managed by array E with $m = \left\lceil \frac{F}{\phi} \right\rceil$ words. The bits in a word are numbered from left to right as $\phi - 1$ to 0, and the words in an array are numbered from $m - 1$ to 0. Hence, the status of OP set q is kept by bit $\omega = q \bmod \phi$ in word E_u, where $u = \left\lfloor \frac{q}{\phi} \right\rfloor$. For example, Fig. 2.19 illustrates the bit-pattern memory management for a frame with $F = 90$ OP sets and $\phi = 32$ bits, where $m = 3$ words E_2, E_1, and E_0 are required to maintain the status of the 90 OP sets. During the initialization of the time-slot scheduling algorithm, all bits of words E_2, E_1, and E_0 are set to 1.

The function *getEmptyOPSet()* returns the reference of the next empty OP set within the frame. The reference of the next desired OP set is given by $q = (p - z) \bmod F$, where q may recirculate to the frame end. If OP set q is empty, q is the returned value. Otherwise, the reference u of the relevant word and the bit number in E_u are computed for OP set q. For instance, we have $u = 2$ and $\omega = 1$ for $q = 65$. Now, $h = u$ is the index of the first word to be evaluated. The next step is to find a word that has at least one set bit. All bits in E_u numbered from ω to $\phi - 1$ are reset to 0 and the result is saved in parameter x. In other words, only bits 0 to $\omega - 1$ in the first word are evaluated to be non-zero. When $x \neq 0$, there are some bits equal to 1 in E_u. Hence, E_u is the answer of the search. Otherwise, the loop searches for a non-zero word, which may be executed up to m times in the worst case. Within the loop, index u is reduced by 1 to point to the next word. However, index u may re-circulate to point to the end of the words. If an index u points to a new word, parameter $x = E_u$ and then its value is compared with 0. If it is still 0, the loop continues. Now, assume index u points to the first word discussed above (i.e., $u = h$). Here, all the bits numbered from 0 to $\omega - 1$ in E_u are reset to zero, and the bits numbered from ω to $\phi - 1$ in E_u are checked for being non-zero. After finding the non-zero word E_u, the position of the most significant set bit in x is found using the function *findPosition()* and saved in parameter y. This function uses low level instructions depending on the processor used in the edge switch, e.g., the BSR function in Intel CPUs. Finally, $y + u \times \phi$ is returned from *getEmptyOPSet()* as the desired OP set reference.

2.1.2.2.3 Traffic Transmission in DA Consider the schedule of a given OP set in a specific time slot. At a suitable time before the time slot beginning (depending on the processing speed), the OBM unit requests the relevant BMPS units (that have scheduling in the OP set) to provide their aggregated packets for transmission. Assume that k optical packets should be sent from Torrent-i in the OP set. Upon request by the OBM, the packet scheduler in BMPS-i provides k aggregated packets and sends them to the OBM. To create an aggregated packet, the relevant PS schedules client packets from the relevant traffic streams and from different classes and aggregates a traffic up to $B_C \times S_T$ bits. Note that an aggregated packet (to be sent in a time slot as an optical packet) must include an integer number of client packets with the total size of less than or equal to $B_C \times S_T$ bits.

For example, consider OP set 2 in Fig. 2.16, where four optical packets are scheduled: two from Torrent-1, one from Torrent-2, and one from Torrent-4. Here, the OBM requests BMPS-1, BMPS-2, and BMPS-4 to provide two, one, and one optical packet(s), respectively. During the header interval or even earlier, the OBM prepares an aggregated header (see Section 1.4.4) based on the optical packets ready for transmission in OP set 2. Then, the OBM sends the aggregated header over the control channel. At the time slot boundary, the OBM sends the ready optical packets to the relevant egress switches on the predetermined fibers/wavelengths.

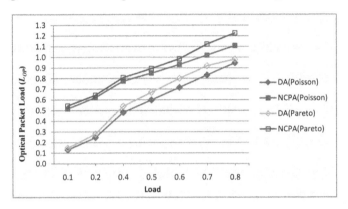

Figure 2.20: Optical packet generation rate over optical packet service rate (L_{OP}) under DA and NCPA

2.1.2.2.4 Advantages of DA   DA has a number of advantages as listed below:

- **DA is a fair bandwidth allocation scheme**: Its fairness can be studied under two cases:

  - By allocating a dedicated buffer to each traffic stream in any class inside a BMPS unit (i.e., using G buffers for G streams in each class) and using fair packet schedulers inside ingress switches to process the traffic streams, all the streams within each class can be fairly processed. Therefore, a bursty source cannot cause high delay and even loss for well-behaved sources within a class. This is unlike the other schemes that save all the same-class client packets from different source routers in a shared FCFS buffer.

  - Each torrent is assigned a number of time slots within a frame period to send its traffic. This bandwidth is guaranteed and no (malicious) torrent can use it. This is a short-term fairness in which all torrents can send their traffic within the frame period. This is unlike the NCPA scheme in which a malicious torrent can generate a large number of optical packets at a very small period of time in such a way that no other well-behaved torrents can send their traffic. Fairness performance is formulated and evaluated in Section 2.3.

- **DA can guarantee a stable edge switch operation:** Define optical packet load L_{OP} to be the optical packet generation rate over optical packet service rate in an ingress switch. In other words, L_{OP} equals average number of optical packets built per time slot divided by $f \times W$. Under DA, the maximum value for L_{OP} happens when $f \times W$ optical packets are scheduled to be transmitted per OP set with $f \times W$ wavelength channels; i.e., $L_{OP} = 1$. By this mechanism, DA guarantees a stable edge switch operation so that the number of optical packets that should be sent in an OP set never exceeds the number of available wavelengths. This is unlike NCPA in which we may have even $L_{OP} > 1$, called unstable edge switch operation, in which the aggregation buffer APTB would overload.

  Figure 2.20 displays average L_{OP} for DA and NCPA when the inter-arrival of client packets follow Poisson and Pareto (with shaping parameter $a = 1.2$) distributions in each ingress switch of an OPS network with eight edge switches. Load L indicates the normalized client traffic load (from source routers) arriving at an edge switch. Traffic load L is normalized with respect to the total bandwidth of each edge switch; i.e., $f \times W \times B_c$. As can be observed, the parameter L_{OP} for DA is always less than for NCPA. Under NCPA, one can observe that $L_{OP} > 1.0$ at $L = 0.6$ under Pareto traffic and also at $L = 0.7$ under Poisson traffic. Therefore, an edge switch becomes unstable in servicing the optical packets waiting in APTB. In other words, the delay of optical packets waiting in APTB can grow to infinity. One can see that L_{OP} never exceeds 1.0 in DA because even when all $f \times W$ time slots in an OP set are occupied, we have at most $L_{OP} = 1.0$. The reason for studying parameter L_{OP} is to show that the number of optical packets that arrive in an OPS network is smaller under DA than NCPA. Therefore, the number of optical packets collisions reduces under DA compared with NCPA, thus decreasing PLR and increasing throughput under DA compared with NCPA as illustrated in [55, 62].

- **DA Can Balance Load on Wavelength Channels Better Than NCPA:** This is evaluated in Section 2.3.

- **DA Can Provide a Smooth Access to the OPS Network:** This is evaluated in Section 2.3.

- **DA Works at a Large Time Granularity:** In terms of operation complexity, on each client packet arrival, NCPA checks the packet size to see whether an optical packet can be created or not. This per-packet operation is a problem at high packet arrival rates. However, DA predicts future traffic and then schedules enough number of time slots appropriately within a frame period. This scheduling is performed only once at each frame interval which can be a much bigger time granularity than the per-packet operation. Since DA operations can be carried out at a large time granularity, its scalability can be assured.

- **The Complexity of DA Is $O(F \times f \times W)$ in the Worst Case:** This is because the complexity components of DA are as the following according to Fig. 2.17:

- The complexity of the $EmptySS()$ procedure and initialization of the OP set *Capacity* array are $O(F \times f \times W)$ and $O(F)$, respectively.

- The algorithm executes the block of codes inside the second loop for at most $\sum_{i=0}^{n_e-1} \Lambda_i \leq F \times f \times W$ times. Hence, there are $F \times f \times W$ runs in the worst case.

- All codes within both loops are executed with complexity $O(1)$. In addition, it can be seen that $getEmptyOPSet()$ has a complexity of almost $O(1)$ since its complexity is equal to $O(m) = O\left(\left\lceil \frac{F}{\phi} \right\rceil\right) \approx O(1)$, especially when $\phi \gg 1.0$ or $F \leq 2 \times \phi$. At $\phi = 128$ in a 128-bit processor, for instance, the complexity is exactly $O(1)$ if $F \leq 128$, and the complexity becomes almost $O(2) \approx O(1)$ at $F \leq 256$.

2.1.2.3 The OPORON Packet Aggregation Approach In Optical Packet switching Over wavelength ROuted Network (OPORON) [64], client packets are aggregated and then sent to an asynchronous OPS network. The OPORON saves client packets as the following. Similar to the ISA1 architecture, there are $n_e = n - 1$ buffer pools in each ingress switch, and M buffers in each buffer pool for saving traffic of M classes of service. A packet classifier unit in an ingress switch determines the egress switch and class of service for any new client packet and then saves it in the relevant buffer. Therefore, if the class of a client packet is c and its egress switch is d, it is saved in class c buffer dedicated to egress switch d. For each buffer, there are two threshold parameters C_{max} and C_{min} on packet aggregation size in bytes, along with two threshold parameters T_{max} and T_{min} on client packets waiting times in queues. Based on these four thresholds, client packet aggregation can be triggered. Considering the incoming traffic characteristics, these four thresholds can be adjusted in a way in order to shape the traffic arrival at the OPS network. It should be noted that the operation of NCPA could be considered as a special case of OPORON in which $C_{max} = C_{min}$ and $T_{max} = T_{min}$.

The OPORON generates an aggregated optical packet to egress switch d in three cases:

- Suppose that an incoming client packet must be buffered in buffer c of buffer pool d, but this buffer is full. In this case, the client packet is stored in a higher class buffer in buffer pool d. If there is no empty buffer at higher classes, the traffic from class c buffer is aggregated and sent to the APTB buffer (similar to ISA1), and then the client packet is buffered in buffer c of buffer pool d.

- When the waiting time in class c exceeds T_{max}, the client packets in this buffer are aggregated and sent to the APTB buffer. Note that on the arrival of a client packet in an empty buffer, the aggregation timer is triggered. The operation here is similar to the PA technique.

- When an incoming client packet is buffered in buffer c of buffer pool d, the traffic from this buffer is aggregated provided that two conditions are satisfied;

the volume of traffic in buffer c is in range $[C_{min}, C_{max}]$ and the waiting time for the first packet in buffer c is in range $[T_{min}, T_{max}]$.

Table 2.3: Aggregation parameters and simulation results for NCPA and OPORON [64]

Trigger zone	C_{min} (bytes)	C_{max} (bytes)	T_{min} (μsec)	T_{max} (μsec)	Hurst parameter (R/S)	Hurst parameter (Time-Variance)	Mean aggregation delay (μsec)	Inter-arrival time (LRD / SRD)
Input traffic	—	—	—	—	0.78	0.81	—	LRD
NCPA	3000	3000	5	5	0.62	0.63	4.8	SRD
OPORON	2000	3000	3	5	0.51	0.50	3.1	SRD

According to the above operation, an optical packet may carry client packets from different classes similar to CPA. Table 2.3 illustrates simulation results and aggregation parameters for NCPA and OPORON packet aggregation mechanisms. In this simulation, the R/S method and the time-variance method have been used for estimating the Hurst parameter H and the degree of self-similarity. The average delay experienced by the client packets in the aggregation mechanism prior to transmission through optical domain can be reduced (by almost 30% according to the performance evaluations). The threshold parameters are obtained from the observation of the input IP traffic profiles, generated by a multi-fractal wavelet model. The performance results (based on the time-variance method) show that the self-similarity of the input traffic generated based on hurst parameter $H = 0.81$ has been decreased to $H = 0.63$ and $H = 0.50$ in the OPS network by using NCPA and OPORON, respectively. The Hurst parameter measured under the R/S method also proves such reduction in decreasing parameter H. Finally, the last column results show that both NCPA and OPORON packet aggregation mechanisms can significantly decrease the Long-Range Dependence (LRD) of client packets inter-arrival times so that the inter-arrival times of transmitted optical packets to the OPS network have Short-Range Dependence (SRD).

2.1.2.4 Network Layer Packet Redundancy The packet redundancy technique discussed for single-class OPS networks in Section 2.1.1.10 can be applied only to HP traffic in order to provide QoS differentiation in a QoS-capable OPS network [65]. This technique avoids resource reservation, but the load of redundant optical packets increases in the network, degrading the total network performance. However, if the redundancy is applied only to HP traffic, the additional load may be negligible since the HP traffic shares a small part of the network traffic compared with the LP traffic.

This technique can support differentiation. However, packet redundancy requires more processing power in the network edge switches compared with the other mentioned techniques. It will also lead to additional delay for the HP client packets because of processing in ingress and egress switches. The more redundant packets, the more processing. Finally, this technique increases network load and PLR as a result.

2.1.2.5 MPLS-Based QoS Differentiation   Quality of service provisioning in OPS networks could be based on MPLS (see 1.4.6.3.1) that offers resource reservation with proper control algorithms to support different classes of service and traffic engineering [66], thus enhancing network efficiency. In the proposed scheme, the connection-oriented forwarding capability of MPLS can be used over connectionless OPS networks by establishing LSPs. This allows optical network resources such as wavelengths and optical buffers to be reserved over predetermined paths in order to provide required QoS for HP optical packets.

The traffic engineering capability of MPLS can route optical packets in an OPS network in a way to balance traffic load on various links and OPS switches. Load balancing obviously can result in reducing PLR. The traffic engineering can also assure a better utilization of available resources for HP optical packets. Once optical packets are classified at ingress switches, specific forwarding rules can be applied in each OPS switch for different class-based optical packets. As a result, PLR of HP optical packets can be minimized. However, implementing MPLS traffic engineering increases complexity of OPS switches. In addition, the MPLS traffic engineering can provide a better reliability for HP optical packets by appropriately routing them around failed links/nodes.

2.1.2.6 Comparison of Software-Based Contention Avoidance Schemes for Multi-Class Traffic   The operations of studied contention avoidance schemes for OPS networks supporting multi-class traffic can be grouped into two categories:

- **No Coordination (NC)**: Ingress switches in some techniques (like CPA, NCPA, DA, OPORON, and network layer packet redundancy) send their optical packets without any coordination with the OPS core network. All the techniques in NC can be easily combined with the network layer packet redundancy technique.

- **Loose Coordination (LS)**: Ingress switches in MPLS-based QoS differentiation are loosely coordinated with the OPS network in order to have a balanced optical packet transmission. This technique can be combined with all the techniques in the NC category.

Table 2.4: Comparison of multi-class software-based contention avoidance schemes

Category	Algorithm	Slotted OPS/ Asynchronous OPS	Single-Hop/ Multi-Hop	Single-Fiber/ Multi-Fiber	Complexity
NC	CPA	both	both	both	moderate
	NCPA	both	both	both	moderate
	DA	slotted OPS	both	both	moderate
	OPORON	both	both	both	moderate
	Network Layer Packet Redundancy	asynchronous OPS	both	single-fiber	moderate
LC	MPLS-based QoS Differentiation	both	muti-hop	single-fiber	moderate

Table 2.4 compares multi-class software-based contention avoidance schemes in terms of suitability for slotted or asynchronous OPS, suitability for single-hop or

multi-hop OPS, suitability for single-fiber or multi-fiber OPS, and operation complexity.

Now, a comparison is made among the features of the software-based contention avoidance schemes proposed for QoS-capable OPS networks:

- The MPLS-based technique deals with traffic at a coarse level in terms of accepting, rejecting, and routing, whereas the packet aggregation, OPORON, and the smooth traffic transmission techniques are bandwidth access techniques that deal with client packets at a finer level. Any other bandwidth access technique can be easily combined with the MPLS-based technique.

- Using network layer packet redundancy increases optical packet load. However, if it is only used for HP traffic, the increase in optical packet load can be ignored since the share of HP traffic is usually small in OPS networks. This technique can be combined with other bandwidth access techniques.

- To support QoS-based traffic under PA, the CPA technique is suggested that can significantly reduce the PLR compared with NCPA. The main idea of CPA is to prevent from transmission of an optical packet that is not completely filled. The OPORON packet aggregation scheme can also improve PLR and burstiness of optical packets compared with NCPA. However, adjusting the parameters of OPORON could be a challenging issue. Using smooth optical packet transmission to an OPS network under DA can resolve the problems stated in Section 2.1.2.1.3 for the PA mechanisms. Unlike NCPA, both CPA and DA can carry client packets from different classes in the same optical packet. The DA in an ingress switch can (1) provide each torrent to have a fairer access to network bandwidth, (2) reduce the average traffic drop rate in the OPS network, (3) balance load on wavelength channels much better, and (4) guarantee a stable edge switch operation in which the optical packet generation rate is lower than the optical packet service rate. In terms of memory requirement, NCPA keeps track of a timer and a threshold for each class as well as an APTB buffer, while DA keeps track of the schedule of all time slots within a frame period (i.e., DA requires more memory than NCPA). Clearly, DA, OPORON, CPA, and NCPA cannot be combined with each other.

The loss and delay performances for different classes are close to each other under CPA. The DA provides lower PLR for both HP/MP traffic at both high and light traffic loads compared with CPA. The CPA can provide lower PLR for HP/MP classes under moderate traffic loads. The DA always provides the worst end-to-end delay for LP traffic, but the best delay for HP/MP traffic, especially at low and moderate traffic loads. Choosing timeout values is important in CPA since lower timeout values may result in infinite delay at heavy traffic loads.

The DA mechanism can provide a stable edge switch operation, and fairness for torrents. It can also reduce collision at the network using smooth traffic transmission and load balancing. Consequently, network throughput can be increased. Furthermore, DA has resolved the problem of choosing a good timeout value required under NCPA because under the proposed architecture the service differentiation is provided by class-based packet scheduling. Finally, there is no optical packet transmission buffer and the optical packets from the torrents are scheduled directly to output wavelength channels in an ingress switch.

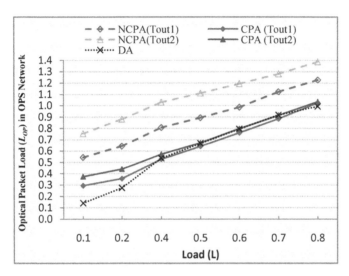

Figure 2.21: Optical packet load in OPS network under DA, CPA, and NCPA bandwidth access schemes. Reproduced from [59] with permission from De Gruyter.

Figure 2.21 displays optical packet load L_{OP} in OPS network under DA, CPA, and NCPA bandwidth access schemes [59]. An eight-node OPS network (i.e., $n = 8$) with $f = 2$, and $W = 4$ is evaluated. Define T_{HP}, T_{MP}, and T_{LP} to be the timeout values set for HP, MP, and LP client packets, respectively. Two different timeout sets are evaluated for CPA and NCPA: (1) Tout1 with [$T_{HP} = 20$ μs, $T_{MP} = 40$ μs, $T_{LP} = 100$ μs] for HP, MP, and LP traffic, respectively; and (2) Tout2 with [$T_{HP} = 15$ μs, $T_{MP} = 25$ μs, $T_{LP} = 50$ μs] for HP, MP, and LP traffic, respectively. Recall DA uses packet scheduling to aggregate client packets from different classes in an optical packet. Hence, performance results for DA are independent of timeout values. Looking at Fig. 2.21 shows that L_{OP} is smaller under DA compared with both NCPA and CPA at light traffic loads and heavy traffic load of 0.8. However, L_{OP} for DA lies between CPA performances at moderate traffic load. One can see that the parameter L_{OP} never exceeds 1.0 in DA because even when all $f \times W$ optical packets in an OP set are occupied we have at most $L_{OP} = 1.0$. Although L_{OP} of CPA and DA are close to each other, DA outperforms CPA. The reason for small L_{OP} under DA is due to its smooth traffic generation. Since L_{OP} is higher under NCPA compared with CPA, the NCPA will have a higher collision rate than CPA in the OPS network. Therefore, CPA can significantly reduce loss rate for all classes compared with NCPA. By reducing timeout values, L_{OP} under NCPA is significantly increased due to the huge number of optical packets generated due to small timeouts.

However, L_{OP} under CPA is slightly increased under the Tout2 set. This is because of the aggregation of client packets from different classes in the same optical packet.

Since $f \times W = 8$ optical packets can be served at each $S_T + S_O = 10$ μs in an ingress switch destined to $n_e = 7$ egress switches, the average optical packet service rate is $\frac{8 \text{ optical packets}}{7 \times 10 \ \mu s} \approx 114,000$ optical packets/s. Define r_i and t_i to be respectively the average traffic arrival rate at class i of a given egress switch in packets/s and timeout assigned to class i, where i = EF, AF, and BE. Under Poisson arrival, an optical packet from class i can be generated when either the timeout event has occurred where $n_1 = 1 + t_i \times r_i$ packets arrive for class i, or when there are $n_2 = \left\lfloor \frac{B_c \times S_T}{L_a} \right\rfloor$ client packets in class i buffer [67] to completely fill an optical packet. Hence, the optical packet generation rate of class i is $\frac{r_i}{n_{min}}$, where $n_{min} = min(n_1, n_2)$.

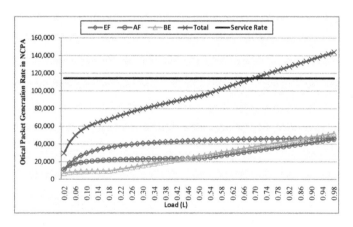

Figure 2.22: Optical packet generation rate under Poisson traffic arrivals in NCPA

Figure 2.22 shows the analytical results in ISA1 under NCPA with Tout1 for optical packet generation rate of EF, AF, and BE classes, and their total rates. The analysis results show that the timeout event that indiscriminately generates optical packets becomes ineffective at $L \geq 0.18$ and $L \geq 0.52$ for the BE and AF traffic, respectively. This is because at these ranges there is enough traffic in BE and AF classes to make a full optical packet before the timeouts relevant to classes BE and AF, respectively, occur. Therefore, the optical packet generation becomes smoother later on. As can be observed, following an average optical packet service rate of 114,000 optical packets/s, the total optical packet generation rate from these three classes still continues to grow, thus leading to $L_{OP} > 1$.

2.2   Hardware-Based Schemes

In this section, the contention avoidance schemes that require additional network hardware for reducing the volume of contending optical packets in an OPS network are detailed. These schemes are suitable both for asynchronous OPS and slotted OPS networks. Contention avoidance in hardware-based approaches is achieved by increasing the hardware components, thus increasing the network cost.

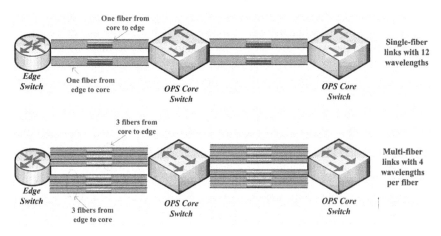

Figure 2.23: OPS multi-fiber links

2.2.1   Designing a Multi-Fiber (MF) OPS Architecture

In this technique [1, 4, 9, 59, 68, 69], each connection link in an OPS network, between any pair of edge switch to OPS switch or any OPS switch to OPS switch, contains f fibers with a smaller number of wavelengths per fiber (see Fig. 1.12) than a single fiber OPS network. For example, if a single-fiber network needs $\omega = 12$ wavelengths per fiber per connection link (see top image in Fig. 2.23) to transport its traffic, then a multi-fiber OPS network with $f = 3$ fibers on each connection link requires $W = \left\lceil \frac{\omega}{f} \right\rceil$ wavelengths per fiber (see bottom image in Fig. 2.23).

Consider that an OPS switch is connected to N input edge/core switches with N input links and N output edge/core switches with N output links. Assume that there are c contending optical packets arriving at the core switch on wavelength λ that compete for output link l at the same time. Let us see what happens in single-fiber and multi-fiber OPS networks:

- In a single-fiber network, the core switch size is $N * N$ and $f = 1$ fiber is used on each input/output link of the OPS switch. Here, only one optical packet can be switched on wavelength λ in output link l, and up to $c - 1$ optical packets may be dropped when there is no means to resolve their contention at the OPS switch. In the worst case, all optical packets coming from all input links may compete for output link l at the same time. In this case, $c = N$ and $N - 1$ optical packets will be dropped, the highest PLR at that moment.

- Now, consider a multi-fiber OPS network with f fibers used on each input/output link, where an input/output link is connected to f input/output ports of the OPS switch. Here, an $N*N$ OPS switch used in a single-fiber network is replaced with an $(N \times f)*(N \times f)$ switch in a multi-fiber architecture. In output link l of the OPS switch, up to f optical packets on the same wavelength λ can be switched on f output ports at the same time. Then, up to $c - f$ optical packets may be dropped, when there is no means to resolve their contention at the OPS switch. In the worst case, all optical packets coming from all input links may compete for output link l at the same time. In this case, we have $c = N \times f$ and $N \times f - f$ optical packets will be dropped, thus leading to the highest PLR at that moment.

Let us show the above-defined c contenting optical packets as c-OP collision. It is proved in Appendix D.4 in [70] for slotted OPS that the role of k-OP collision in dropping optical packets is larger than the role of all c-OP collision where $c > k$. For example, by considering $k = 2$, PLR due to 2-OP collision is larger than sum of PLR due to 3-OP collision, PLR due to 4-OP collision, PLR due to 5-OP collision, and so on. In other words, if a device is used to resolve the collision of two optical packets, the PLR will significantly reduce. However, this significance is exponentially reduced when k increases. Therefore, to achieve a very low PLR, so many contention resolution devices must be used. One of these devices is to use multi-fiber OPS network. By using f fibers on each connection link, it is possible to have f wavelength channels of the same wavelength in each output link, and therefore up to f optical packets (on the same wavelength) can be switched successfully at the same time. By this scheme, PLR due to 2,3,...,f-OP collision are all removed.

A multi-fiber network can result in a smaller number of collisions and a lower number of optical packet drops than can a single-fiber OPS network. The requirement for a large switch size (the switch size is increased by a factor of f), however, could be a disadvantage for the multi-fiber architecture. The advantages of multi-fiber networks have been stated in Section 1.3. It should be mentioned that the deployment of multi-fiber architecture can be easily provided in a metro area as there are carrier companies in many metro areas.

Figure 2.24 depicts PLR in a single-hop slotted OPS network with an OPS switch with size 30 * 30 (i.e., $N = 30$) and when there is no contention resolution mechanism used at the optical switch. Fixed-length client packets with Poisson inter-arrival times are generated so that a desired load on wavelength channels can be achieved, and each optical packet carries only one client packet. Here, each fiber has four wavelengths and each wavelength carries the same traffic load. Different loads and a different number of fibers are investigated. The upper diagram shows PLR for the case that all edge switches are connected to the core OPS switch with $f = 1$ fiber. For example, PLR increases from 21% to 33.5% when load changes from 0.5 to 0.9.

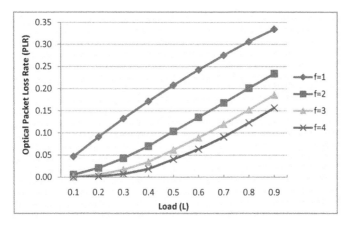

Figure 2.24: The effect of using multi-fiber OPS switch on reducing PLR at $N = 30$

The diagram at $f = 2$ shows a better gain in PLR reduction. As mentioned, for the most part, the collisions are due to the collision of two optical packets. A two-fiber architecture leads to the PLR of almost 10% to 23% for different loads in range 0.5 and 0.9 (11% less drop compared with $f = 1$). A network with three fibers results in PLR of 6% to 18.5% for $L = 0.5$ to 0.9. As can be observed, the PLR reduction (almost 5%) is smaller compared with $f = 2$ (11%). The four-fiber network has also reduced the PLR (almost by 3%) compared with $f = 3$. Now the performance results show that in a four-fiber network and at high load of $L = 0.9$, the contention resolution schemes at the core switch must resolve contention of just 15.5% instead of 33.5% in a single-fiber network.

2.2.2 Using Additional Wavelengths (AW) to Carry the Same Traffic

An equivalent effect to lowering traffic load in order to reduce PLR is to use Additional Wavelengths (AW) to transmit the same traffic on each connection link [4, 31, 69]. Recall that L_{OP} is the normalized load of optical packet arrival at the OPS network and L is the normalized traffic load of client packet arrivals in ingress switches. Let an ingress switch with W wavelengths in a slotted OPS has $N_p \leq W$ optical packets to transmit. One can write $N_p = L \times W$. In this case, by using all W available wavelengths, a wavelength is occupied with probability $\frac{N_p}{W}$, and therefore the normalized traffic load on each wavelength channel equals $L_{OP} = \frac{N_p}{W} = L$, provided that the traffic distribution on the wavelengths is uniform. Now, consider that the same ingress switch uses W_a additional wavelengths to transmit the same number of optical packets. By distributing N_p optical packets uniformly over $W + W_a$ wavelengths, a wavelength is occupied with probability $L_{OP} = \frac{L \times W}{W + W_a}$, which is obviously less than $\frac{N_p}{W}$. Since the normalized traffic load on a wavelength is reduced, PLR decreases in the OPS network. In a window of m time slots on a wavelength channel, we expect $E = m \times (1 - L_{OP})$ empty time slots. By increasing E, PLR reduces as a result.

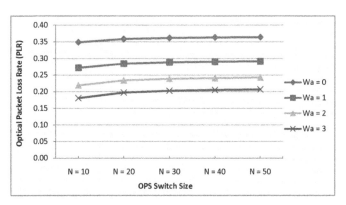

Figure 2.25: The effect of using additional channels vs. different number of OPS switch sizes under a network with $W = 3$ wavelengths

Figure 2.25 depicts the PLR from simulation results in a single-hop slotted OPS network with an $N * N$ OPS switch at $N = 10, 20, ..., 50$, where the OPS switch uses no contention resolution mechanism. Fixed-length client packets with Poisson inter-arrival times are generated so that a desired load L on wavelength channels can be achieved, and each optical packet carries only one client packet. Here, we have $W = 3$ wavelengths and the number of additional channels W_a varies from 0 to 3. Assume that all ingress switches send their optical packets with the same load $L = 1.0$ (i.e., full traffic load). For $N = 10$, using one, two, and three additional channels leads to the PLR of 27.1%, 21.9%, and 18.1%, respectively. As can be seen, the role of using more additional channels decreases when W_a tends to W. As the diagram shows, by using one additional channel, PLR reduces by almost 20% for all cases. The second and the third additional channels reduce the PLR by almost 35% and 45%, respectively. The same situation can be observed for OPS switches with $N > 10$. Note that to use additional wavelengths, more transceivers must be employed in ingress and egress switches. This clearly results in a high network cost.

2.2.3 Combined Hardware-Based Schemes

The aforementioned hardware-based contention avoidance schemes can be combined with each other so that in each fiber of a multi-fiber network, W_a additional wavelengths can be used in order to significantly reduce the PLR. The combined Multi-Fiber (MF) and Additional Wavelength (AW) channels is called MFAW. However, the cost-effective combination of fiber and additional wavelengths channels should be considered [69].

Let ω be the number of required wavelengths to carry full traffic load of a single-fiber OPS network. In a multi-fiber architecture with f fibers per connection link, ω may not be always divisible by $f \times W$. In this case, one should use $W = \left\lceil \frac{\omega}{f} \right\rceil$ wave-

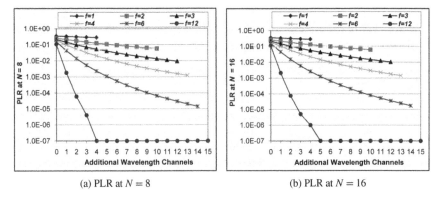

(a) PLR at $N = 8$ (b) PLR at $N = 16$

Figure 2.26: PLR at the core switch for different architectures at $\omega = 12$

lengths per fiber, where each fiber would have $f \times \left\lceil \frac{\omega}{f} \right\rceil - \omega$ additional wavelengths channels on each connection link.

In MFAW, an ingress switch has $f \times (W + W_a)$ available wavelength channels, where the optical packets departing the ingress switch must be uniformly distributed over these wavelength channels. As computed in Section 2.2.2, a wavelength is occupied by probability $L_{OP} = \frac{L \times W}{W + W_a}$. However, note that traffic load L should be updated to $L \times \frac{\omega}{f \times W}$ in multi-fiber network since ω may not be always divisible by $f \times W$. Having $f \times W \geq \omega$ has the same effect as using additional wavelength channels. Therefore, we have $L_{OP} = L \times \frac{\omega}{f \times W} \times \frac{W}{W + W_a}$. Note that without using additional wavelength channels we have $L_{OP} = L \times \frac{\omega}{f \times W}$. Clearly, the load of optical packets in MFAW is less than the load of optical packets in multi-fiber networks. Therefore, PLR reduces under MFAW when $W_a > 0$. Note that the analytical equation for computing PLR in a multi-fiber slotted OPS switch with size N, f fibers on each connection link, and optical packet load L_{OP} is given by [68]

$$PLR(N, f, L_{OP}) = \frac{\displaystyle\sum_{i=f+1}^{N \times f} \binom{N \times f}{i} \left(\frac{L_{OP}}{N}\right)^i \left(1 - \frac{L_{OP}}{N}\right)^{N \times f - i} (i - f)}{f \times L_{OP}}. \quad (2.5)$$

Note that to compute the convolution used in Eq. (5) in [68], one should recall that the r-fold discrete convolution of the binomial distribution $\binom{m}{k} p^k (1-p)^{m-k}$ is $\binom{m \times r}{k} p^k (1-p)^{m \times r - k}$. Equation (2.5) proves that PLR decreases by reducing L_{OP}. Since the load of transmitted traffic is assumed to be the same on all wavelength channels of each ingress switch, PLR is independent of W or W_a and is the same on all wavelength channels.

Figures 2.26a and 2.26b depict PLR (in a logarithmic scale) at the core switch of a single-hop OPS network under MFAW for two network scenarios (with core switch size $N = 8$ and $N = 16$) at $\omega = 12$ and $L = 1.0$. Different network architecture at $f = 1, 2, 3, 4, 6, 12$ fibers on each connection link are evaluated. In this scenario, ω is always divisible by f. Since maximum number of wavelength channels on a fiber (W_a) is limited to $W_{max} = 16$ in this scenario, the number of additional channels in a fiber cannot exceed $W_{max} - \left\lceil \frac{\omega}{f} \right\rceil$. With switch size of $N = 8$ and $f = 1$ fiber, PLR starts from 34.3% at $W_a = 0$ and reduces to 27.3% at $W_a = 4$. As another example, with $N = 8$ and $W_a = 1$, PLR reduces from 32.3% at $f = 1$ to 0.17% at $f = 12$. With $N = 16$ edge switches, the same trend can be observed. However, under the same number of fibers and additional channels, PLR for $N = 16$ is a little higher than for $N = 8$. Choosing $f = 2$ or $f = 3$ has much more significant reduction in PLR when compared with $f > 3$. This is also true for additional wavelengths so that one can obtain considerable reduction in PLR when we have smaller number of additional wavelengths such as $W_a = 1$, $W_a = 2$.

A desired PLR may be obtained under different hardware components. For example, in both scenarios, PLR < 0.01 is obtained at $(f, W_a) = [(4, 7), (6, 3), (12, 1)]$, but some of them may not be cost-effective. To achieve a very low PLR, a relatively high number of fibers or additional wavelengths must be used, so that for PLR $< 10^{-5}$, we can only rely on the network architecture based on $f = 12$. PLR is almost zero at ($N = 8$, $f = 12$, $W_a \geq 4$) and ($N = 16$, $f = 12$, $W_a \geq 5$).

Additional performance evaluations show that be decreasing ω, PLR increases. For example, at $\omega = 6$ and $N = 8$, the smallest PLR cannot be lower than 2.74×10^{-7} (at $N = 8, f = 6, W_a = 15$). This level is higher than the one observed for the $\omega = 12$.

2.2.4 Comparison of Hardware-Based Schemes

The hardware-based contention avoidance schemes reduce PLR by increasing the OPS network resources without decreasing load of optical packet arrival in the OPS network. All hardware-based contention avoidance schemes can be easily combined with every software-based contention avoidance scheme reviewed for single-class and multi-class OPS networks.

The use of additional wavelengths in a fiber provides more bandwidth to carry the same traffic. However, this requires more transceivers at ingress and egress switches and more devices to handle a high number of wavelengths on a fiber [71].

Compared with the single-fiber OPS architecture, however, the number of wavelengths is the same in a multi-fiber architecture and only the number of fibers and the size of each OPS switch must be increased in the network. Since a multi-fiber architecture with a small number of wavelengths per fiber is more cost-effective than a single-fiber architecture with a large number of wavelengths per fiber [71], the former technique seems to be much suitable to be applied to OPS networks (see Section 1.3 for additional advantages of multi-fiber networks). For example, consider two scenarios: The first scenario has one fiber and h wavelengths inside the fiber, and the second scenario has h fibers and each fiber has one wavelength. Note that the latter scenario has two noticeable advantages: network survivability and less PLR. Moreover, the cost of the second scenario is less than the architecture that uses the first scenario plus wavelength converters at the core switch to achieve the same PLR reduction as the first scenario [72].

Finally, as depicted, MFAW gains the advantages of both using additional channels and multi-fiber architecture and can provide the least PLR compared with its basic mechanisms. However, for choosing an appropriate number of fibers f and number of additional channels W_a, one needs to run an optimization algorithm to make a tradeoff between network cost and desired PLR level, where the optimum point of this optimization results in the appropriate f and W_a [69].

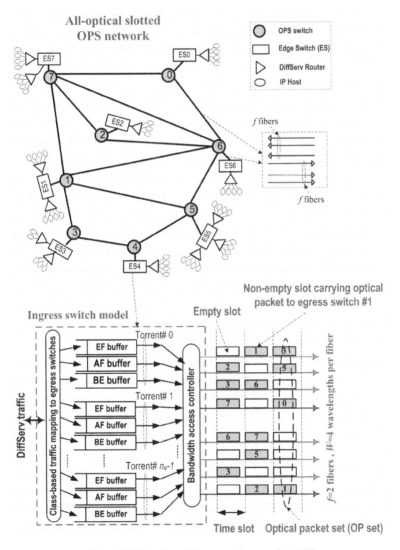

Figure 2.27: Network model for even packet transmission [73]

2.3 Formulation of Even Traffic Transmission in Slotted OPS

In even traffic transmission, it is desirable for an ingress switch to evenly (smoothly) transmit its optical packets to an OPS network during frame periods and among wavelengths and fibers so that arrivals at each intermediate OPS switch and egress switch can be even (smooth). Even optical packet transmission can reduce the burstiness of optical packet arrival in the OPS network, thus reducing PLR.

In [73], some formulations have been provided for even optical packet transmission in synchronized multi-fiber multi-wavelength OPS networks with f fibers per connection link and W wavelengths per fiber. By these formulations, one can quantitatively compute evenness parameters for bandwidth access schemes and then compare the superiority of one bandwidth access scheme with another one. To do this, a figure of merit called "distribution index" is defined to compare different bandwidth access schemes.

Figure 2.27 depicts the network model for even optical packet transmission. Within a time slot, an integer number of client packets of any type and class (at least one client packet) can be carried. Recall that the packet traffic carried in a time slot is referred to as an optical packet. An empty time slot therefore has no traffic to any egress switch. On the other hand, a non-empty time slot carries traffic toward a given egress switch. In this figure, a number written inside a gray time slot indicates an egress switch address. Each ingress switch can transmit at the same time up to $f \times W$ simultaneous optical packets, called an optical packet (OP set), on its $f \times W$ wavelengths. In Fig. 2.27, we have $f = 2$ and $W = 4$. Before transmitting each OP set and during the header interval, an OP set header is sent over the control channel in each fiber (but not displayed in Fig. 2.27) based on the aggregated header discussed in Section 1.4.4. Each header includes the traffic information for each optical packet in the relevant OP set, such as the egress switch address and the number of client packets in each class carried in the optical packet. An electronic buffer is used at the output of each ingress switch for possible aggregation of class-based traffic packets from their hosts (e.g., legacy networks) before transmitting them in fixed optical time slots.

We shall use $|a|$, $\lceil a \rceil$, $\lfloor a \rfloor$, and $a \bmod b$ to denote respectively the absolute value of a, the ceiling value of a, the floor value of a, and the remainder of a divided by b. Note that when a is an integer number, we assume $\lceil a \rceil = \lfloor a \rfloor = a$. Let a torrent be all client packets going to the same egress switch. In other words, in an ingress switch the traffic of Torrent-i is destined to egress switch i.

2.3.1 Evenness Index Parameters

The evenness is computed on the distribution of optical packets within a frame period snapshot. For this, a Gauging Frame (GF) with size τ OP sets is defined in an ingress switch. There are all together $\tau \times f \times W$ optical packets that can be sent in a GF. Let Z_i be the number of optical packets (out of $\tau \times f \times W$) that Torrent-i uses within a GF. Considering $n_e = n - 1$ torrents in a given ingress switch, total number of optical packets leaving the ingress switch in a GF is given by

$$Z_t = \sum_{i=0}^{n_e-1} Z_i \leq \tau \times f \times W . \tag{2.6}$$

To provide evenness indexes, the Even Assignment Problem (EAP) is first formulated. Consider two sets A and B with α and β items, respectively. The objective of EAP is to assign all the items in Set B to as many items as possible in set A so that no item in Set A is "overwhelmed". Ideally, $\frac{\beta}{\alpha}$ items from Set B should be assigned to each item in Set A. Since β may not be divisible by α, the desired assignment $\frac{\beta}{\alpha}$ may not always be practical. Let m_j be an integer number of items from Set B that must be assigned to Item j in Set A. The assignment index X for EAP is defined as

$$X = \sum_{j=0}^{\alpha-1} \left| m_j - \frac{\beta}{\alpha} \right| - X^* , \tag{2.7}$$

where the summation calculates how close is the assignment of the items in Set B to the items in Set A to $\frac{\beta}{\alpha}$, and X^* is the optimum assignment index.

Parameter X^* in Eq. (2.7) is necessary because $\frac{\beta}{\alpha}$ may not always result in an integer value. Using X^* guarantees $X = 0$ under an optimum distribution. Parameter X^* is computed by the following mechanism. When β is not divisible by α, we have two cases:

- $\left\lfloor \frac{\beta}{\alpha} \right\rfloor$ items in Set B must be assigned to each one of $k = \alpha \times \left\lceil \frac{\beta}{\alpha} \right\rceil - \beta$ items in Set A.

- $\left\lceil \frac{\beta}{\alpha} \right\rceil$ items in Set B must be assigned to each one of the $\alpha - k$ items in Set A.

  Based on the above assignment, all the items in Set B are mapped to Set A because we always have $k \times \left\lfloor \frac{\beta}{\alpha} \right\rfloor + (\alpha - k) \times \left\lceil \frac{\beta}{\alpha} \right\rceil = \beta$.

As stated, we have $X = 0$ under an optimum distribution. Therefore, we can find X^* by

$$X^* = \sum_{j=0}^{\alpha-1} \left| m_j - \frac{\beta}{\alpha} \right| = k \times \left(\frac{\beta}{\alpha} - \left\lfloor \frac{\beta}{\alpha} \right\rfloor \right) + (\alpha - k) \times \left(\left\lceil \frac{\beta}{\alpha} \right\rceil - \frac{\beta}{\alpha} \right) , \tag{2.8}$$

which is simplified as

$$X^* = 2 \left(\alpha \times \left\lceil \frac{\beta}{\alpha} \right\rceil - \beta \right) \times \left(\frac{\beta}{\alpha} - \left\lfloor \frac{\beta}{\alpha} \right\rfloor \right) . \tag{2.9}$$

Note that when β is divisible by α, we have $k = 0$ and $\left\lceil \frac{\beta}{\alpha} \right\rceil = \left\lfloor \frac{\beta}{\alpha} \right\rfloor = \frac{\beta}{\alpha}$. Hence, all items in Set B can be equally assigned to each one of the α items in Set A. In this case, we have $X^* = 0$.

In general, even traffic transmission can be categorized into four types of distributions as discussed in the following. Each category can be mapped to the EAP problem with different α and β parameters.

2.3.1.1 Distance-Based Even Distribution

Under this distribution, optical packets sent to a given egress switch are equally spaced from each other within a GF frame so that a deterministic optical packet service rate is provided for each torrent within the frame. This method of even distance-distribution through GF is denoted as EDF_{di}.

Let $\Phi_{i,N_i} = \left\{ (x_j, y_j) | 0 \le j \le N_i - 1, 0 \le x_j \le \tau - 1, 0 < y_j \le \min(f \times W, Z_i) \right\}$ be the distribution of Z_i time slots allocated to Torrent-i in N_i OP set positions. Here, x_j refers to the OP set number within the GF, and y_j (as density) shows the number of optical packets destined to egress switch i at OP set x_j. Note that for the remaining $\tau - N_i$ OP sets, we have $y_j = 0$. For instance, in $\Phi_{2,3} = \{(1,2),(5,1),(7,4)\}$, we have $x_0 = 1, x_1 = 5, x_2 = 7$ with the densities of $y_0 = 2, y_1 = 1, y_2 = 4$ optical packets, respectively. Here, $Z_2 = y_0 + y_1 + y_2 = 7$ time slots are allocated to Torrent-2 within the GF and they are distributed in only $N_2 = 3$ OP set positions. Define the difference between two OP set positions x_j and x_{j-1} that carry traffic for Torrent-i to be $d_j = x_j - x_{j-1}$, where $j \in (0, 1, .., N_i - 1)$ and $x_{-1} = x_{N_i-1} - \tau$.

The distance-distribution index $X_{DI,i}$ for Torrent-i is computed by

$$X_{DI,i} = \sum_{j=0}^{N_i-1} \left| d_j - \frac{\tau}{N_i} \right| - X_{DI,i}^* , \tag{2.10}$$

where the summation computes how close are the optical packets distributed for Torrent-i to the ideal distance of $\frac{\tau}{N_i}$. Since the ideal distance of $\frac{\tau}{N_i}$ may not always be an integer number, the bias parameter $X_{DI,i}^*$ should be used in order to guarantee $X_{DI,i} = 0$ under an optimum distance-distribution. When $\frac{\tau}{N_i}$ is an integer value, we have $X_{DI,i}^* = 0$ since the optical packets can be distributed exactly in equal time slot distances from each other. Note that $X_{DI,i}$ is sensitive to the difference between the time slot positions assigned to optical packets within the GF, not the positions of the optical packets.

In an optimum distribution for N_i optical packets, k OP sets must have a distance of $k_1 = \left\lfloor \frac{\tau}{N_i} \right\rfloor$ time slots from each other, and $N_i - k$ OP sets must have a distance of $k_2 = \left\lceil \frac{\tau}{N_i} \right\rceil$ time slots from each other. Consider set $\{x^*, y^*\}$ to be an optimum distance distribution for a set with N_i optical packets. Then, we should have $X_{DI,i} = 0$, and therefore we have

$$X_{DI,i}^* = \sum_{j=0}^{N_i-1} \left| d_j^* - \frac{\tau}{N_i} \right|, \tag{2.11}$$

where $d_j^* = x_j^* - x_{j-1}^*$. Considering $k \times k_1 + (N_i - k) \times k_2 = \tau$, parameter $X_{DI,i}^*$ can be expanded to

$$X_{DI,i}^* = k \times \left(\frac{\tau}{N_i} - \left\lfloor \frac{\tau}{N_i} \right\rfloor \right) + (N_i - k) \times \left(\left\lceil \frac{\tau}{N_i} \right\rceil - \frac{\tau}{N_i} \right), \tag{2.12}$$

which is simplified as

$$X_{DI,i}^* = 2 \left(N_i \left(\left\lfloor \frac{\tau}{N_i} \right\rfloor + 1 \right) - \tau \right) \left(\frac{\tau}{N_i} - \left\lfloor \frac{\tau}{N_i} \right\rfloor \right). \tag{2.13}$$

Finally, define X_{DI} to be the average of EDF_{di} indices among all torrents within a GF as $X_{DI} = \frac{\sum_i X_{DI,i}}{n_e}$. The index X_{DI} can be used to relatively compare the distance-distribution performances of different bandwidth allocation schemes. Note that an optimum distribution within a frame for all torrents is obtained when $X_{DI} = 0$.

2.3.1.2 Density-Based Even Distribution

Here, optical packets of the same torrent are distributed with the same density among all OP sets within the GF frame so that an equal number of optical packets from each torrent can be served in time slots. This method of even density distribution through the GF is denoted as EDF_{de}. For example, consider there are 23 optical packets to be sent toward a given egress switch. These optical packets can be scheduled to be transmitted in a set of 6 OP sets. In this case, equal distribution of optical packets means that five OP sets will carry 4 optical packets and one OP set will carry 3 optical packets (difference of 1) to the egress switch.

Considering the parameters defined in Section 2.3.1.1, the density-distribution index $X_{DE,i}$ for Torrent-i can be computed by mapping the EDF_{de} problem to the EAP problem by setting $\beta = Z_i$ and $\alpha = \tau$:

$$X_{DE,i} = \sum_{j=0}^{\tau-1} \left| y_j - \frac{Z_i}{\tau} \right| - X_{DE,i}^*, \tag{2.14}$$

where $X_{DE,i}^*$ is computed as

$$X_{DE,i}^* = 2 \left(\tau \times \left\lceil \frac{Z_i}{\tau} \right\rceil - Z_i \right) \left(\frac{Z_i}{\tau} - \left\lfloor \frac{Z_i}{\tau} \right\rfloor \right). \tag{2.15}$$

Finally, the density-distribution index X_{DE} within a GF is defined as the average of EDF_{de} indices among all torrents in an ingress switch $X_{DE} = \frac{\sum_i X_{DE,i}}{n_e}$. The index X_{DE} can be used to relatively compare the density-distribution performances of different bandwidth allocation schemes. Note that an optimum distribution within a frame for all torrents is achieved when $X_{DE} = 0$.

2.3.1.3 Fair Dissemination (FD) Distribution The same opportunity should be provided for different torrents to access to an OPS network under the FD distribution. The FD deals with the fairness issue and computes how uniform are the egress switches of the optical packets going to n_e egress switches within a GF. In simple words, each ingress switch should send an almost equal number of optical packets to each egress switch within the τ OP sets interval; i.e., a symmetric traffic transmission to the network. The FD problem can be mapped to the EAP problem, where Set B has $\beta = Z_t$ items and Set A has $\alpha = n_e$ items. The FD index X_{FD} for all torrents within the GF is computed by

$$X_{FD} = \sum_{j=0}^{n_e-1} \left| Z_j - \frac{Z_t}{n_e} \right| - X_{FD}^* , \qquad (2.16)$$

where the first term computes how close are the distribution of the optical packets for each egress switch to the fair distribution of $\frac{Z_t}{n_e}$, and X_{FD}^* is the optimum fairness dissemination index. Since $\frac{Z_t}{n_e}$ may not always be an integer value, the parameter X_{FD}^* is required as a bias in order to provide $X_{FD} = 0.0$ under an optimum FD distribution. This parameter can be computed by

$$X_{FD}^* = 2 \left(n_e \times \left\lceil \frac{Z_t}{n_e} \right\rceil - Z_t \right) \left(\frac{Z_t}{n_e} - \left\lfloor \frac{Z_t}{n_e} \right\rfloor \right) . \qquad (2.17)$$

The index X_{FD} can be used to relatively compare fairness performances of different bandwidth access techniques with each other. Note that an optimum distribution is obtained when $X_{FD} = 0$.

2.3.1.4 Load Balanced (LB) Distribution Under LB distribution, transmission of optical packets should be in a way to provide load balancing among different wavelengths/fibers in a given time slot. The aim of LB is not to overwhelm wavelengths/fibers in time slots. Unlike the density distribution, the LB distribution tries to evenly distribute the optical packets to different egress switches among the wavelengths/fibers at each OP set within a GF. Let set $\{\theta_j | j \in (0, \tau - 1)\}$ be the distribution of optical packets on τ OP sets within a GF, where $\theta_k \in (0, f \times W)$ denotes the total number of distributed optical packets on OP set k, where each optical packet may be destined to a different egress switch. Mapping the LB problem to the EAP problem by $\beta = Z_t$ and $\alpha = \tau$, the LB index X_{LB} is defined by

$$X_{LB} = \sum_{j=0}^{\tau-1} \left| \theta_j - \frac{Z_t}{\tau} \right| - X_{LB}^* , \tag{2.18}$$

where the summation computes how close is the load balancing of optical packets distributed at each OP set to the desired density of $\frac{Z_t}{\tau}$. The optimum load-balanced distribution index X_{LB}^* is necessary since $\frac{Z_t}{\tau}$ may not always lead to an integer value. This guarantees $X_{LB} = 0$ under an optimum load-balanced distribution. Similar to the previous distribution categories, X_{LB}^* is given by

$$X_{LB}^* = 2 \left(\tau \times \left\lceil \frac{Z_t}{\tau} \right\rceil - Z_t \right) \left(\frac{Z_t}{\tau} - \left\lfloor \frac{Z_t}{\tau} \right\rfloor \right). \tag{2.19}$$

The index X_{LB} is used to relatively compare the load-balanced distribution performances of different bandwidth allocation schemes in a synchronized OPS network. Note that an optimum distribution is obtained when $X_{LB} = 0$.

Figure 2.28: Three types of optical packet distributions at $\tau = 10$ OP sets [73]

2.3.1.5 Benefits of Even Optical Packet Transmission

Figure 2.28 shows three distributions within a GF of 40 time slots at $\tau = 10$ OP sets, $f = 1$, and $W = 4$ (i.e., each OP set can carry at most four optical packets). The number displayed inside a time slot indicates a torrent number (i.e., the relevant egress switch address). In Fig. 2.28a, optical packets are evenly distributed in terms of both distance and density. For instance, Torrent-6 optical packets (i.e., the optical packets going to egress switch 6) have a distance of two time slots from each other. In addition, under Distribution-a, the density of Torrent-6 optical packets in different OP sets is even (i.e., always 1). However, Distribution-b has an uneven distribution. For instance, the distance between two Torrent-6 optical packets is not even (i.e., either one or seven time slots apart). Also, there is an uneven density among the optical packets distributed for Torrent-6 (namely, 3, 2, 1, and 1 optical packet(s) in the first four OP sets within the GF). As seen, the density distribution is the worst for Torrent-7 optical packets with 1, 3, and 1 optical packet(s) in the last three OP sets of the GF. This "uneven-ness" gives rise to high latency variation and low bandwidth utilization, which may not be desirable for many traffic applications.

In general, the following advantages can be counted for even traffic transmission:

- **Less Jitter:** Using even distance-distribution, optical packet transmission can be smoothed and jitter can be reduced due to the deterministic service rate within a frame. For example, in Distribution-b of Fig. 2.28, Torrent-6 optical packets are sent in the first part of the frame. Thus, Torrent-6 optical packets experience less delay and jitter in the first part of the frame since they are all sent back-to-back. Until the end of the frame, no optical packet is sent from Torrent-6. Let another frame (say Distribution-c in Fig. 2.28) immediately follows the frame of Distribution-b. In Distribution-c, Torrent-6 optical packets are transmitted at the end of the frame (i.e., in OP sets 8 and 9). Therefore, jitter of Torrent-6 optical packets will be high because of the large distance between the transmissions of Torrent-6 optical packets in these two consecutive frames. It is easy to see that Torrent-6 can send its optical packets with a better jitter or delay in Distribution-a than in Distribution-b.

- **Fairness in Traffic Transmission:** Different torrents can find short-term fair chances to access to an OPS network bandwidth when both distance-based and density-based distributions are respected.

- **Less Buffering:** Even optical packet transmission can reduce the amount of buffering required for client packets arriving at ingress switches.

- **Better Utilization of Network Bandwidth:** Here, optical packets can be transmitted in appropriate time slots that may match the volume of traffic arrival for a torrent. For example, in Distribution-b of Fig. 2.28, five optical packets of Torrent-6 will be transmitted in the first three time slots within the frame. However, it is possible that during the transmission of these optical packets, less client packets have already arrived for Torrent-6. Hence, bandwidth may be wasted. On the other hand, in Distribution-a of Fig. 2.28 the optical packets are transmitted for Torrent-6 once every two time slots. Therefore, better bandwidth utilization could be obtained, especially under multimedia traffic with regular traffic arrival rates.

- **Reduction of Burst Generation:** The arrival of regular traffic to each egress switch is obtained by even traffic transmission. For example, Distribution-b in Fig. 2.28 generates a burst of seven optical packets to egress switch 6 in four consecutive time slots, while Distribution-a sends the same traffic with no burst, and this leads to the arrival of regular traffic at egress switch 6.

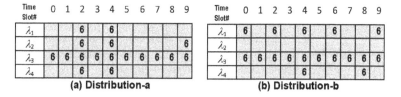

(a) Distribution-a   (b) Distribution-b

Figure 2.29: Two types of optical packet distributions at $\tau = 10$ OP sets for distance and density evaluations [73]

Example 2.1 *Figure 2.29 shows two distributions for the scheduling of 17 optical packets allocated to Torrent-6. For $Z_6 = 17$ and $\tau = 10$ OP sets, we can obtain $X_{DE,6}^* = 4.2$ and $X_{DI,6}^* = 0$. Using Eq. (2.14), one can compute $X_{DE,6} = 5.6$ and $X_{DE,6} = 1.4$, respectively, for distributions a and b. This shows that Distribution-b is better than Distribution-a in terms of density. On the other hand, using Eq. (2.10), we obtain $X_{DI,6} = 0$ for both distributions that shows both distribution are the best in terms of distance. This example shows that both distributions have the same EDF_{di}, but their evenness of density (EDF_{de}) are different. Therefore, evaluation of both distance and density distributions must be performed for bandwidth access schemes.*

Time Slot#	0	1	2	3	4	5	6	7	8	9
λ_1	1	3	3				1	2	1	2 3
λ_2	1	4	5	1	3	0		0		4
λ_3	0		1	3	2	1	3	2	5	1

(a) Distribution-a

Time Slot#	0	1	2	3	4	5	6	7	8	9
λ_1	3	5	2	4	2	3	4		4	0
λ_2	1		3	0	4	4	0	1		3
λ_3		0	5	1	5	1	2		2	5

(b) Distribution-b

Time Slot#	0	1	2	3	4	5	6	7	8	9
λ_1	1	2				2	5	4		0 0
λ_2	0	0	1	4	1		2	1	4	2
λ_3	2	4	4	0	4	4	0		5	1

(c) Distribution-c

Figure 2.30: Three types of optical packet distributions at $\tau = 10$ OP sets for fairness evaluation [73]

Example 2.2 *Figure 2.30 shows three sets of distributions at $n_e = 6, f = 1, W = 3$ channels, and $\tau = 10$. For all these distributions, we have $Z_t = 25$ and $X_{FD}^* = 1.67$. Then, the FD index for Distribution-a, Distribution-b, and Distribution-c are $X_{FD} = 9.67$, $X_{FD} = 0.00$, and $X_{FD} = 11.00$, respectively. Hence, among these three distributions, Distribution-b has the best and Distribution-c has the worst fairness dissemination distributions.*

Example 2.3 *Consider $f = 3$ fibers, $W = 4$ wavelengths, and $\tau = 10$ OP sets in a slotted OPS network. For $Z_5 = 26$ optical packets of Torrent-5 (destined to egress switch 5), consider three different distributions: $\Phi_{5,6}^1 = \{(1,3), (2,5), (6,1), (7,4), (8,7), (9,6)\}$ as the first distribution, $\Phi_{5,6}^2 = \{(1,4), (2,8), (3,6), (4,1), (5,2), (6,5)\}$ as the second distribution, and $\Phi_{5,8}^3 = \{(1,2), (2,3), (3,4), (4,3), (5,2), (6,5), (8,2), (9,5)\}$ as the third distribution. Note that for the first two distributions, we have $N_5 = 6$, whereas for the third distribution we have $N_5 = 8$. For these distributions, we have $X_{DE,5}^* = 4.8$. For the first two distributions, we have $X_{DI,5}^* = 2.67$, whereas we obtain $X_{DI,5}^* = 3.0$ for the third distribution. For the first distribution, we obtain $X_{DE,5} = 19.2$ and $X_{DI,5} = 2.7$; for the second distribution we compute $X_{DE,5} = 20.4$ and $X_{DI,5} = 4.0$; and for the third distribution we calculate $X_{DE,5} = 9.2$ and $X_{DI,5} = 0.0$. Therefore, among these distributions, the third distribution has the best distribution in terms of both distance and density, but the second distribution has the worst optical packet distribution within the GF frame.*

2.3.2 Normalized Evenness Index Parameters

The evenness parameters stated in Section 2.3.1 are single metrics where their values could be larger than 1.0. For better comparison among bandwidth access algorithms, normalized evenness index parameters are obtained in the following. It should be noted that the normalized value of index X is obtained by $X^{Norm} = \frac{X}{X_{max}}$, where X_{max} is the largest value of index X obtained under the worst distribution. A lower normalized index parameter is equivalent to a better distribution. In other words, the best distribution leads to $X^{Norm} = 0$, and the worst distribution results in $X^{Norm} = 1$.

2.3.2.1 Normalized Distance-Based Even Distribution To compute normalized EDF_{di} index $X_{DI,i}^{Norm}$ for Torrent-i, we need to compute maximum of Eq. (2.10). For this, Eq. (2.10) can be expanded to

$$X_{DI,i} = \sum_{if\,d_j > \frac{\tau}{N_i}} \left(d_j - \frac{\tau}{N_i} \right) + \sum_{if\,d_j \le \frac{\tau}{N_i}} \left(\frac{\tau}{N_i} - d_j \right) - X_{DI,i}^* \,. \qquad (2.20)$$

Equation (2.20) is maximized when both summations are maximized. This case only happens when all optical packets are distributed consecutively within a GF frame. The first summation is maximized at $d_j = \tau - N_i + 1$ for one term, and the second summation is maximized when $d_j = 1$ (i.e., $N_i - 1$ terms with distance of 1 inside the second summation). Therefore, we have

$$X_{DI,i}^{max} = \left(\tau - N_i + 1 - \frac{\tau}{N_i} \right) + (N_i - 1) \times \left(\frac{\tau}{N_i} - 1 \right) - X_{DI,i}^* \,, \qquad (2.21)$$

which is simplified to

$$X_{DI,i}^{max} = 2(N_i - 1) \left(\frac{\tau}{N_i} - 1 \right) - X_{DI,i}^* \,. \qquad (2.22)$$

Therefore, normalized index value of EDF_{di} for Torrent-i within a GF frame is given by $X_{DI,i}^{Norm} = \frac{X_{DI,i}}{X_{DI,i}^{max}}$. Finally, normalized index value of EDF_{di} for all torrents within the GF frame is defined by $X_{DI}^{Norm} = \frac{\sum_{i=0}^{n_e-1} X_{DI,i}^{Norm}}{n_e}$.

Note that at $N_i = 1$, $N_i = \tau - 1$ and $N_i = \tau$, we can define the normalized value for Torrent-i to be 0. This is because at $N_i = 1$ and $N_i = \tau$, we can compute $X_{DI,i}^* = 0$, and therefore $X_{DI,i}^{max} = 0$ according to Eq. (2.22). In addition at $N_i = \tau - 1$, we can compute $X_{DI,i}^* = 2\frac{\tau-2}{\tau-1}$. This results in $X_{DI,i}^{max} = 0$ according to Eq. (2.22). For these cases, we also have $X_{DI,i} = 0$. Hence, for these three special cases, we have $X_{DI,i}^{Norm} = 0$.

2.3.2.2 Normalized Density-Based Even Distribution The maximum value of EDF_{de} index parameter for Torrent-i within a GF frame can be obtained under two cases as

- When Z_i is not divisible by $(f \times W)$: The summation in Eq. (2.14) is maximized when for Torrent-i, (1) $c_1 = \left\lfloor \frac{Z_i}{f \times W} \right\rfloor$ OP sets carry $f \times W$ optical packets, (2)

one OP set carries $Z_i mod(f \times W)$ optical packets and (3) $c_2 = \tau - \left\lceil \frac{Z_i}{f \times W} \right\rceil$ OP sets carry no optical packets. In this case, we have

$$X_{DE,i}^{max} = c_1 \times \left| f \times W - \frac{Z_i}{\tau} \right| + \left| Z_i \bmod (f \times W) - \frac{Z_i}{\tau} \right| + c_2 \times \frac{Z_i}{\tau} - X_{DE,i}^*. \quad (2.23)$$

- When Z_i is divisible by $(f \times W)$: The summation in Eq. (2.14) is maximized when for Torrent-i, (1) $\frac{Z_i}{f \times W}$ OP sets carry $f \times W$ optical packets and (2) $\tau - \frac{Z_i}{f \times W}$ OP sets carry no optical packets. Now, we have

$$X_{DE,i}^{max} = \frac{Z_i}{f \times W} \left| f \times W - \frac{Z_i}{\tau} \right| + \left(\tau - \frac{Z_i}{f \times W} \right) \frac{Z_i}{\tau} - X_{DE,i}^*. \quad (2.24)$$

Therefore, normalized index value of EDF_{de} for Torrent-i within a GF frame is given by $X_{DE,i}^{Norm} = \frac{X_{DE,i}}{X_{DE,i}^{max}}$. Finally, normalized index value of EDF_{de} for all torrents within the GF frame is defined by $X_{DE}^{Norm} = \frac{\sum_{i=0}^{n_e-1} X_{DE,i}^{Norm}}{n_e}$.

In special cases of $Z_i = 1$, $Z_i = \tau \times f \times W - 1$, and $Z_i = \tau \times f \times W$, we always have $X_{DE}^{Norm} = 0$. This is because for both $Z_i = 1$ and $Z_i = \tau \times f \times W - 1$, we have $X_{DE,i}^* = 2\left(1 - \frac{1}{\tau}\right)$ according to Eq. (2.15). In addition, we obtain $X_{DE,i}^* = 0$ for $Z_i = \tau \times f \times W$. By substituting these values in Eq. (2.23) and Eq. (2.24), we can obtain $X_{DE,i}^{max} = 0$ for all three special cases. For these cases, we also have $X_{DE,i} = 0$. Therefore, for these three special cases, we have $X_{DE,i}^{Norm} = 0$.

2.3.2.3 Normalized Fair Dissemination (FD) Distribution

It is easy to see that the summation in Eq. (2.16) is maximized when all Z_t optical packets within a GF frame are destined to the same egress switch and no optical packet is carried to any of $n_e - 1$ other egress switches in the network. Therefore, we can compute X_{FD}^{max} by

$$X_{FD}^{max} = \left(Z_t - \frac{Z_t}{n_e} \right) + (n_e - 1) \left(\frac{Z_t}{n_e} \right) - X_{FD}^*. \quad (2.25)$$

or

$$X_{FD}^{max} = 2\left(Z_t - \frac{Z_t}{n_e} \right) - X_{FD}^*. \quad (2.26)$$

Therefore, we calculate normalized value of fair dissemination by $X_{FD}^{Norm} = \frac{X_{FD}}{X_{FD}^{max}}$. In special case of $Z_t \leq 1$, we have both $X_{FD}^{max} = 0$ and $X_{FD} = 0$, and therefore $X_{FD}^{Norm} = 0$.

2.3.2.4 Normalized Load Balanced (LB) Distribution The maximum of the LB index parameter X_{LB}^{max} can be defined under two cases as

- When Z_t is not divisible by $(f \times W)$: The summation in Eq. (2.18) is maximized when (1) $c_1 = \left\lfloor \frac{Z_t}{f \times W} \right\rfloor$ OP sets carry $f \times W$ optical packets for any torrent; (2) one OP set carries $Z_t \bmod (f \times W)$ optical packets for any torrent; and (3) $c_2 = \tau - \left\lceil \frac{Z_t}{f \times W} \right\rceil$ OP sets carry no optical packets. In this case, X_{LB}^{max} is obtained from Eq. (2.18) as

$$X_{LB}^{max} = c_1 \times \left| f \times W - \frac{Z_t}{\tau} \right| + \left| Z_t \bmod (f \times W) - \frac{Z_t}{\tau} \right| + c_2 \times \frac{Z_t}{\tau} - X_{LB}^*. \quad (2.27)$$

- When Z_t is divisible by $(f \times W)$: In this case, the summation in Eq. (2.18) is maximized when (1) $\frac{Z_t}{f \times W}$ OP sets carry $f \times W$ optical packets for any torrent; and (2) $\tau - \frac{Z_t}{f \times W}$ OP sets within the GF frame carry no optical packets. Now, we have

$$X_{LB}^{max} = \frac{Z_t}{f \times W} \left| f \times W - \frac{Z_t}{\tau} \right| + \left(\tau - \frac{Z_t}{f \times W} \right) \frac{Z_t}{\tau} - X_{LB}^*. \quad (2.28)$$

The normalized index value for LB within a GF frame is now defined by $X_{LB}^{Norm} = \frac{X_{LB}}{X_{LB}^{max}}$. In special cases of $Z_t = 1$, $Z_t = \tau \times f \times W - 1$, and $Z_t = \tau \times f \times W$, we have both $X_{LB}^{max} = 0$ and $X_{LB} = 0$, and therefore $X_{LB}^{Norm} = 0$.

2.3.3 Hybrid Index Parameters

Three types of hybrid indices can be defined as follows:

- **Hybrid Index $H1$**: This index considers time domain distributions at the same time, and is defined by $H1 = \frac{X_{DI}^{Norm} + X_{DE}^{Norm}}{2}$.

- **Hybrid Index $H2$**: This index considers time domain and wavelength/fiber domain distributions at the same time, and is computed by $H2 = \frac{X_{DI}^{Norm} + X_{DE}^{Norm} + X_{LB}^{Norm}}{3}$.

- **Hybrid Index $H3$**: This index considers all distributions at the same time, and is calculated by $H3 = \frac{X_{DI}^{Norm} + X_{DE}^{Norm} + X_{LB}^{Norm} + X_{FD}^{Norm}}{4}$.

Note that a lower hybrid index is equivalent to a better distribution. Clearly, $H3$ could be the best index for comparing different bandwidth access schemes. It should be mentioned that $H3$ should be used when traffic transmission in each ingress switch

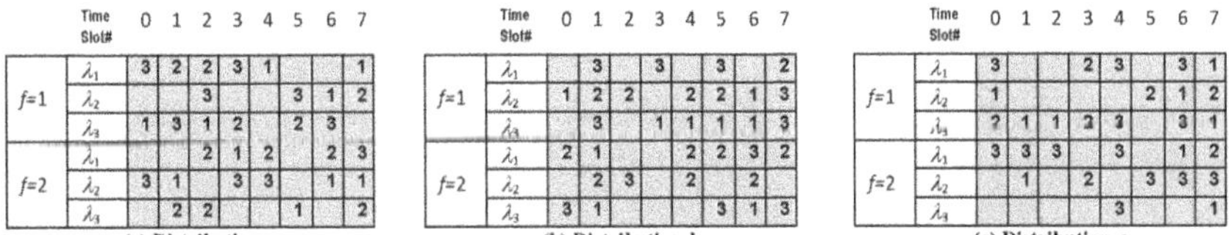

(a) Distribution-a (b) Distribution-b (c) Distribution-c

Figure 2.31: Three distributions at $\tau = 8$ OP sets, $n_e = 3$ edge switches (and 3 Torrents), $f = 2$, and $W = 3$ for hybrid index evaluations [73]

to different egress switches is almost symmetric. When traffic transmission is asymmetric, $H2$ and $H1$ could be used to compare different bandwidth access schemes in a slotted OPS network.

Example 2.4 *Three distributions in ingress switch 0 are illustrated in Fig. 2.31 within a GF of $\tau = 8$ OP sets in a network with $n_e = 3$ egress switches (i.e., a network with $n = 4$ edge switches in total). Table 2.5 compares normalized and hybrid indices $H1$, $H2$, and $H3$ for these distributions. Among these three distributions, Distribution-a with $H3 = 0.058$ is the best distribution and Distribution-c with $H3 = 0.336$ is the worst distribution. A similar decision can be made according to the $H1$ and $H2$ indices.*

Table 2.5: Computation of normalized and hybrid indices [73]

Index	Distribution-a	Distribution-b	Distribution-c
X_{DI}	0.000	0.000	1.333
X_{DI}^{Norm}	0.000	0.000	0.444
X_{DE}	0.667	2.583	5.083
X_{DE}^{Norm}	0.052	0.206	0.465
X_{LB}	3.500	10.000	5.250
X_{LB}^{Norm}	0.179	0.500	0.292
X_{FD}	0.000	2.000	5.333
X_{FD}^{Norm}	0.000	0.048	0.143
$H1$	0.026	0.103	0.455
$H2$	0.077	0.235	0.400
$H3$	0.058	0.189	0.336

2.3.4 Even Optical Packet Transmission under DA and NCPA

In this section, the DA and NCPA bandwidth access schemes are compared in terms of optical packet transmission by computing the $X_{DI}, X_{DE}, X_{LB}, X_{FD}, H1, H2$, and $H3$ in a random network topology with $n = 8$ nodes. Note that all the "evenness" measurements are performed only in ingress switches during simulation, and therefore the results have no dependency on the network topology. A number of immediate routers that support some host networks (as traffic generator) are connected to each edge switch. Each OPS core switch uses the shortest number of hops to route its optical packets between an ingress–egress pair. Consider that the network has the following characteristics: $f = 2$ fibers per link, $W = 4$ wavelengths per fiber, and $\tau = 19$ OP sets. Recall that in an (ideal) optimal bandwidth access scheme, we must have $X_{DI} = 0, X_{DE} = 0, X_{LB} = 0, X_{FD} = 0, H1 = 0, H2 = 0, H3 = 0$.

The frame length used in DA is set to 100 OP sets. Timeout values in NCPA are set to 20 μs, 40 μs, and 100 μs for EF, AF, and BE classes, respectively. Besides Poisson traffic modeling, the Pareto probability density function $p(t) = a \times b^a \times t^{-a-1}$ with shape parameter $a = 1.2$ is used to model the inter-arrival times of client packets. In each ingress switch, traffic in any class is uniformly distributed to each one of the egress switches.

Two network scenarios are studied in the following. Scenario 1 uses the packet length distribution in [74]. The routers connected to an edge switch generate multiclass traffic with EF: 25%, AF: 35%, and BE: 40% of total network traffic. In Scenario 2, the lengths of client packets are exponentially distributed with mean 512 bytes and routers generate EF: 20%, AF: 30%, and BE: 50% of total network traffic. Clearly, under these two models, generated optical packets will encounter different distribution indices under DA and NCPA. In the following, DA1 and NCPA1 mean DA and NCPA, respectively, under Scenario 1. Whereas DA2 and NCPA2 mean DA and NCPA, respectively, under Scenario 2.

The OPNET simulation software has been used to develop the simulation models in this section and 95% confidence intervals are found to be within 1% of the mean values displayed. Each evenness index is obtained by averaging the optical packet distribution indices computed through the simulation period among eight edge switches.

The average X_{FD} and X_{LB} indices are compared in Fig. 2.32 under Poisson and Pareto traffic arrivals. For Poisson traffic, Scenario 1 has only been evaluated (see Fig. 2.32a). Whereas both Scenario 1 and Scenario 2 have been studied under Pareto traffic (see Fig. 2.32b and Fig. 2.32c). Until load 0.5, load balancing is better under DA than NCPA because of the scheduling nature of DA. However, NCPA has a much lower LB index at high loads than DA because almost all the wavelength channels in each OP set are full under NCPA. Fairness in dissemination of optical packets is always better (lower) for DA than for NCPA under both traffic arrivals and both Scenario 1 and Scenario 2.

Average values of X_{DI} and X_{DE} indices within simulation period are compared in Fig. 2.33 under Scenario 1 and Scenario 2. DA has a lower X_{DI} compared with NCPA. Under DA, the X_{DI} index decreases when traffic load increases because op-

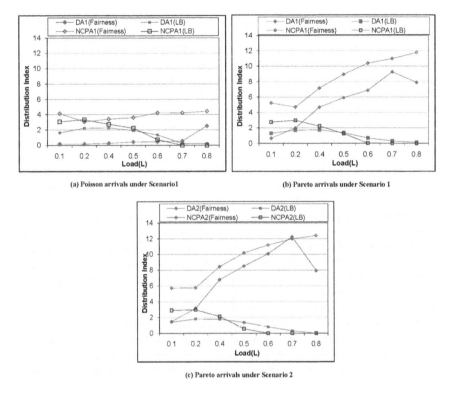

Figure 2.32: Average LB and FD distribution indices [73]

tical packets belonging to each torrent appear with a high probability in all OP sets within each frame of DA. In addition, DA has a very lower X_{DE} compared with NCPA. The DA experiences $X_{DE} = 0$ until load of 0.7 under Poisson traffic and until load of 0.4 under Pareto traffic.

Figure 2.34 displays the performance of $H1, H2,$ and $H3$ indices under Scenario 1. As it can be seen, DA can always provide the best distribution index compared with NCPA.

In short, DA is much better than NCPA to be used in slotted OPS networks for accessing to network bandwidth. One should note that DA has the nature of client packet scheduling in order to create optical packets and transmit them in a smooth way to the OPS network. This is why better transmission indices can be achieved under DA compared with NCPA.

2.4   Summary

Contention avoidance schemes to reduce the contention problem in OPS networks have been detailed in this chapter. Their operations, advantages, and disadvantages have been stated. In addition, the capability of each technique to be combined with

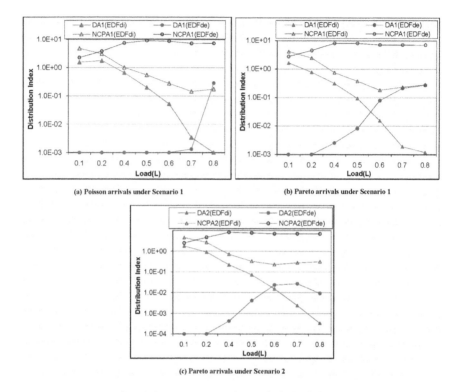

(a) Poisson arrivals under Scenario 1

(b) Pareto arrivals under Scenario 1

(c) Pareto arrivals under Scenario 2

Figure 2.33: Average EDF_{di} and EDF_{de} distribution indices [73]

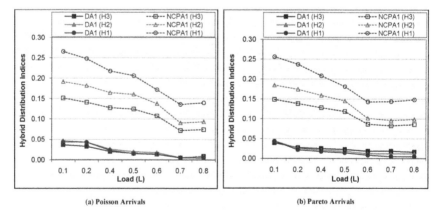

(a) Poisson Arrivals

(b) Pareto Arrivals

Figure 2.34: Average hybrid distribution indices under Scenario 1 [73]

other techniques has been detailed. Finally, the approaches proposed for reducing contention in QoS-capable OPS networks have been investigated in this chapter.

It should be stated that a contention avoidance scheme alone cannot effectively reduce PLR to a very small value. Hence, besides contention avoidance, contention resolution schemes (detailed in Chapter 3) must be used in OPS core switches in order to reduce PLR to the desired value. When less traffic collides in OPS core switches, one may even decide to use less contention resolution hardware in the core switched, which leads to network price reduction.

REFERENCES

1. A. G. Rahbar. PTES: A new packet transmission technique in bufferless slotted OPS networks. *IEEE/OSA Journal of Optical Communications and Networking (JOCN)*, 4(6):490–502, June 2012.

2. A. G. Rahbar and O. Yang. CST: A new contention reduction scheme in slotted all-optical packet switched networks. *Elsevier Performance Evaluation*, 67(5):361–375, May 2010.

3. A. G. Rahbar and O. Yang. Reducing loss rate in slotted optical networks: a lower bound analysis. In *IEEE ICC*, Istanbul, Turkey, June 2006.

4. A. G. Rahbar and O. Yang. Contention avoidance in slotted optical networks. In *International conference on Optical Communication Systems and Networks, SPIE Photonics North*, volume 5970, Toronto, Canada, Sept. 2005.

5. P. Zhou and O. Yang. How practical is optical packet switching in core networks? In *IEEE Globecom*, San Francisco, 2003.

6. M. Klinkowski, F. Herrero, D. Careglio, and J. Sol-Pareta. Adaptive routing algorithms for optical packet switching networks. In *IFIP Conference on Optical Network Design and Modelling (ONDM)*, Milan, Italy, Feb. 2005.

7. L. Mason, A. Vinokurov, N. Zhao, and D. Plant. Topological design and dimensioning of agile all photonic networks. *Computer Networks, Special issue on Optical Networking*, 50(2):268–287, Feb. 2006.

8. M. Jin and O. W. W. Yang. A TDM solution for all-photonic overlaid-star networks. In *Conference on Information Sciences and Systems (CISS)*, Princeton University, Princeton, NJ, Mar. 2006.

9. A. G. Rahbar. EBvN: efficient BvN in multi-fiber/multi-wavelength overlaid-star optical networks. *Springer Journal of Annals of Telecommunications*, 67(11-12):575–588, Nov. 2012.

10. C. S. Chang, W. J. Chen, and H. Y. Huang. Birkhoff-von Neumann input buffered crossbar switches. In *IEEE Infocom*, Mar. 2000.

11. N. McKeown. The iSLIP scheduling algorithm for input-queue switches. *IEEE Transactions on Networking*, 7(2):188–201, Apr. 1999.

12. D. Liu, N. Ansari, and E. hou. A novel fairness criterion for input queued switches. In *IEEE MILCOM*, McLean, VA, Oct. 2001.

13. A. Bianco, M. Guido, and E. Leonardi. Incremental scheduling algorithms for WDM/TDM networks with arbitrary tuning latencies. *IEEE Journal on Selected Areas in Communications*, 51(3):464–475, Mar. 2003.

14. C. S. Chang, W. J. Chen, and H. Y. Huang. On service guarantees for input buffered crossbar switches: a capacity decomposition approach by Birkhoff and von Neumann. In *IEEE/IFIP Seventh International Workshop on Quality of Service (IWQoS'99)*, London, UK, June 1999.

15. C. H. Papadimitriou and K. Steiglitz. *Combinatorial Optimization: Algorithms and Complexity*. Prentice Hall, 1982.

16. T. Anderson, S. Owicki, J. Saxe, and C. Thacker. High speed switch scheduling for local area networks. *ACM Transactions on Computer Systems*, 11(4):319–352, Nov. 1993.

17. I. Keslassy, M. Kodialam, T. V. Lakshma, and D. Stiliadis. On guaranteed smooth scheduling for input-queued switches. *IEEE/ACM Transactions on Networking*, 13(6):1364–1375, Dec. 2005.

18. K. Kar, T. V. Lakshman, D. Stiliadis, and L. Tassiulas. Reduced complexity input buffered switches. In *Hot Interconnects VIII*, Palo Alto, CA, Aug. 2000.

19. C. S. Chang, D. S. Lee, and Y. J. Shih. Mailbox switch: a scalable two-stage switch architecture for conflict resolution of ordered packets. In *IEEE Infocom*, Hong Kong, Mar. 2004.

20. D. Careglio, A. Rafel, J. Sol-Pareta, S. Spadaro, A. M. Hill, and G. Junyent. Quality of service strategy in an optical packet network with a multi-class frame-based scheduling. In *IEEE International Workshop on High Performance Switching and Routing (HPSR 2003)*, Torino, Italy, June 2003.

21. A. Bianco, P. Giaccone, E. Leonardi, and F. Neri. A framework for differential frame-based matching algorithms in input-queued switches. In *IEEE Infocom*, Hong Kong, 2004.

22. C. S. Chang, W. J. Chen, and H. Y. Huang. Birkhoff–von Neumann input-buffered crossbar switches for guaranteed-rate services. *IEEE Transactions on Communications*, 49:1145–1147, 2001.

23. X. Ming, C. Minghua, C. Hongwei, and X. Shizhong. Contention resolution through network global control in optical packet switching networks. In *INFOCOM'09 IEEE International conference on Computer Communications Workshops*, Rio de Janeiro, Brazil, 2009.

24. Z. Lu, D. K. Hunter, and I. D. Henning. Contention resolution scheme for slotted optical packet switched networks. In *International conference on Optical Network Design and Modelling (ONDM)*, Milan, Italy, Feb. 2005.

25. Z. Lu, D. K. Hunter, and I. D. Henning. Contention reduction in core optical packet switches through electronic traffic smoothing and scheduling at the network edge. *IEEE Journal of Lightwave Technology*, 24(12):4828–4837, Dec. 2006.

26. Z. Lu, D. K. Hunter, and I. D. Henning. Congestion control scheme in optical packet switched networks. In *IEEE/OSA OFC*, Anaheim, CA, Mar. 2006.

27. S. Yao, F. Xue, B. Mukherjee, S. J. B. Yoo, and S. Dixit. Electrical ingress buffering and traffic aggregation for optical packet switching and their effect on TCP-level performance in optical mesh networks. *IEEE Communications Magazine*, 40(9):66–72, 2002.

28. A. Rostami and A. Wolisz. Virtual optical bus: an efficient architecture for packet-based optical transport networks. *IEEE/OSA Journal of Optical Communications and Networking*, 2(11):901–914, Nov. 2010.

29. L. Tancevski, S. Yegnanarayanan, G. A. Castaon, L. Tamil, F. Masetti, and T. McDermont. Optical routing of asynchronous, variable length packets. *IEEE Journal of Selected Areas on Communications*, 18:2084–2093, Oct. 2000.

30. H. Øverby and N. Stol. Effects of bursty traffic in service differentiated optical packet switched networks. *Optics Express*, 12(3):410–415, 2004.

31. F. Xue, Z. Pan, Y. Bansal, and J. Cao. End-to-end contention resolution schemes for an optical packet switching network with enhanced edge routers. *IEEE Journal of Lightwave Technology*, 21(11):2595–2604, Nov. 2003.

32. F. Xue and S. J. B. Yoo. Self-similar traffic shaping at the edge router in optical packet-switched networks. In *IEEE International Conference on Communications*, Apr. 2002.

33. C. Raffaelli and P. Zaffoni. TCP performance in optical packet-switched networks. *Photonic Network Communications*, 11(3):243–252, 2006.

34. C. Raffaelli and P. Zaffoni. Packet assembly at optical packet network access and its effects on TCP performance. In *IEEE High Performance Switching and Routing (HPSR)*, Torino, Italy, June 2003.

35. L. Zhaobiao, P. Yong, Z. Min, and Y. Peida. Comparison study of length-threshold, timer-based and hybrid algorithm for optical packet assembly. In *Proceedings of SPIE, Volume 5626*, Feb. 2005.

36. J. Ramamirtham and J. Turner. Time sliced optical burst switching. In *IEEE Infocom*, San Francisco, Mar. 2003.

37. F. Xue, S. Yao, B. Mukherjee, and S. J. B. Yoo. The performance improvement in optical packet-switched networks by traffic shaping of self-similar traffic. In *IEEE OFC*, Anaheim, CA, Mar. 2002.

38. M. OMahony, D. Simeonidou, D. Hunter, and A. Tzanakaki. The application of optical packet switching in future communications networks. *IEEE Communications Magazine*, 39(3):128–135, Mar. 2001.

39. G. Hu, K. Dolzer, and C. Gauger. Does burst assembly really reduce the self-similarity. In *IEEE/OSA OFC*, Atlanta, 2003.

40. A. Todani, S. Choudhury, and A. Bhowmik. A contention avoidance scheme for optical packet switched networks. *International Journal of Innovative Research in Science, Engineering and Technology*, 2(5):1417–1424, May 2013.

41. H. Øverby. Network layer packet redundancy in optical packet switched networks. *OSA Optics Express*, 12(20):4881–4895, Oct. 2004.

42. V. Sivaraman, D. Moreland, and D. Ostry. Ingress traffic conditioning in slotted optical packet switched networks. In *Australian Telecommunication Networks and Applications Conference (ATNAC)*, Sydney, Dec. 2004.

43. F. Xue and S. J. B. Yoo. TCP-aware active congestion control in optical packet-switched networks. In *IEEE Optical Fiber Communications (OFC)*, Mar. 2003.

44. V. Sivaraman, D. Moreland, and D. Ostry. A novel delay-bounded traffic conditioner for optical edge switches. In *IEEE HPSR*, Hong Kong, May 2005.

45. V. Sivaraman, H. Elgindy, D. Morel, and D. Ostry. Packet pacing in short buffer optical packet switched networks. In *IEEE Infocom*, 2006.

46. V. Sivaraman, H. Elgindy, D. Morel, and D. Ostry. Packet pacing in small buffer optical packet switched networks. *IEEE/ACM Transactions on Networking (TON)*, 17(4):1066–1079, Aug. 2009.

47. W. Wei, C. Wang, and X. Liu. Adaptive IP/Optical OFDM networking design. In *IEEE Optical Fiber Communication (OFC)*, San Diego, CA, Mar. 2010.

48. W. Wei, J. Hu, C. Wang, T. Wang, and C. Qiao. A programmable router interface supporting link virtualization with adaptive Optical OFDMA transmission. In *IEEE OFC/NFOEC*, 2009.

49. W. Wei, C. Wang, J. Yu, N. Cvijetic, and T. Wang. Optical orthogonal frequency division multiple access networking for the future Internet. *IEEE Journal of Optical Communications and Networking*, 1:236–246, 2009.

50. N. Andriolli, P. Castoldi, H. Harai, J. Buron, and S. Ruepp. GMPLS-based packet contention resolution in connection-oriented OPS networks. *Journal of Optical Networking*, 8(1):14–25, Jan. 2009.

51. H. Øverby. An adaptive service differentiation algorithm for optical packet switched networks. In *International Conference on Transparent Optical Networks (ICTON)*, Warsaw, Poland, Jul. 2003.

52. J. V. Gontan, P. P. Marino, J. V. Alonso, and J. G. Haro. A feasibility study of GMPLS extensions for synchronous slotted optical packet switching networks. In *WGN5: Workshop in G/MPLS Networks*, Girona, Spain, Mar. 2006.

53. I. Kotuliak. Optical networks access node performance. In *International Symposium Electronics in Marine (ELMAR)*, Zadar, Croatia, June 2004.

54. D. Zhang, M. Zhang, and P. Ye. Traffic shaping at the edge node in synchronous optical packet switched networks. *Photonics Networks Communications*, 13:103–110, 2007.

55. A. G. Rahbar and O. Yang. A new bandwidth access framework in slotted-OPS networks. In *IEEE Conference on Local Computer Networks (LCN)*, Tampa, Florida, Nov. 2006.

56. A. G. Rahbar and O. Yang. Contention avoidance and resolution schemes in bufferless all-optical packet-switched networks: a survey. *IEEE Communications Surveys and Tutorials*, 10(4):94–107, Dec. 2008.

57. D. Careglio, J. S. Pareta, and S. Spadaro. Optical slot size dimensioning in IP/MPLS over OPS networks. In *IEEE International Conference on Telecommunications (ConTEL)*, Zagreb, Croatia, June 2003.

58. C. Raffaelli and P. Zaffoni. Effects of slotted optical packet assembly on end-to-end performance. In *Lecture Notes in Computer Science, Springer*, volume 3079, June 2004.

59. A. G. Rahbar. Impairment-aware merit-based scheduling in QoS-capable transparent multi-fiber optical packet switched networks. *Journal of Optical Communications*, 33(2):103–121, June 2012.

60. A. G. Rahbar and O. Yang. Contention avoidance by composite slot assembling. In *IEEE/OSA OFC*, Anaheim, CA, Mar. 2006.

61. M. Casoni, E. Luppi, and M. L. Merani. Impact of assembly algorithms on end-to-end performance in optical burst switched networks with different QoS classes. In *International Workshop on Optical Burst Switching*, San Jose, CA, Oct. 2004.

62. A. G. Rahbar and O. Yang. Distribution-based bandwidth access scheme in slotted all-optical packet-switched networks. *Elsevier Computer Networks*, 53(5):744–758, Apr. 2009.

63. A. G. Rahbar and O. Yang. OCGRR: a new scheduling algorithm for differentiated services networks. *IEEE Transactions on Parallel and Distributed Systems*, 18(5):697–710, May 2007.

64. R. Nejabati, G. Zervas, D. Simeonidou, M. J. OMahony, and D. Klonidis. The OPORON project: demonstration of a fully functional end-to-end asynchronous optical packet-switched network. *IEEE Journal of Lightwave Technology*, 25(11):3495–3510, Nov. 2007.

65. S. Bjørnstad and H. Øverby. Quality of service differentiation in optical packet/burst switching: A performance and reliability perspective. In *International Conference on Transparent Optical Networks*, Barcelona, Spain, Jul. 2005.

66. S. Bjørnstad, M. Nord, D. R. Hjelme, and N. Stol. Optical burst and packet switching: node and network design, contention resolution and quality of service. In *IEEE International Conference on Telecommunications (ConTEL)*, Zagreb, Croatia, June 2003.

67. T. Khattab, A. Mohamed, A. Kaheel, and H. Alnuweiri. Optical packet switching with packet aggregation. In *IEEE SoftCom*, Ancona, Venice, Italy, Oct. 2002.

68. Y. Li, G. Xiao, and H. Ghafouri-Shiraz. On the benefits of multifiber optical packet switch. *Microwave and Optical Technology Letters*, 43(5):376–378, Dec. 2004.

69. A. G. Rahbar and O. Yang. Fiber-channel tradeoff for reducing collisions in slotted single-hop optical packet-switched (OPS) networks. *OSA Journal of Optical Networking*, 6(7):897–912, Jul. 2007.

70. A. G. Rahbar and O. Yang. *OPS Networks: Bandwidth Management & QoS*. VDM Verlag, Germany, 2009.

71. O. Gerstel and H. Raza. Merits of low-density WDM line systems for long-haul networks. *IEEE Journal of Lightwave Technology*, 21:2470–2475, 2003.

72. A. K. Somani, M. Mina, and L. Li. On trading wavelengths with fibers: a cost performance based study. *IEEE/ACM Transactions on Networking*, 12:944–951, 2004.

73. A. G. Rahbar and O. Yang. Even slot transmission in slotted optical packet-switched networks. *IEEE/OSA Journal of Optical Communications and Networking*, 1(2):A219–A235, Jul. 2009.

74. Caida. Packet size reports, http://www.caida.org/research/traffic-analysis/fix-west-1998/packetsizes/. Technical report, 1998.

CHAPTER 3

CONTENTION RESOLUTION IN OPS NETWORKS

In this chapter, contention resolution schemes proposed for both asynchronous OPS and synchronous (slotted) OPS networks are studied. Each core switch in the network first receives the information of optical packets (carried in their headers) in its input ports. Then, the core switch evaluates the potential contention, resolves the contention, and finally makes its switching modules ready to switch the optical packets toward the next hop core switches that lead to their desired egress switches. The core switch may make a new header for the optical packets departing from each output link. In slotted-OPS, all the mentioned operations must be performed within the slot offset (see Section 1.4.4).

As mentioned when optical packets arrive at a core switch, there may be a collision in any output port of the core switch. A contention resolution scheme should be provided in the core switch to successfully switch some of the optical packets and drop the rest due to the lack of resources. The core switch needs to resolve the contention at all its output links.

In general, contention resolution schemes are mainly grouped into two hardware-based and software-based schemes. In another categorization, contention in OPS can be resolved in four domains that will be detailed in the following sections [1, 2]:

Quality of Service in Optical Packet Switched Networks, First Edition.
By Akbar Ghaffarpour Rahbar Copyright © 2015 IEEE. Published by John Wiley & Sons, Inc.

- **Wavelength Domain**: The wavelength domain uses WDM feature to solve contention. It uses tunable wavelength converters to convert contending arriving optical packets to different wavelengths for solving contention.

- **Time Domain**: For time domain, optical buffering using FDLs is used in which contending optical packets are delayed in OPS core switches and then transmitted at appropriate times toward their destinations.

- **Space Domain**: For space domain, deflection routing is used where a contending optical packet is routed through a different route toward its destination.

- **Retransmission Domain**: Retransmission of optical packets in the optical domain can achieve a very low PLR. Here, the higher layer nodes are not aware of loss of optical packets in the optical domain.

3.1 Hardware-Based Contention Resolution Schemes

Here, the hardware-based contention resolution schemes that use additional hardware for resolving contention are studied. Although these schemes can resolve contention in OPS core switches, they impose an additional hardware cost on the network implementation. The hardware-based contention resolution schemes are studied for single-class traffic without QoS differentiation and multi-class traffic with QoS differentiation.

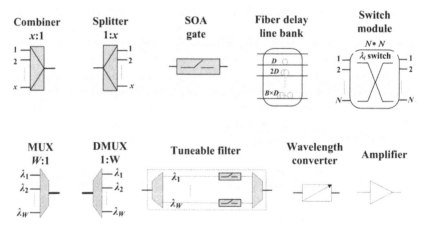

Figure 3.1: Symbols used in OPS core switch architectures

Figure 3.1 shows different symbols used in OPS core switch architectures in the following subsections.

3.1.1 Single-Class Traffic

In this section, hardware-based contention resolution schemes developed for single-class traffic for both asynchronous OPS and slotted OPS networks are studied.

3.1.1.1 Optical Buffers In electronic packet switching networks, contention in an output port of an electronic switch is resolved by storing contending packets in a queue until the output port becomes available again. In fact, buffering provides some delay for a contending packet until its switching becomes possible. Similar to the electronic networks, buffering in an OPS core switch provides delays for a contending optical packet until the core switch finds the possibility of switching of the optical packet. Buffering in optical networks can be implemented in three ways, each having its own advantages and disadvantages:

- **Buffering in the Electronic Domain:** In opaque optical switches, an OPS core switch first converts an optical packet to the electronic domain, stores it in an electronic buffer, and then transmits it in the optical form. However, this type of buffering is not acceptable since the speed of OPS networking is limited by the speed of electronic memories, especially when there are many optical packets to be buffered. In addition, it leads to high price and complexity of core switches [1].

- **Slow Light Buffering:** In slow light buffering, decreasing the velocity of an optical packet propagation increases the delay for the optical packet. For slowing an optical signal, there are two main techniques: Electromagnetically Induced Transparency (EIT) and Coupled Resonant Structure (CRS). The EIT uses resonances between electromagnetic field and polarizable medium and CRS uses resonances between electromagnetic waves. However, both mechanisms have limitation such as dispersion and absorption losses and cannot be used for high data rates [3].

  Variable slow light buffers using tunable-coupled resonator rings can also be used [4], where a buffer is made up of micro-resonator rings coupled with a bus waveguide. Choosing different power couplings (strength of resonance) between bus and rings can produce different delays for different wavelengths. Therefore, it can produce arbitrary distinct delay for each wavelength.

- **Fiber Delay Lines (FDLs):** In an all-optical OPS core switch, optical queuing can be realized with some fiber with a given length. Assuming that the light propagation speed in a fiber is $v = 2 \times 10^5$ km/s, an FDL with length l km can provide a delay equal to $\frac{l}{v} = 5 \times l$ μs for an optical packet passing through it. It is possible to pass multiple optical packets coming on different wavelengths through a single FDL at the same time, thus decreasing the need for FDL by factor W.

  The FDL buffering has a major difference from a conventional electronic buffering. In an electronic queue, a packet may be stored for a period as long as it needs, and then it can be accessed at any time. However, random access memory is not feasible in all-optical OPS switches because an FDL delaying is fixed and inflexible. This means that when an optical packet is sent to the FDL, it cannot be retrieved before it reaches the end of the FDL. In other words, the optical packet can only be in the delay line for a time equal to the propagation delay of the delay line. This issue adds more limitations than electronic RAM.

Many all-optical OPS architectures rely on FDL buffers to resolve contention. However, this optical buffering technology is immature because it still relies on bulky optical FDLs [5]. An OPS switch also needs a large number of hardware components and complex scheduling algorithms to realize FDL buffering [6]. In addition, an optical signal is usually degraded in FDLs [6]. To compensate the signal degradation, optical amplifiers are often used along the data path. However, the cascaded amplifiers also accumulate noise that can severely limit the network size at very high data rates, unless expensive signal regeneration units are employed [6].

In slotted OPS, FDL lengths are selected as an integer number of time-slot size. For example, suppose two optical packets have arrived at the same time slot at a core switch on the same wavelength λ_i. The core switch forwards one of those optical packets to its desired output port and sends the other optical packet on wavelength λ_i of an FDL with one time-slot delay. In the next time slot, the second optical packet exits the FDL, and then it can be scheduled through its output port without contention [1].

In asynchronous OPS, two methods can be used for FDLs. In the first method, the FDL size must be larger than the maximum optical packet size in the network [7]. The second method uses different FDLs with different lengths to provide different amounts of delay for contending optical packets. Using FDLs for contention resolution is easier and more effective in slotted OPS than in asynchronous OPS [7].

The architecture of FDLs could be categorized as either fixed delay or variable delay [8].

- **Fixed Delay FDL Architecture:** Figure 3.2 displays an optical buffer module with B fiber delay lines. The module provides delays from 0 to $B \times D$, where D is a constant delay time. An optical packet can enter at different inputs of the buffer module to experience a delay in range $[0, B \times D]$. In slotted OPS networks, we have $D = S_{ts}$, where S_{ts} is the time-slot length in seconds. In asynchronous OPS networks, D should be adjusted as a function of the average optical packet length.

- **Variable Delay FDL Architecture:** In this architecture, an FDL has multiple stages of buffering structure and 2*2 switching elements in order to retrieve an optical packet at the end of the stages. By this, very large buffers are also realized by cascading FDLs. Figure 3.3 illustrates a d-stage optical buffer with $(d+1)$ 2*2 switching elements, where each switch acts as either in cross or bar state. By adjusting the states of $d+1$ switches, an optical

Optical buffer module

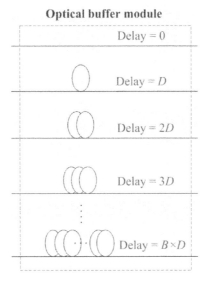

Figure 3.2: An optical buffer module

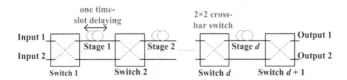

Figure 3.3: A d-stage optical buffer with $d+1$ cross-bar switching elements [5]

packet entering an input could experience a delay from 0 to d time slots [5, 9, 10].

3.1.1.1.1 FDL Buffer Architectures

Using FDLs in an OPS switch has an important role in contention resolution and their capacities provide great impact on the switch performance. The location of FDLs in a core switch affects the contention problem as well. There are three main architectures for employing FDLs in an OPS core switch [13]:

- **Feed-Forward Output Buffering**: In this architecture, an optical packet directly goes toward its destination after passing through an FDL; i.e., it is only buffered once. For example, the architecture for multi-stage buffering displayed in Fig. 3.3 acts like the feed-forward mechanism. In this buffering, FDLs are located after the switching module in a core switch. For contention resolution of contending optical packets, the switch controller decides which optical packets on which wavelengths should be delayed [1, 14–16].

Feed-forward output buffering can be built in one of three architectures [16]:

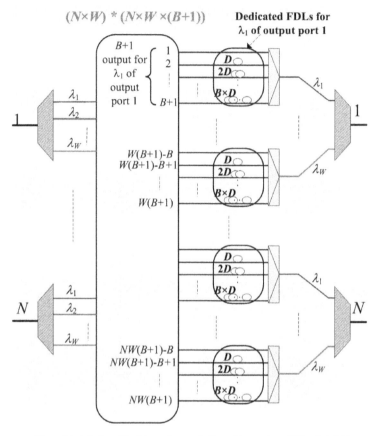

Figure 3.4: A dedicated feed-forward output buffer architecture in an OPS switch [11]

- **Feed-Forward Dedicated Buffering (FF-DB):** Here, a set of B FDLs are dedicated to each output wavelength channel. Therefore, an output port with W wavelengths needs $W \times B$ FDLs. This architecture also needs $W \times (B+1)$ output ports from switch space in an OPS core switch as depicted in Fig. 3.4. In this case, the switch space size should be $(N \times W) * (N \times W \times (B+1))$ and the number of required FDLs is $N \times W \times B$. In FF-DB, at each FDL bank dedicated to wavelength λ_i of an output port, at most one FDL can be used to delay only one optical packet on wavelength λ_i. Although this architecture needs a large number of ports and FDLs, and most of the FDLs are unused, it leads to the lowest PLR [11].

- **Feed-Forward Shared-per-Port (FF-SP) Buffering:** Here, a set of B FDLs are shared among several wavelengths on an output port. Figure 3.5 and Figure 3.6 display two shared-per-output buffering architectures in an OPS core switch.

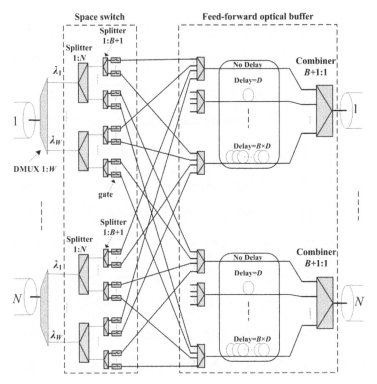

Figure 3.5: A feed-forward shared-per-output buffer architecture in an OPS switch (Architecture 1) [1]

* In the first architecture, after switching of optical packets, they enter one of the $B + 1$ combiners located before the FDL bank at a
giver output port. The FDL bank provides delay from 0 to $B \times D$,
where D is a constant delay value. Delay 0 is for the optical packets that do not require any buffering. The optical packets passing
through the FDL bank enter a combiner that provides output wavelengths λ_1 to λ_W for the output port.

* The space switch in the second architecture needs $W + B$ ports for
each output port, where the optical packets that do not need buffering are switched on W ports, and the optical packets that require
buffering are switched on B ports. In each output port, there is
a $B * W$ space switch to direct the delayed optical packets to the
appropriate 2:1 combiner. The output of each combiner is a wavelength that goes toward a multiplexer. Clearly, the space switch
size in FF-SP is much smaller than in the FF-DB architecture, but
at the expense of increasing PLR compared with FF-DB.

– **Feed-Forward Shared-per-Node (FF-SN) Buffering**: In FF-SN, a set of B
FDLs are shared among N output ports (see Fig. 3.7). The first switch space

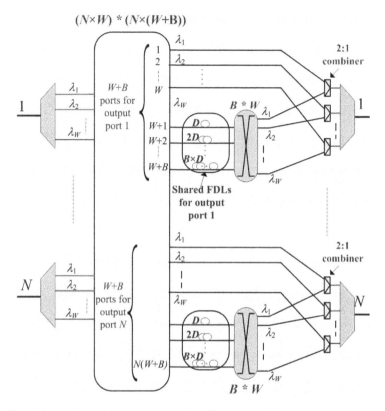

Figure 3.6: A feed-forward shared-per-output buffer architecture in an OPS switch (Architecture 2) [12]

is a $(N \times W) * (N \times W + B))$ switch with W ports for every output ports and B ports for the FDL bank that can delay at most B optical packets. The small space switch at the output of the FDL bank is a $B * (N \times W)$ switch that directs the delayed optical packets toward their desired output ports. This architecture reduces the number of FDLs required in an OPS switch.

- **Feed-Forward Input Buffering**: Similar to output buffering, an optical packet will not return back to the optical buffer and is only buffered for once in feed-forward input buffering. However, unlike the output buffering, optical packets are buffered before the switching module. Figure 3.8 shows the architecture of this class of buffering, where a demultiplexing module separates the optical packets from each incoming fiber on W wavelengths. All optical packets coming on wavelength λ_i from different input ports are all routed to switching module i. There are W switching modules in the OPS switch, where the figure shows only switching modules 1 and W. In each switching module, there are N FDL banks, N splitters, and a switch fabric used for each wavelength with size $(N \times (B + 1)) * N$. In switching module i, each wavelength is split into $B + 1$

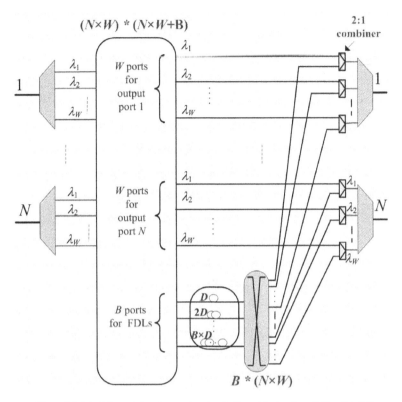

Figure 3.7: A feed-forward shared-per-node output buffer architecture in an OPS switch [17]

fibers, where one of them is connected to the switch fabric without delay and the rest of them are connected to the switch fabric after emerging the delaying fibers. The switch controller decides which copy of delayed signal should be switched to the output port. As can be observed, this class of OPS switches needs FDL bank for each wavelength. In addition, the number of switching ports is increased based on the dimension of FDL banks.

- **Recirculation Buffering (Feed-Backward)**: An optical packet can recirculate in an FDL buffer for more than once to obtain additional delays. In this architecture, contention of optical packets are solved at the input and output sides of the core switch all together. There are two architectures proposed for using recirculation FDLs in an OPS switch as [16]:

  - **Feed-Backward Shared-per-Node (FB-SN)**: Figure 3.9 displays a shared-per-node recirculation buffer architecture with B FDL ports shared among all ports, where D is an amount of fixed delay. There are $B \times W$ ports to delay at most $B \times W$ optical packets in the FDLs, where each FDL can delay at most W optical packets at the same time. Here, the switch size is $(W \times (N+B)) * (W \times (N+B))$. A contending optical packet on wavelength

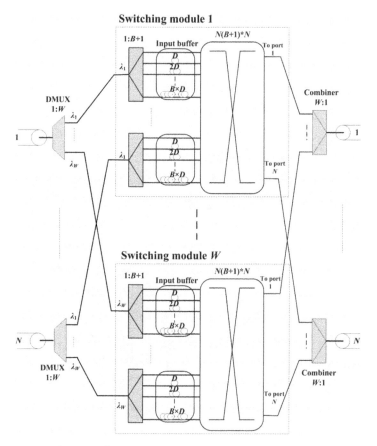

Figure 3.8: A feed-forward input buffer FDL architecture in an OPS switch

λ_i that cannot be switched and needs a delay of $k \times D$ is forwarded to the k-th output port relevant to FDLs. After recirculation, the optical packet finds an opportunity to be switched toward its desired output port. If the optical packet cannot still be switched, it may be forwarded to one of the FDLs again [11, 18].

Of course, there could be another scenario similar to Fig. 3.9 in which there are only B ports for delaying at most B optical packets. Clearly, this scenario does not need any multiplexers and demultiplexers in the FDLs section. In this scenario, an FDL can delay only one optical packet. Here, the space switch size reduces to $(N \times W + B) * (N \times W + B)$. One can use as many output–input FDLs since lower PLR can be achieved by increasing the number of FDLs, but at the expense of increasing the cost and bulkiness of switch.

- **Feed-Backward Shared-per-Port (FB-SP)**: In FB-SP, there are N banks of FDLs, each with size B FDLs, to delay the optical packets of the relevant

$$(N \times W + B \times W) * (N \times W + B \times W)$$

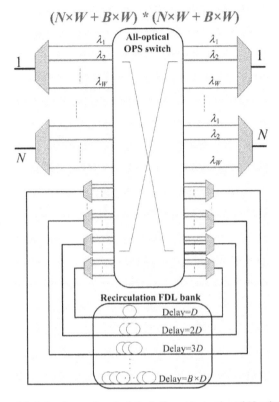

Figure 3.9: A shared-per-node recirculation buffer architecture in an OPS switch [11]

ports, where the contending optical packets at output port k are only sent to FDL bank k. The switch size here is $(N \times (W + B)) * (N \times (W + B))$ (see Fig. 3.10) [19].

The feed-forward FDL architecture has no limitation on optical packet length, but it needs FDLs for each port and requires a high number of components. Thus, an OPS switch using feed-forward FDLs has a large volume. Although the number of FDL banks are reduced in the recirculation FDL architecture compared with the feed-forward FDL architecture, the drawback of recirculation is the need for additional ports on a switch fabric. This limits switch fabric for very large interconnection networks. In addition, the recirculation architecture may increase signal impairments (such as attenuation, crosstalk, and dispersion). Hence, the feed-forward FDL architecture may be preferred in practice. [3, 14, 16, 20–22].

3.1.1.1.2 Issues on Fiber Delay Lines There are some drawbacks for using FDLs in an OPS switch:

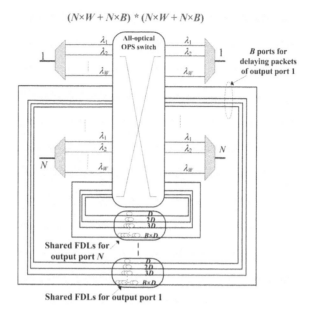

Figure 3.10: A shared-per-port recirculation buffer architecture in an OPS switch

- FDLs require additional ports on switch fabric which. This increases price and limits switch extension.

- The capacity of FDL buffering is small since a limited number of optical packets can be buffered in an FDL.

- It is difficult to maintain FDLs when temperature changes [7].

- Since FDLs delay optical packets by propagating them inside fixed length optical fibers, they cause physical layer impairments, signal distortions, and power loss, thus increasing bit error rate [7].

- The physical size and bulkiness of FDLs limits the usage of FDLs [7].

- Although having a large number of FDLs can reduce PLR, however, using additional FDLs would not improve the PLR performance when it reaches a limit [7].

- Optical packets may be reordered after passing through FDLs, which is usually an undesirable issue.

As a result, the design of optical buffers that mitigates the effects of the aforementioned limitations has emerged as an important research area for OPS. The important issues to be considered in designing FDL architectures include packet loss, cost, control complexity, packet reordering, and signal loss along the FDL. In short, usage of FDLs should be minimized [22].

3.1.1.1.3   *Wavelength and Delay Selection (WDS) Techniques*   When contention occurs in an OPS switch, the control unit of the switch detects contention. After determining the output fiber for a given optical packet by the routing strategy in a core switch, the core switch controller must execute a two-dimensional scheduling problem involving the choice of a suitable wavelength on which to transmit the optical packet; in case this wavelength is currently busy, a delay should be assigned to that optical packet. This problem is called the Wavelength and Delay Selection (WDS) problem [23–25].

Consider a core switch in an asynchronous connectionless OPS that uses B FDLs at each output port with lengths D_j (where $j = 1$ to B) for providing delays in range D to $B \times D$ in order to resolve the contention of contending optical packets. Assume that a new optical packet has arrived at the core switch at time t_0 on wavelength λ_i. However, since it cannot be scheduled on the desired output port on wavelength λ_i, it is considered as a contending optical packet.

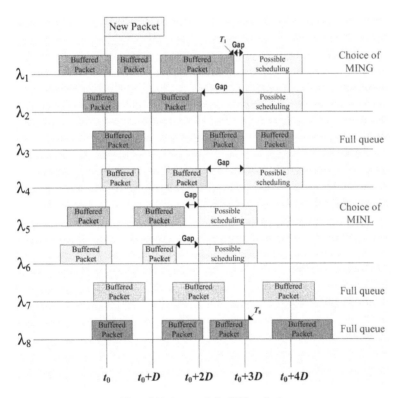

Figure 3.11: An example for WDS mechanism

Define T_w to be the time when the last bit of the last optical packet currently buffered on wavelength λ_w will leave the core switch. For instance, in a network with eight wavelengths, Fig. 3.11 shows the departure times of previously buffered optical packets on wavelengths λ_1 and λ_8 with T_1 and T_8, respectively. Consider

$B = 4$ for delaying of at most $4 \times D$ in this figure. There are some wavelengths (i.e., wavelengths λ_3, λ_7, and λ_8) that are called full queues. This is because the new optical packet cannot be scheduled on these wavelengths at all as they are occupied even at time $t_0 + 4 \times D$.

If there is a wavelength λ_w so that $t_0 \geq T_w$, then the new optical packet can be immediately transmitted on wavelength λ_w after wavelength conversion from wavelength λ_i. In case $t_0 < T_w$, the following steps should be executed to schedule the new optical packet on an FDL in a connectionless asynchronous OPS network:

1. Search for the set of wavelengths Λ such that for each wavelength $\lambda_w \in \Lambda$, we have $T_w - t_0 \leq D_B$. In case $\Lambda = \{\varnothing\}$, all queues are full and the new optical packet must be dropped.

2. Select a suitable wavelength λ_w from set Λ to transmit the new optical packet.

3. Select delay D_j such that $t_0 + D_j \geq T_w$, and then send the new optical packet on FDL j.

Step 2 above has a strong influence on the PLR performance and can be implemented by different techniques such as:

- **Random Choice Among Neither Empty Nor Full Queues (RNENF):** RNENF selects randomly wavelength λ_w from set Λ excluding full queues (i.e., the queues with $T_w - t_0 > D_B$). This technique is aware of the congestion state of the output queues. When all wavelengths are full, the optical packet is lost. For example, in Fig. 3.11, all wavelengths, except wavelengths λ_3, λ_7, and λ_8 which denote to full queues, can be chosen to schedule the new optical packet. The possible scheduling times for the new optical packet have been displayed in the figure with "Possible scheduling".

- **RANDOM:** This is the pure random mechanism that only selects a wavelength without having any information from the congestion state of the output queues.

- **Minimum Length Queue (MINL):** Under MINL, the shortest queue is selected. Here, wavelength λ_w is chosen such that $T_w \leq T_k$ for all $k \neq w$. If there are more than one shortest queue, the one with minimum gap is selected. For example, in Fig. 3.11, two wavelengths λ_5 and λ_6 have the minimum queue length for the new optical packet as it can be scheduled on both wavelengths at time $t_0 + 2 \times D$. However, MINL chooses λ_5 since the possible scheduling time of the new optical packet with the last buffered optical packet on wavelength λ_5 has the smaller gap than the gap on wavelength λ_6 as depicted in the figure. Therefore, the new optical packet is wavelength converted from wavelength λ_i to wavelength λ_5 and then transmitted on the FDL with $2 \times D$ buffering. Then, at time $t_0 + 2 \times D$, the new optical packet starts leaving the FDL and going toward the next core switch on wavelength λ_5.

- **Minimum Gap Queue (MING):** The queue with the smallest gap between the possible scheduling time of the new optical packet and the last buffered optical packet is chosen under MING. For example, in Fig. 3.11, wavelength λ_1 is chosen by MING since it provides the smallest gap as displayed in the figure. Therefore, the new optical packet is wavelength converted from wavelength λ_i to wavelength λ_1 and then transmitted on the FDL with $3 \times D$ buffering. If there are more than one queue with the smallest gap, the one with the smallest queue length is selected.

In connection-oriented OPS networking implemented with MPLS, one possibility is to have a static LSP-to-wavelength assignment performed at the connection set-up phase, maintained over the whole connection holding time. However, the static wavelength assignment cannot provide acceptable performance results since it does not regard the congestion state of the wavelengths. The connectionless WDS algorithms that work on per-packet basis may be applied at connection level to design a dynamic LSP-to-wavelength assignment technique. In connection-oriented OPS, the following heuristic wavelength allocation techniques can be used under WDS aiming at maximizing the performance [23]:

- **Round-Robin Wavelength Selection (RRWS)**: Assume that an incoming optical packet (following an LSP) must be delivered to a congested wavelength. In this case, RRWS performs the simple round-robin search for finding a non-congested wavelength on the same fiber. If such a wavelength is available, the optical packet is sent to it and the forwarding table is updated accordingly in order to guarantee that all the subsequent optical packets on that LSP will follow the same route. The RRWS can significantly improve the PLR performance with respect to the pure static wavelength allocation. However, it provides a non-optimal allocation strategy based on the round-robin search.

- **Minimum Queue length Wavelength Selection (MQWS)**: The output wavelength selection is similar to MINL. When the desired wavelength to which an incoming optical packet has to be scheduled is congested, MQWS chooses a wavelength (among all the non-congested wavelengths) that provides the smallest waiting queue. Then, all optical packets on the same LSP are rerouted to this wavelength until a new congestion occurs. This approach is smarter than RRWS since it optimizes the utilization of queuing resources.

- **Empty Queue Wavelength Selection (EQWS)**: When an optical packet following an LSP arrives and has to be delivered to a congested wavelength, EQWS searches for a wavelength with an empty queue. Then, the LSP is switched to that wavelength as long as needed; i.e., until either congestion arises in the new wavelength or the congestion of the previous wavelength disappears. In the latter case, the LSP is switched back to the original wavelength.

Figure 3.12 depicts PLR as a function of D normalized to the average packet length for different connectionless wavelength allocation algorithms in WDS at $N = 4$, $W = 16$, $L_{OP} = 0.8$, and $B = 8$ FDLs. Since RNENF is aware of the congestion state of the output queues, it provides performance improvement of about two orders

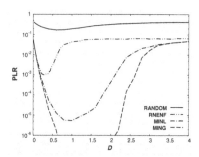 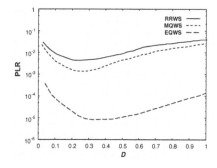

Figure 3.12: PLR under different wavelength selection schemes in WDS suitable for connectionless OPS. Reprinted from *Computer Networks*, Vol. 44, F. Callegati, W. Cerroni, C. Raffaelli, and P. Zaffoni, Wavelength and time domain exploitation for QoS management in optical packet switches, 569–582, copyright 2004, with permission from Elsevier.

Figure 3.13: PLR under different wavelength selection schemes in WDS for connection-oriented OPS. Reprinted from *Computer Networks*, Vol. 44, F. Callegati, W. Cerroni, C. Raffaelli, and P. Zaffoni, Wavelength and time domain exploitation for QoS management in optical packet switches, 569–582, copyright 2004, with permission from Elsevier.

of magnitude compared with the pure random mechanism. The intelligent MING provides better performance than MINL. This is because the limitation of unused resources due to the minimization of the gaps between optical packets is more important than the minimization of the queue length. The observed behavior of the curves is due to the influence of the granularity D on the buffer capacity, where the position of the minimum PLR depends on the adopted wavelength selection strategy. At $D \ll 1$, the buffering space becomes negligible, and the system becomes almost bufferless and the four techniques become equivalent. At $D \gg 1$, the gaps inside the buffer are so large that the buffer itself becomes useless and the system acts as there is no buffer at all, resulting in the same situation as above. In short, the best PLR performance is obtained first by the MING followed by the MINL and random mechanisms. Figure 3.13 shows PLR as a function of D for different connection-oriented wavelength allocation algorithms in WDS under the same parameters. Here, EQWS outperforms both RRWS and MQWS due to the exploitation of unused queuing space [23].

3.1.1.2 Wavelength Converters In this section, contention resolution using different types of wavelength converters are discussed for single-class OPS networks. Wavelength conversion can play an important role in improving the utilization of the available wavelengths in an OPS network. An OPS core switch can use Wavelength Converters (WCs) to resolve the blocking of optical packets by transmitting a contending optical packet on another wavelength. In other words, using wavelength conversion, optical packet loss due to the wavelength continuity problem can be reduced. Obviously, avoiding blocking reduces PLR.

Compared with optical buffers, wavelength converters have the benefits of small size, no additional delay, no jitter, and no optical packet reordering problems [26]. However, hardware cost of wavelength conversion devices is high because of using expensive components [27, 28]. This is why many researchers employ partial wavelength conversion techniques (say shared-per-node conversion, shared-per-link conversion, etc.) in order to reduce the wavelength conversion cost in the network. In addition, decreasing the number of required WCs, while keeping PLR at an acceptable level, is the goal of many contention resolution algorithms. As illustrated in [7], increasing the number of WCs improves throughout of the network. However, marginal improvement is achieved with increasing the number of WCs too much.

For example, consider an output port of a core switch with $W = 8$ wavelengths, where at a given time it is going to switch three optical packets on wavelength λ_2 and four optical packets on wavelength λ_6; i.e., totally seven optical packets are going to be switched. Without using WCs, two of the seven optical packets will be switched through their wavelengths (i.e., two wavelengths are only used), and five optical packets will be dropped. Having enough WCs that can be used at the output port, all the optical packets can be switched by sending the five contending packets, after wavelength conversion, on available unused wavelengths at the output port.

3.1.1.2.1 Types of Wavelength Converters

Wavelength conversion can be categorized as Fixed Wavelength Conversion (FWC) and Tunable Wavelength Conversion (TWC). In FWC, an input wavelength λ_i can only be converted to a fixed defined wavelength λ_j. However, in TWC, an input wavelength λ_i can be converted to a set of output wavelengths. Tunable wavelengths converters are scalable, but they are more expensive than fixed wavelength converters [13, 29].

In the following, r_{wc} denotes the number of wavelength converters used in a converter bank in an OPS switch, and N_{WC} refers to total number of wavelength converters used in the OPS switch. Note that we always have $N_{WC} \geq r_{wc}$. There are five types of TWCs as follows [30]:

- **Full-Range Wavelength Converter (FRWC):** A WC with type FRWC can convert any input wavelength to any arbitrary wavelength in the spectrum range, e.g., [33, 34]. A network with FRWC can achieve the lowest PLR. However, full-range wavelength converters are expensive and difficult to implement due to the technological limitations. In addition, under actual traffic conditions, full-range wavelength conversion capability may not always be necessary, thus making OPS unnecessarily expensive [7, 35–37].

- **Limited-Range Wavelength Converter (LRWC):** Since FRWC is expensive, Limited-Range Wavelength Converters (LRWCs) can be used in OPS switches, where an input wavelength can only be converted to a limited set of adjacent wavelengths, e.g., in slotted OPS [38, 39] and in asynchronous OPS [29, 37, 40, 41]. In limited-range conversion with conversion degree d_c, an optical packet arriving on input wavelength λ_i can only be converted to d_c output wavelengths above and below λ_i. Although using LRWC can result in a cheaper core switch design than by using FRWC, it may need a complex scheduling algorithm at the output port in order to find the best match between input wavelengths and the converters.

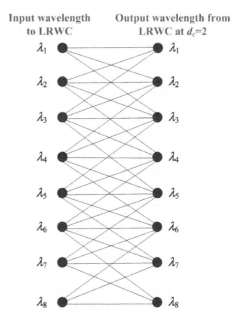

Figure 3.14: A non-circular LRWC example at $W = 8$ and $d_c = 2$

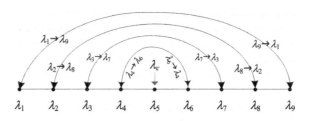

Figure 3.15: Parametric wavelength converter at pump wavelength λ_5 under NRP [31, 32]

There are three conversion policies for LRWC [41]: (1) Far conversion policy chooses the farthest idle output wavelength with respect to the input wavelength; if there exist two such wavelengths one of them is selected randomly, (2) Near conversion policy selects the nearest free output wavelength with respect to the input wavelength; if there exist more than one such wavelengths one of them is chosen randomly, and (3) Random conversion policy picks a random available output wavelength.

LRWC can be designed in one of two types:

- **Non-circular**: If wavelengths are numbered from 1 to W and the input wavelength number is i, the output wavelength can be any of $\min(1, i - d_c)$ to $\max(i + d_c, W)$ wavelengths. Figure 3.14 illustrates the conversion range for a non-circular LRWC at $W = 8$ and $d_c = 2$. As can be observed, each wavelength can only be converted to two above and two below wavelengths.

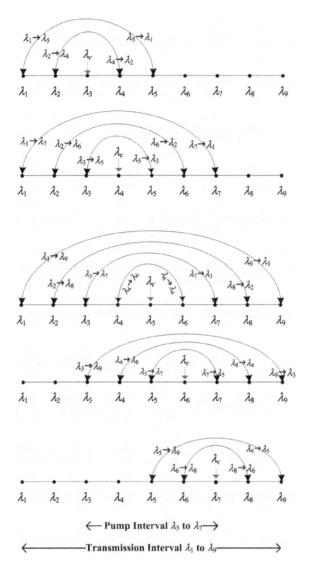

Figure 3.16: Parametric wavelength converter at different pump wavelengths under VRP [31, 32]

However, wavelength λ_1 can only be converted to λ_2 and λ_3, while wavelength λ_8 can only be converted to λ_6 and λ_7.

– **Circular**: Using circular LRWC [42], all wavelengths can be converted to $2 \times d_c + 1$ output wavelengths, even the border wavelengths. For example, wavelength λ_1 at $d_c = 2$ can be converted to output wavelengths $(\lambda_{W-1}, \lambda_W, \lambda_1, \lambda_2, \lambda_3)$. Similarly, wavelength λ_W can be converted to output wavelengths $(\lambda_{W-2}, \lambda_{W-1}, \lambda_W, \lambda_1, \lambda_2)$.

- **Fixed-Input/Tunable-Output WC (FTWC):** A WC with type FTWC can convert fixed wavelength λ_k to any output wavelength from λ_1 to λ_W. This type of converter is cheaper than a Tunable-input/Tunable-output WC (TTWC) [43].

- **Tunable-Input/Fixed-Output WC (TFWC):** A WC with type TFWC can convert any input wavelength from λ_1 to λ_W to a fixed output wavelength λ_k. This type of converter is less expensive than both TTWC and FTWC [43, 44].

- **Parametric Wavelength Converter (PWC):** A PWC converts multiple input wavelengths to another set of wavelengths simultaneously in the same device [31, 37, 45]. In PWC, an input wavelength is converted to an output wavelength which is symmetric with respect to a pump wavelength. If the pump is positioned on wavelength λ_c, then either wavelength λ_i is converted to wavelength $\lambda_{2 \times c - i}$ or wavelength $\lambda_{2 \times c - i}$ is converted to wavelength λ_i, but not both conversions at the same time. Putting the pump wavelength on the central wavelength of the spectrum leads to high switching performance improvement because this configuration provides maximum number of wavelength conversion pairs in a PWC device. Note that the pump wavelength can be set either at a transmission wavelength (say on wavelength channel λ_5 as in Fig. 3.15) or at the middle of two adjacent transmission wavelength channels (say between wavelength channel λ_5 and λ_6).

Figure 3.15 illustrates a PWC device with nine wavelengths, where a pump wavelength has been set at $\lambda_c = \lambda_5$. Here, there are four conversion pairs as $\lambda_1 \leftrightarrow \lambda_9$, $\lambda_2 \leftrightarrow \lambda_8$, $\lambda_3 \leftrightarrow \lambda_7$, and $\lambda_4 \leftrightarrow \lambda_6$. However, at the same time, only four conversions can happen; i.e., the PWC device cannot convert both $\lambda_2 \rightarrow \lambda_8$ and $\lambda_8 \rightarrow \lambda_2$ at the same time. Of course, the pump wavelength can be placed on another wavelength, thus creating a different set of conversion patterns. It should be mentioned that a PWC may need wavelength a guard band around the pump wavelength due to crosstalk effect, and therefore all wavelengths in the guard band could not be used for conversion purposes [36]. For example, in Fig. 3.15, if λ_4 and λ_6 are within the guard band, we can convert neither λ_4 to λ_6 nor λ_6 to λ_4.

The conversion set of wavelengths at a PWC with a given pump wavelength is called a conversion pattern. Obviously, changing pump wavelength leads to different patterns. There are mainly two policies to select pump wavelength in a PWC device as [31, 32]:

- **Number-Rich Policy (NRP):** In all PWC devices, the pump wavelength is always set at the middle of lowest and highest channel wavelengths (as in Fig. 3.15). This leads to maximum number of conversion pairs, but a given wavelength can only be converted to another given wavelength in all PWCs, say $\lambda_2 \rightarrow \lambda_8$ is possible in every PWC used in a core switch. Clearly, this limits the conversion flexibility in PWCs as all conversion patterns are not available in the core switch (e.g., $\lambda_2 \rightarrow \lambda_7$ is not possible in any PWC). Therefore, some wavelength conversion requests for optical packets may be blocked, even if the desired output wavelength is free.

- **Variety-Rich Policy (VRP)**: Each PWC device in an OPS core switch has a different pump wavelength which provides different conversion pairs in the core switch. However, to obtain maximum possible conversion pairs counts, the pump wavelengths should be chosen around the central wavelength. Let there be W transmission wavelengths (i.e., λ_1 to λ_W). Then, the place for positioning a pump wavelength could be within the pump interval λ_l to λ_h, where the center of this interval is located at the center of transmission wavelengths; i.e., $\frac{l+h}{2} = \frac{1+W}{2}$. Figure 3.16 depicts five positions for placing pump wavelengths within pump interval λ_3 to λ_7. In an OPS switch using K PWCs, pump wavelengths should be placed in equal distances of $\delta = \frac{h-l}{K-1}$ from λ_l to λ_h. In other words, pump wavelength λ_{p_m} for the m-th PWC is located at $\lambda_{p_m} = \lambda_{l+(m-1)\times\delta}$. In a slotted OPS, a core switch can choose pump wavelengths dynamically. The switch controller evaluates the status of contending optical packets arriving at a time slot and provides the best conversion pattern. Then, it adjusts the pump wavelength of each PWC in order to satisfy maximum wavelength conversion pair requests [32].

Compared with single-channel converters, PWC devices can reduce cost and operational power. This is because a single PWC device can provide simultaneous wavelength conversions, thus reducing the number of required WCs. In addition, since only one PWC device performs the conversions, total power consumption is expected to decrease. Although PWC devices require a high-power pump, the total number of active pumps at the same time is less than the case in which single-channel converters are used and each WC needs a separate pump wavelength [45].

3.1.1.2.2 Architectures of Wavelength Converters in OPS Switches Different architectures have been proposed to use WCs in core switches:

- **Single-Per-Channel (SPC)**: At each output port of a core switch, there is one WC per output wavelength, where each WC is a FRWC; i.e., it can convert any input wavelength to any output wavelength, e.g., [33, 34]. Here, we have $r_{wc} = W$ and $N_{WC} = N \times r_{wc} = N \times W$. There are two architectures proposed for SPC:

  - SPC Architecture 1: Figure 3.17 illustrates SPC Architecture 1. In space switch A, there are W space switches, each with N inputs and $N + N \times W$ outputs. In space switch i that switches the optical packets coming on wavelength λ_i, the optical packets that do not need wavelength conversion are directed through N switch outputs toward their output links. Optical packets going on different wavelengths through different paths toward output link j are combined together and sent to the output link fiber using a passive combiner device. On the other hand, an optical packet destined to output link j that requires wavelength conversion is forwarded through $N \times W$ switch outputs toward the combiners located at the inputs of wavelength converter bank j dedicated to output link j. Note that only one optical packet must

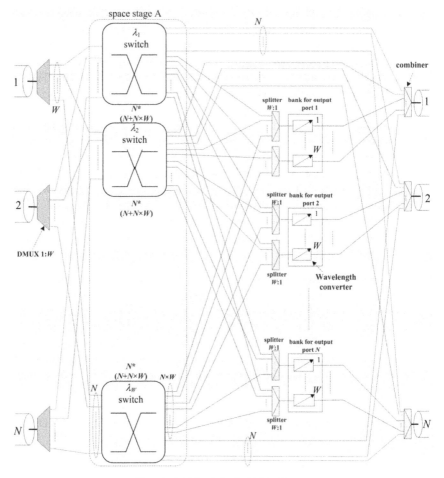

Figure 3.17: OPS switch with SPC Architecture 1

arrive through the combiner at a given WC in bank j. After conversion, the optical packet is sent to the combiner of output link j.

- SPC Architecture 2: Figure 3.18 displays SPC Architecture 2 that includes a space switch for each output link, a number of SOA gates, tunable filters, multiplexers, demultiplexers, combiners, and splitters. The optical signal coming from each input link is split into N signals and directed toward N output links. Each split signal is further split again into W signals and enter W tunable filters. A tunable filter includes a 1:W DMUX, a W:1 MUX, and W SOA gates. It selects one wavelength that can be converted by a WC in the relevant converter bank. Inside a tunable filter, only one of the gates must be connected (i.e., on state) and the remaining $W - 1$ gates must be disconnected (i.e., off state). All gates in Fig. 3.18 are depicted in off state.

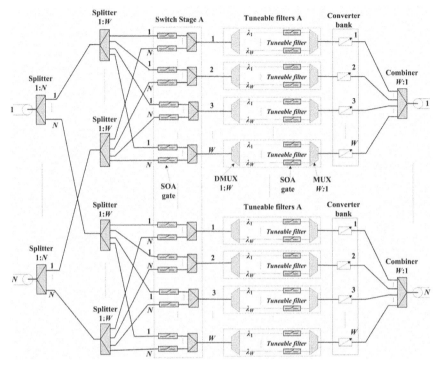

Figure 3.18: OPS switch with SPC Architecture 2 [33]

After conversion, all W signals carried on W wavelengths are combined and sent to the relevant output link.

Using SPC, any optical packet loss due to wavelength continuity can be removed. However, full wavelength converters are expensive and difficult to implement due to the technological limitations [35]. In addition, not all the optical packets arriving at the output port need wavelength conversion (especially at low traffic loads) [46], and therefore such expensive devices may be wasted. Clearly, when more than W optical packets are destined to the same output link, only W of them can be switched and the remaining will be lost.

▪ **Sparse Nodal Conversion:** This design reduces wavelength conversion cost by giving full conversion capabilities only to a given number of core switches in an OPS network. Having carefully selected the nodes to be equipped with WCs is an important optimization issue by which most of the benefits of equipping all core switches in the network with full wavelength conversion can be obtained.

▪ **Shared Wavelength Conversion:** In this technique, WCs are shared in an OPS core switch in order to reduce the core switch cost by the following architectures. Shared WCs are collected in a place called converter bank.

$(N \times W + r_{cw}) * (N \times W + r_{cw})$

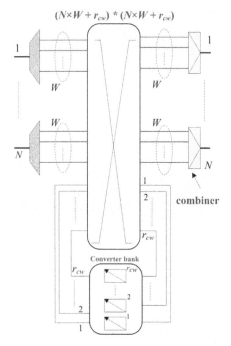

Figure 3.19: OPS switch with recirculating SPN WC architecture [47, 48]

- **Shared-Per-Node (SPN) wavelength conversion:** In this case, either a number of full-range WCs [33, 35, 46, 49] or a number of limited-range WCs (e.g., [50]) are collected in a converter bank and shared among all the output ports of a core switch. Under SPN, we have $N_{WC} = r_{wc}$. The SPN architecture shares converter banks in either recirculating mode (depicted in Fig. 3.19) [47, 48] or feed-forward mode [29]. In the recirculating mode, WCs could be either FRWC or LRWC. After wavelength conversion, optical packets re-enter in the switch and then are scheduled to their desired output ports. There are two architectures proposed for the feed-forward SPN:

  * SPN Architecture 1: There are two sets of switching modules in SPN Architecture 1 as depicted in Fig. 3.20, namely space switch A and space switch B, where all the switching modules in these stages are strictly non-blocking. In space switch A, there are W space switches, each with N inputs and $N + r_{wc}$ outputs. As depicted in Fig. 3.20, the controller of the core switch directs only those optical packets that require conversion to the converter bank. A converted optical packet is then switched to its desired output port using the space switch B. For this purpose, the space switch B uses N SOA gates for each converter output; i.e., totally $N \times r_{wc}$ gates [33, 43].

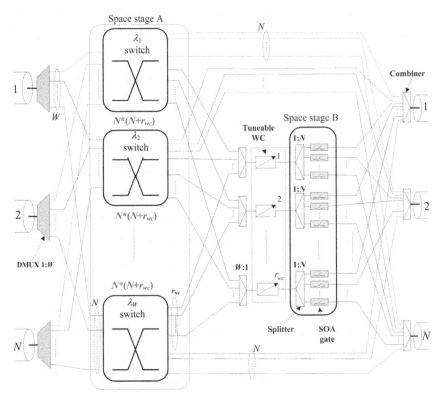

Figure 3.20: OPS switch with feed-forward SPN Architecture 1. Reprinted from *Optics Communications*, Vol. 54, N. Akara, C. Raffaellib, M. Savib, and E. Karasana, Shared-per-wavelength asynchronous optical packet switching: a comparative analysis, 2166–2181, copyright 2010, with permission from Elsevier.

* SPN Architecture 2: In SPN Architecture 2, displayed in Fig. 3.21, the core switch controller forwards the optical packets not needing wavelength conversion toward their output ports through switch space A and tunable filters A. Note that inside a tunable filter, only one SOA gate can be set on (i.e., connected) and the remaining SOA gates must be set off (i.e., disconnected). On the other hand, those optical packets that need wavelength conversion are forwarded through switch space B and tunable filters B toward the converter bank with r_{wc} converters. After conversion, optical packets are directed toward their output ports through switch stage C [33, 49].

The SPN architecture can even use PWCs in its converter bank using two different architectures in an OPS switch with N input ports and N output ports [31, 51]:

* Wavelength-Blocking OPS (WB-OPS): As Fig. 3.22a shows, the output of each PWC can be connected directly to a single input port

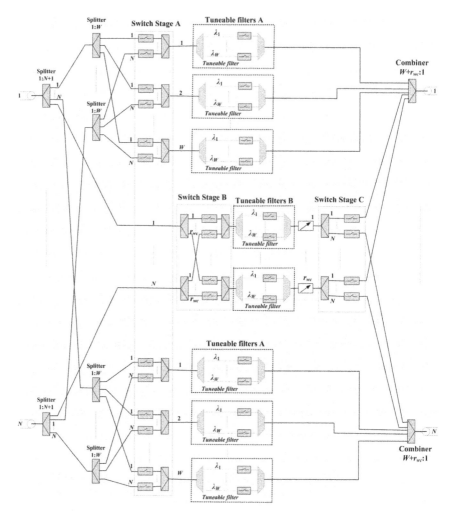

Figure 3.21: OPS switch with feed-forward SPN Architecture 2 [33]

of an OPS switch. This means that all converted wavelengths at the output of the PWC arrive at the switch input port, and therefore they all must be switched to the same output fiber, thus limiting the utilization of PWCs. The switch fabric size is $(N \times W + N_{WC}) * (N \times W + N_{WC} \times W)$, where N_{WC} is the number of shared PWCs used in the switch.

* Non-Wavelength-Blocking OPS (NWB-OPS): As Fig. 3.22b illustrates, the output of a PWC can be demultiplexed to at most W wavelengths and then enter the switch fabrics. In this case, the converted wavelengths from a single PWC can be switched to different output fibers, thus increasing the utilization of PWCs. The

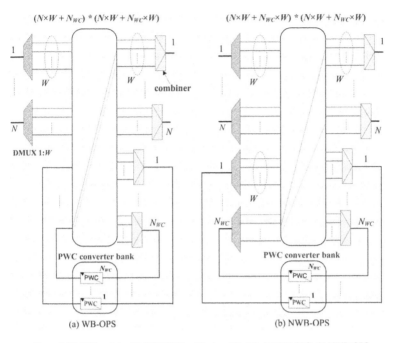

Figure 3.22: OPS switch with SPN-PWC architecture [31, 51]: (a) WB-OPS, (b) NWB-OPS

switch fabric size is $(N \times W + N_{WC} \times W) * (N \times W + N_{WC} \times W)$. Thus, the switch fabric in NWB-OPS needs more ports than WB-OPS.

As can be seen, both of the WB-OPS and NWB-OPS architectures use only one PWC for converting the wavelength of an optical packet. This is called Non-Recursive Parametric Wavelength Conversion (NRPWC). Under NR-PWC, there may be some idle wavelengths and some PWCs conversion pairs may be available, but input optical packets could not be converted to the output idle wavelengths due to the lack of appropriate conversion patterns.

Figure 3.24 compares PLR in a single 16*16 NWB-OPS switch and single 16*16 WB-OPS switch with $W = 32$ wavelengths per fiber, when using N_{WC} PWC converters. Using the VRP in the NWB-OPS results in a lower PLR than the NRP for $N_{WC} = \{1, 3, 5, 7\}$, where each of them has the corresponding minimum pump interval size according to N_{WC}. Moreover, for the VRP under NWB-OPS, PLR decreases as N_{WC} increases. In contrast, PLR under NRP in the NWB-OPS remains almost unchanged as N_{WC} changes. This is because when N_{WC} increases, the variety of conversion pairs remains the same. In the WB-OPS, either changing N_{WC} or using a different policy results in almost the same PLR. The figure also shows that the NWB-OPS has a lower PLR than the WB-OPS. Figure 3.25 compares PLR for the WB-

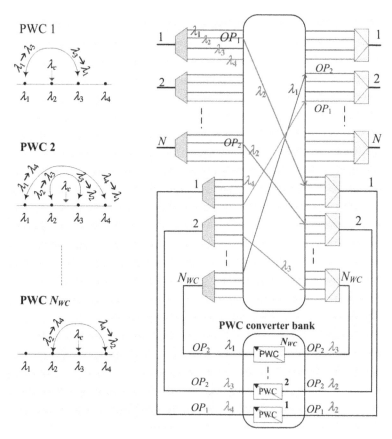

Figure 3.23: An OPS switch with SPN architecture with recursive PWC [52]

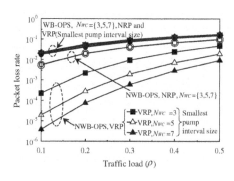

Figure 3.24: PLR of the NRR and VRP at $N = 16$ and $W = 32$ [31]

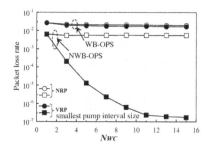

Figure 3.25: PLR of NRP and VRP in the WB-OPS and the NWB-OPS at $N = 16$, $W = 32$, and $L_{OP} = 0.1$ [31]

OPS and the NWB-OPS using the NRP and VRP at optical packet load $L_{OP} = 0.1$. The horizontal axis shows the number of PWCs (i.e., N_{WC})

used in the switch. The PLR in the WB-OPS, for both the NRP and VRP with the smallest pump interval size, is higher than that in the NWB-OPS. The PLR saturates at $N_{WC} = 2$ in the NRP and VRP in the WB-OPS and the NRP in the NWB-OPS, while the VRP in the NWB-OPS saturates at $N_{WC} = 13$. This shows that the performance of the VRP in the NWB-OPS is much better than the others. In short, using VRP PWCs under NWB-OPS can provide the lowest PLR.

Unlike NRPWC, Recursive Parametric Wavelength Conversion (RPWC) uses more than one PWC in a cascaded manner for converting the wavelength of an optical packet. The RPWC cascades available patterns to obtain more conversion patterns. It is proved that at most three stages of cascaded PWCs are enough to provide significant PLR improvement over NRPWC at any number of switch sizes. For example, consider in Fig. 3.23 that four optical packets are destined to output port 2, where two of them can be directly switched on wavelengths λ_2 and λ_3 without using any wavelength conversion (but not displayed in the figure). Wavelength conversion patterns are displayed in the figure, where, for example, there are two conversion patterns for PWC 2 as $\lambda_1 \leftrightarrow \lambda_4$ and $\lambda_2 \leftrightarrow \lambda_3$. However, optical packet OP_1 should be wavelength converted through PWC 1 from λ_2 to λ_4 toward output port 2. To schedule optical packet OP_2 coming on λ_2 from input port N, two cascaded wavelength conversions are required: first from λ_2 to λ_3 through PWC 2 and then from λ_3 to λ_1 through PWC N_{WC}. The RPWC architecture significantly reduces PLR compared with NRPWC [52, 53].

– **Shared-Per-Link (SPL) wavelength conversion:** In Shared-Per-Link (SPL) architecture, converter banks are placed at each output port; therefore, WCs are shared among the wavelengths of an output port. In SPL, we have $N_{WC} = N \times r_{wc}$. There are two approaches proposed for SPL architecture as:

* **Shared-Per-Output-Link (SPOL) wavelength conversion:** Here, a bank of WCs is dedicated to each output link of a core switch. The converters are placed after the switching stage in the core switch. The converter bank of a given output link can only be used by those optical packets that are directed to this output link. If an optical packet needs wavelength conversion, it is sent to the output branch in which there are WCs. Otherwise, the optical packet is forwarded to the output branch without wavelength conversion. The converters could be either full-range, e.g., [33] or limited-range, e.g., [54, 55]. There are two architectures proposed for SPOL:

· SPOL Architecture 1: According to Fig. 3.26, space stage A uses W switches, each with N inputs and $N + N \times r_{wc}$ outputs, to forward optical packets toward their output ports either with wavelength conversion (through $N \times r_{wc}$ paths) or without wavelength conversion (through N paths).

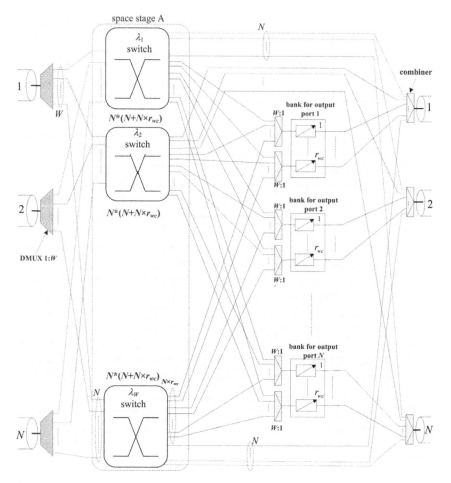

Figure 3.26: OPS switch with SPOL Architecture 1. Reprinted from *Optics Communications*, Vol. 54, N. Akara, C. Raffaellib, M. Savib, and E. Karasana, Shared-per-wavelength asynchronous optical packet switching: a comparative analysis, 2166–2181, copyright 2010, with permission from Elsevier.

- SPOL Architecture 2: According to Fig. 3.27, space stage A and tunable filters are used to direct optical packets toward their output ports.

* **Shared-Per-Input-Link (SPIL) wavelength conversion:** Here, a bank of WCs dedicated to each input link is placed before the switching stage of a core switch (see Fig. 3.28). In a wavelength converter bank, an optical packet incoming from its relevant input link on a wavelength is converted to one of the wavelengths used within the switching fabric, before reaching the desired output link. An arriving optical packet is first wavelength filtered using tunable filters depicted in Fig. 3.28. Then, its wavelength is converted if

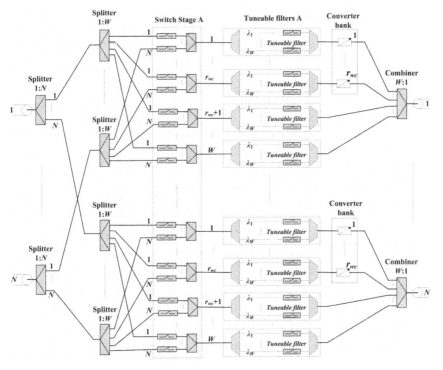

Figure 3.27: OPS switch with SPOL Architecture 2 [33]

needed, and next it is forwarded to the desired output link. For this purpose, the SOA gates used inside the tunable filters and switching stage A must be appropriately switched on/off [56–59].

- **Shared-per-input-wavelength (SPIW) wavelength conversion**: In SPIW configuration, a pool of WCs is dedicated to each wavelength in such a way that all converters in the pool have the same input wavelength, but they can provide different output wavelengths (i.e., fixed-input/tunable-output WCs) [43, 60–64]. In SPIW, we have $N_{WC} = W \times r_{wc}$. Figure 3.29 displays an $N * N$ OPS switch with SPIW capability. There are two sets of switching modules in Fig. 3.29, namely space switch A and space switch B. All the switching modules in these stages are strictly non-blocking. In space switch A, there are W space switches, each with N inputs and $N + r_{wc}$ outputs. After demultiplexing of input wavelengths from N input fibers, the optical packets coming on the same wavelength λ_i are directed to space switch i. The space switch i forwards the optical packets not needing conversion directly toward their output ports, and it forwards the optical packets needing wavelength conversion toward the converter bank dedicated to λ_i. There are W converter banks for wavelengths λ_1 to λ_W, where each bank has r_{wc} full-range fixed-input/tunable-output WCs (i.e., FTWCs). The FTWCs are

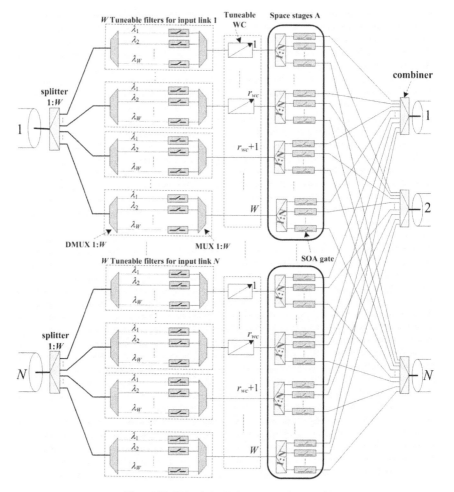

Figure 3.28: OPS switch with SPIL WC architecture [56]

expected to be less costly than tunable-input/tunable-output WCs since an FTWC needs a fixed laser pump while a TTWC needs a tunable laser pump which is expensive and hard to implement. For example, the inputs of all r_{wc} converters in bank 2 relevant to λ_2 are fixed to λ_2, but their outputs can be any wavelength from λ_1 to λ_W. Finally, space switch B forwards the optical packets after wavelength conversion toward their desired output fibers. In this switch, there are $W \times r_{wc}$ of $1 : N$ splitters, $W \times r_{wc} \times N$ of Semiconductor Optical Amplifier (SOA) on/off gates, and N of $W \times r_{wc} : 1$ combiners.

In Fig. 3.29, an optical packet may be lost because of either output blocking (i.e., there are not enough number of wavelengths at the destination output port to accommodate all the optical packets addressed to it) or the lack of

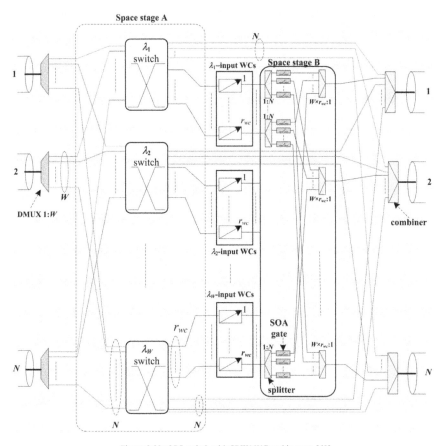

Figure 3.29: OPS switch with SPIW WC architecture [60]

wavelength conversion in a bank (i.e., when more than r_{wc} optical packets that need wavelength conversion arrive on the same wavelength at the same time) [60].

– **Shared-per-output-wavelength (SPOW) wavelength conversion**: Under SPOW configuration, a pool of WCs is dedicated to each wavelength in such a way that all converters in the pool have the same output wavelength, but they have different input wavelengths (i.e., tunable-input/fixed-output WCs) [43]. In SPOW, we have $N_{WC} = W \times r_{wc}$. Figure 3.30 depicts an $N * N$ OPS switch with SPOW capability. There are two sets of switching modules in Fig. 3.30, namely space switch A and space switch B. All the switching modules in these stages are strictly non-blocking. In space switch A, there are W space switches, each with N inputs and $N + (W - 1) \times r_{wc}$ outputs. After demultiplexing of input wavelengths from N input fibers, the optical packets coming on the same wavelength λ_i are directed to space switch i. The space switch i forwards the optical packets not needing conversion di-

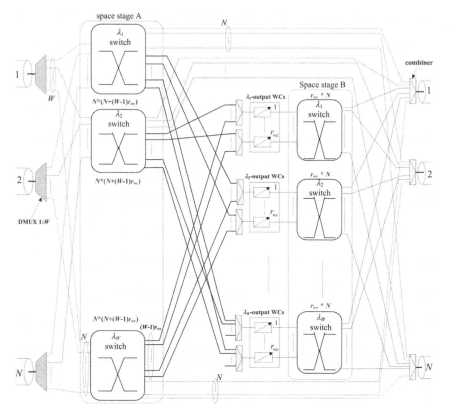

Figure 3.30: OPS switch with SPOW WC architecture. Reprinted from *Optics Communications*, Vol. 54, N. Akara, C. Raffaellib, M. Savib, and E. Karasana, Shared-per-wavelength asynchronous optical packet switching: a comparative analysis, 2166–2181, copyright 2010, with permission from Elsevier.

rectly toward their desired output ports, and it forwards the optical packets requiring wavelength conversion to λ_i toward the wavelength converter bank with output λ_i. There are W wavelength converter banks for wavelengths λ_1 to λ_W, where each bank uses r_{wc} full-range tunable-input/fixed-output WCs, where TFWCs are least complex and cheap WCs. An incoming optical packet that needs wavelength conversion finds the permission to access to any of the $(W - 1) \times r_{wc}$ converters dedicated to $W - 1$ wavelengths, except the one that it is coming from. For instance, the outputs of all r_{wc} converters in bank 2 relevant to λ_2 are fixed to λ_2, but their inputs can be any wavelength from λ_1 to λ_W. Finally, space switch B forwards the optical packets after wavelength conversion toward their desired output fibers. In this stage, there are W switches since WCs on the same bank are fixed-output, where each switch has the size $r_{wc} * N$ because the optical packets on the same wavelength are directed to different output ports.

In SPOW, an optical packet can be forwarded on a given wavelength λ_i after conversion if and only if that wavelength is free at the destination output port and there is at least one free WC in the corresponding bank i. When an optical packet cannot be forwarded on a given wavelength due to the lack of free WC for that wavelength, another wavelength should be checked until a free wavelength on the output port and an available WC in the relevant converter bank is found. According to Fig. 3.30, an optical packet can use more than r_{wc} converters under SPOW. However, in SPIW, an optical packet coming on wavelength λ_i can only use the r_{wc} converters available in bank i.

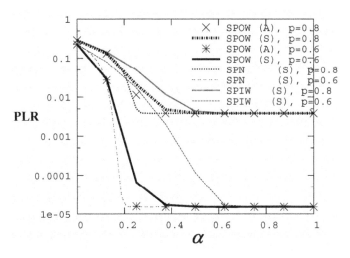

Figure 3.31: PLR of SPOW, SPIW, and SPN switches as a function of α and varying load at $N = 8$ and $W = 48$ [65]

Define conversion ratio α to be the total number of WCs normalized by total number of wavelength channels available in a core switch $\alpha = \frac{N_{WC}}{N \times W}$, where α is a discrete value. In SPIW, for example, one can compute conversion ratio in an OPS switch by $\alpha = \frac{r_{wc} \times W}{N \times W}$ or $r_{wc} = N \times \alpha$.

Figure 3.31 plots PLR of the SPOW obtained by analysis (denoted by "A") and simulation (displayed as "S"), SPIW, and SPN as a function of α and traffic load $p = \{0.6, 0.8\}$, at $N = 8$, $W = 48$ in slotted OPS. Bernoulli arrivals are assumed with probability p in each time slot. It can be seen that the SPOW switch performs very close to the SPN in the region with limited number of WCs (i.e., smaller α), where the performance of SPIW is very poor. This figure also shows that SPOW can perform closely to SPN even when W is high, thus facilitating the implementation of DWDM switches. The SPIW provides good performance when N is high and W is low, while it performs worse when N is low and W is high (as observed in Fig. 3.31) [65].

Figure 3.32 compares the PLR performances of SPIL, SPOL, and SPN architectures in terms of number of WCs. The number of WCs needed by the SPOL

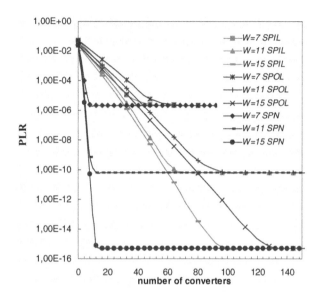

Figure 3.32: PLR under the SPIL, SPOL, and SPN architectures at $L_{OP} = 0.8$, $N = 16$, and $W = 7, 11, 15$ [56]

architecture to reach the PLR saturation are 64, 96, 128 for $W = 7, 11, 15$, respectively. Instead, for the SPIL architecture, the threshold values are 48, 64, 96. This leads to the percentage gains equal to 25%, 33%, and 25% with respect to SPOL architecture. The better performance obtained by SPIL is due to the more efficient sharing of WCs, where optical packets directed to a given output port can use all of the WCs of the core switch, thus sharing them with optical packets destined to other output ports. Under SPOL, however, optical packets can use only the WCs placed at the output port which they are directed to [56].

3.1.1.2.3 Complexity Comparison of Wavelength Converters

To implement an optical space switch, a high number of SOA Gates (OGs) are required. Table 3.1 illustrates the number of OGs required in switching stages, along with the minimum number of WCs required to obtain asymptotic PLR (i.e., PLR of very close to 0) at traffic load 0.25 [43]. Note that the implementation of space switches are assumed to be based on the cross-bar switch architecture, where a space switch with a inputs and b outputs (i.e., $a * b$) requires $a \times b$ OGs. Note that according to the proposed number of WCs in the last column, the required number of OGs is computed at $N = W = 16$ (see the penultimate column). According to this table, SPN needs the least number of WCs. The SPL switch not only needs the highest number of WCs, but also requires the highest number of OGs. On the other hand, the SPIW switch requires a quite high number of WCs, but the least number of OGs compared with the others. This is due to the partitioning of the WCs among the wavelengths which provides significant cost savings. Finally, SPOW switch needs a larger number of WCs compared with SPN and therefore needs a quite higher number of OGs. However, it should be stated that the type of WCs used in SPOW are fixed-output which are less expensive than TTWCs and FTWCs [44]. In addition, performance evaluation results show that SPOW can be equipped with the same number of WCs as the SPN, but at the expense of slightly increasing PLR. In this case, SPOW requires the same number of OGs and WCs as the SPN, but the WCs are less expensive fixed-output converters [43]. It should be noted that in the SPIL architecture illustrated in Fig. 3.28, there are totally $N \times W^2$ gates inside the tunable filters, along with $W \times N^2$ gates inside the switch space stage A; totally $N \times W \times (N + W)$. In addition, there are $N \times r_{wc}$ converters in total inside an OPS switch.

Table 3.1: Complexity comparison of shared WCs in an OPS switch

Sharing Arch.	Number of OGs in Space Stage A	Number of OGs in Space Stage B	Number of OGs in Space Stage C	Number of OGs in Tunable Filters	Total Number of OGs in OPS Switch	Number of OGs at $N =$ $W = 16$	Total Number of WCs (N_{WC}) to Achieve PLR $\approx$ 0
SPIW	$W \times N \times (N + r_{wc})$	$W \times N \times r_{wc}$	—	—	$W \times N \times (N + 2 \times r_{wc})$	8,704	144
SPOW	$W \times N \times (N + (W-1) \times r_{wc})$	$W \times N \times r_{wc}$	—	—	$W \times N \times (N + W \times r_{wc})$	20,480	64
SPN (Arch.1)	$W \times N \times (N + r_{wc})$	$N \times r_{wc}$	—	—	$W \times N^2 + N \times r_{wc} \times (W+1)$	17,152	48
SPN (Arch.2)	$W \times N^2$	$N \times r_{wc}$	$N \times r_{wc}$	$N \times W^2 + W \times r_{wc}$	$W \times N^2 + N \times W^2 + r_{wc}(2N+W)$	10,496	48
SPOL (Arch.1)	$W \times N^2 (1 + r_{wc})$	—	—	—	$W \times N^2 (1 + r_{wc})$	57,344	208
SPOL (Arch.2)	$W \times N^2$	—	—	$N \times W^2$	$W \times N^2 + N \times W^2$	8,192	—
SPIL	$W \times N^2$	$N \times W^2$	$N \times W \times (N+W)$	—	—	8,192	—
SPC (Arch.1)	$W \times N^2 (1+W)$	—	—	—	$W \times N^2 (1+W)$	69,632	·
SPC (Arch.2)	$W \times N^2$	—	—	$N \times W^2$	$W \times N^2 + N \times W^2$	8,192	—

3.1.1.2.4 Scheduling of Wavelength Converters in OPS Switches

To correctly resolve contention and manage forwarding of optical packets in an OPS switch that uses WCs, an appropriate Contention Resolution Scheduling Algorithm (CRSA) is required. These algorithms should consider switching matrix characteristics and the switching context (slotted or asynchronous). They should typically minimize the number of wavelength conversions, thus maximizing the number of optical packets forwarded. Scheduling is especially critical in slotted OPS, where a decision for a number of optical packets arriving at the same time slot must be made within a very small time within the time slot, while in the asynchronous OPS a decision will be made for a single optical packet arriving at the switch input port [43].

In the following, list $S_{k,i,m}$ denotes the set of m contending optical packets indexes incoming on wavelength λ_i destined to output port k of an OPS switch, N_{WC} refers to total number of WCs used in an OPS core switch, and U keeps the list of unused wavelengths when scheduling optical packets without wavelength conversion. Parameters i and l denote wavelengths and C keeps the number of used WCs. It is assumed that WCs in the following CRSA algorithms are LRWC with conversion

Algorithm: Phase 1 of Contention Resolution under any Type of WC Sharing
Architecture

1: **for** $k = 0$ to $N - 1$ {scan all N output ports} **do**
2:    **for** $i = 0$ to $W - 1$ {scan all wavelengths on port k} **do**
3:       Make set $S_{k,i,m}$ {prepare the set of contending optical packets on λ_i for output port k}
4:       **if** $m = 0$ {no optical packet contending for λ_i on port k} **then**
5:          Add pair (i,k) to list U {save unused wavelength i in list U}
6:       **else**
7:          {there are m optical packets contending for wavelength i at output port k}
8:          Remove randomly one optical packet OP from $S_{k,i,m}$
9:          Schedule OP on λ_i at output port k
10:         Set $m = m - 1$ in $S_{k,i,m}$ {update m in list $S_{k,i,m}$}
11:       **end if**
12:    **end for**
13: **end for**
14: END

Figure 3.33: Pseudo code of Phase 1 of contention resolution scheduling in a core switch

degree d_c for SPN, SPIL, and SPOL. Clearly, when using FRWC in converter banks, it is enough to set $d_c = W$. However, for the SPN architecture, CRSA for PWCs are also described.

In the CRSA of all types of wavelength conversion sharing architectures, the first phase (called Phase 1) is the same. Figure 3.33 shows Phase 1 in which all output ports and all wavelengths on each output port are scanned in order to schedule the optical packets that do not need wavelength conversion. First, for each output port k and wavelength λ_i, set $S_{k,i,m}$ is created that includes the set of m contending optical packets destined to output port k on wavelength λ_i. After scheduling of an optical packet from set $S_{k,i,m}$ at output port k on wavelength λ_i, this set is updated appropriately; where updating means removing the scheduled optical packet index from this set and reducing the number of items m in the set by 1. In addition, when there is an unused wavelength on output link k, the information of this wavelength including output link k and wavelength number i is added to list U.

- **CRSA in SPN**: In short, when an optical packet arrives at a core switch on wavelength λ_i, the CRSA first checks if wavelength λ_i is free on the corresponding output port. If yes, the optical packet is directly forwarded to the output port (see Figure 3.33). Otherwise, the CRSA picks a free wavelength at the destination output port and sequentially checks the available shared WCs in the OPS switch, until a free WC is found. After then, the optical packet is forwarded on the selected wavelength. The optical packet is lost when there is no free wavelength at the destination output port (i.e., output blocking) or no WC is available [43].

Figure 3.34 displays the pseudo code for Phase 2 of CRSA at a core switch with SPN limited range wavelength converters for a slotted OPS. Phase 2 deals with converting the optical packets that need wavelength conversion. First, a randomizing function is applied on list U to randomly change the positions of the elements in this list. This randomizing makes a fair allocation of WCs to different output ports. Then, the elements in list U are evaluated sequentially, where two parameters (i,k) are obtained for each element of this list. Then, list

```
Algorithm:   Phase 2 of Contention Resolution in SPN
```

1: $C = 0$ {initialize the number of used WCs in core switch}
2: Randomize list U {randomize list U}
3: **for** $j = 1$ to $sizeof(U)$ {scan all unused wavelengths in list U} **do**
4:     **if** $C = N_{WC}$ **then**
5:         Exit loop {end of phase 2 if all WCs are used}
6:     **end if**
7:     Get pair (i,k) from $U(j)$ {get the information of an unused output wavelength}
8:     **for** $l = 0$ to $W - 1$ {scan all incoming wavelengths destined to output port k} **do**
9:         Consider list $S_{k,l,m}$
10:         **if** $m > 0$ AND $|l - i| \le d_c$ **then**
11:             Remove randomly one optical packet OP from $S_{k,l,m}$ incoming on λ_l
12:             Schedule OP on λ_i at output port k {use one WC to convert OP from incoming wavelength λ_l to outgoing wavelength λ_i}
13:             Set $C = C + 1$ {update the number of used WCs}
14:             Set $m = m - 1$ in $S_{k,l,m}$ {reduce the number of contending optical packets}
15:             Exit loop {evaluate the next unused wavelength}
16:         **end if**
17:     **end for**
18: **end for**
19: Drop all optical packets remaining in sets $S_{k,i,m}$ when $m > 0$
20: END

Figure 3.34: Pseudo code of Phase 2 of contention resolution scheduling in a core switch with SPN converters

$S_{k,l,m}$ is found randomly in such a way that $m > 0$. In this case, a wavelength converter is allocated to output link k in order to schedule an optical packet from list $S_{k,l,m}$ for transmission on wavelength λ_i after wavelength conversion from wavelength λ_l. After each scheduling, set $S_{k,l,m}$ is updated appropriately. Phase 2 ends whenever the number of used WCs (i.e., C) equals N_{WC} [66]. The remaining optical packets in sets $S_{k,l,m}$ are finally lost.

In the following, the CRSA of Phase 2 for two WB-OPS and NWB-OPS architectures that use PWCs in SPN architecture are detailed. In both cases, the PWCs are saved in converter bank B. Pump wavelengths can be located in PWCs based on either the Number-Rich policy or the Variety-Rich policy. For each PWC, a set of conversion pairs is assigned. For example, if a PWC has conversion pair $\lambda_1 \leftrightarrow \lambda_5$ and $\lambda_2 \leftrightarrow \lambda_4$, the conversion pair set $\{(1,5),(2,4),(4,2),(5,1)\}$ is assigned to the PWC. More details are as follows:

- Figure 3.35 illustrates the pseudo code in Phase 2 of CRSA for the WB-OPS architecture depicted in Fig.3.22a. Here, all output ports are evaluated one by one. For each output port k, its unused wavelengths are extracted from list U and saved in list OUT. Then, list IN is created from the index ind of contending optical packets in set $S_{k,l,m}$ destined to output port k, but only one optical packet from each $S_{k,l,m}$ since there is only one λ_l at the input of each PWC. Then, CRSA starts finding the best PWC that can convert most of the optical packets from IN to OUT. Consider that function $Match(IN, OUT, z)$ returns the list of possible wavelength conversions. After evaluating all PWCs, consider that PWC x can have the most conversion possibility and conversion pairs are kept in $BestMatch$. PWC x is first removed from set B. Then, for each item (ind, l, i) in $BestMatch$, the relevant

Algorithm: Contention Resolution in SPN with PWC under WB-OPS Architecture

1: B = set of N_{WC} PWC converters {save bank of converters in B}
2: $v = N_{WC}$ {save number of PWCs in the converter bank in v}
3: **for** $k = 0$ to $N - 1$ {for all N output ports} **do**
4:    **if** $v = 0$ **then**
5:       Exit loop {if all PWCs used}
6:    **end if**
7:    Extract free wavelengths at output port k from U and append to list OUT
8:    **if** list $OUT \neq \varnothing$ **then**
9:       **for** $l = 0$ to $W - 1$ {scan all wavelengths destined to output port k} **do**
10:          Consider list $S_{k,l,m}$
11:          **if** $m > 0$ **then**
12:             Get index ind of one optical packet from $S_{k,l,m}$ and append (ind,l) to list IN
13:          **end if**
14:       **end for**
15:       {start to find the best match}
16:       $BestMatch = \varnothing$ {Initialize $BestMatch$}
17:       **for** every PWC z in B {scan all v available PWCs in the PWC bank B} **do**
18:          $ML = Match(IN, OUT, z)$ {list of conversions from IN to OUT through PWC z}
19:          **if** $size(ML) > size(BestMatch)$ **then**
20:             $BestMatch = ML$ {save the best match until now}
21:             $x = z$ {save the index of best PWC until now}
22:          **end if**
23:       **end for**
24:       Exclude PWC x from set B
25:       $v = v - 1$
26:       **for** every item (ind,l,i) in $BestMatch$ **do**
27:          Remove optical packet OP with index ind from list $S_{k,l,m}$
28:          Schedule OP on λ_i at output port k {convert wavelength λ_l of OP to outgoing wavelength λ_i}
29:          Set $m = m - 1$ in set $S_{k,l,m}$ {reduce the number of contending optical packets}
30:       **end for**
31:    **end if**
32: **end for**
33: Drop all optical packets remaining in sets $S_{k,l,m}$ when $m > 0$
34: END

Figure 3.35: Pseudo code of contention resolution scheduling in a core switch with SPN with PWC in WB-OPS

optical packets are scheduled from sets $S_{k,l,m}$. Finally, the optical packets remaining in sets $S_{k,l,m}$ that cannot be converted are finally lost.

- Figure 3.36 illustrates the pseudo code in Phase 2 of CRSA for the NWB-OPS architecture illustrated in Fig.3.22b. After randomizing list U, each free wavelength λ_i at output port k from list U is evaluated. Assume that a contending optical packet coming on wavelength λ_l is destined to output port k. All available PWCs are searched for finding a conversion pair from λ_l to λ_i. Let PWC z be able to perform this conversion. First, both conversion pairs (λ_l, λ_i) and (λ_i, λ_l) are removed from the conversion pair set belonging to PWC z. Then, the contending optical packet is scheduled on this conversion, and set $S_{k,l,m}$ is updated appropriately. Next, if all the conversion pairs in PWC z become scheduled, it is removed from B. Finally, the CRSA should exit both loops and evaluate the next unused wavelength in list U. The algorithm is over when set B becomes empty. The remaining optical packets in sets $S_{k,l,m}$ are finally dropped.

- **CRSA in SPOL**: When an optical packet arrives at a core switch on wavelength λ_i, the CRSA first checks if wavelength λ_i is free on the corresponding output

Algorithm: Contention Resolution in SPN with PWC under NWB-OPS Architecture

1: B = set of N_{WC} PWC converters {save bank of converters in B}
2: $v = N_{WC}$ {save number of PWCs in converter bank in v}
3: Randomize list U {randomize list U}
4: **for** $j = 1$ to $sizeof(U)$ {scan all unused wavelengths in list U} **do**
5:     **if** $v = 0$ **then**
6:         Exit loop {if all PWCs used}
7:     **end if**
8:     Get pair (i,k) from $U(j)$ {get the information of an unused output wavelength}
9:     **for** $l = 0$ to $W - 1$ {scan all incoming wavelengths destined to output port k} **do**
10:        Consider list $S_{k,l,m}$
11:        **if** $m > 0$ **then**
12:            **for** every PWC z in B {scan all v available PWCs in the PWC bank B} **do**
13:                **if** conversion pair (l,i) is available in PWC z **then**
14:                    Remove conversion pair (l,i) and (i,l) from PWC z {both (l,i) and (i,l) must be removed since conversion is possible in one direction}
15:                    Remove randomly an optical packet OP from list $S_{k,l,m}$
16:                    Schedule OP on λ_i at output port k through PWC z {convert OP from incoming wavelength λ_l to outgoing wavelength λ_i}
17:                    Set $m = m - 1$ in set $S_{k,l,m}$ {reduce the number of contending optical packets}
18:                    **if** all conversion pairs in PWC z scheduled **then**
19:                        Exclude PWC z from set B
20:                        $v = v - 1$
21:                    **end if**
22:                    Exit both loops to evaluate the next unused wavelength
23:                **end if**
24:            **end for**
25:        **end if**
26:    **end for**
27: **end for**
28: Drop all optical packets remaining in sets $S_{k,i,m}$ when $m > 0$
29: END

Figure 3.36: Pseudo code of contention resolution scheduling in a core switch with SPN with PWC in NWB-OPS

Algorithm: Phase 2 of Contention Resolution in SPOL

1: **for** $k = 0$ to $N - 1$ {for all N output ports} **do**
2:     $C_k = 0$ {initialize the number of used WCs in output port k}
3: **end for**
4: **for** $j = 1$ to $sizeof(U)$ {scan all unused wavelengths in list U} **do**
5:     Get pair (i,k) from $U(j)$ {get the information of an unused output wavelength}
6:     **if** $C_k < r_{wc}$ **then**
7:         **for** $l = 0$ to $W - 1$ {scan all incoming wavelengths destined to output port k} **do**
8:             Consider list $S_{k,l,m}$
9:             **if** $m > 0$ AND $|l - i| \le d_c$ **then**
10:                Remove randomly one optical packet OP from $S_{k,l,m}$ incoming on λ_l
11:                Schedule OP on λ_i at output port k {use one WC to convert wavelength λ_l of OP to outgoing wavelength λ_i}
12:                Set $C_k = C_k + 1$ {update the number of used WCs at output port k}
13:                Set $m = m - 1$ in $S_{k,l,m}$ {reduce the number of contending optical packets}
14:                Exit loop {evaluate the next unused wavelength}
15:            **end if**
16:        **end for**
17:    **end if**
18: **end for**
19: Drop all optical packets remaining in sets $S_{k,i,m}$ when $m > 0$
20: END

Figure 3.37: Pseudo code of Phase 2 of contention resolution scheduling in a core switch with SPOL converters

port. If yes, the optical packet is forwarded according to the code in Fig. 3.33. Otherwise, the CRSA randomly selects a wavelength on the corresponding output port and checks whether there is a free WC among the shared WCs on the corresponding converter bank [43]. This is performed using the code depicted in Fig. 3.37 for limited range wavelength converters, where there are r_{wc} converters for each output port. Here, C_k keeps the number of used WCs in output port k. By scanning all unused wavelengths from list U, the CRSA schedules the incoming optical packets that can be converted to unused outgoing wavelengths at their desired output ports. For each unused wavelength λ_i at output port k, CRSA tries to find a contending optical packet coming on wavelength λ_l and destined to output port k in order to convert to wavelength λ_i using a WC in the converter bank allocated to output port k. Those optical packets that cannot be shifted in wavelength are finally dropped.

```
Algorithm:  Contention Resolution in SPC

 1: for k = 0 to N − 1 {for all N output ports} do
 2:    for i = 0 to W − 1 {scan all wavelengths destined to output port k} do
 3:        Consider list S_{k,i,m}
 4:        if m > 0 then
 5:            Remove randomly one optical packet from S_{k,i,m} incoming on λ_i for scheduling on λ_i
 6:            Set m = m − 1 in S_{k,i,m} {reduce the number of contending optical packets}
 7:        else
 8:            {either no packet coming on λ_i or all packets in set S_{k,i,m} have already been scheduled}
 9:            for l = 0 to W − 1 {scan all incoming wavelengths destined to output port k} do
10:                Consider list S_{k,l,m}
11:                if m > 0 AND |l − i| ≤ d_c then
12:                    Remove randomly one optical packet OP from S_{k,l,m} incoming on λ_l
13:                    Schedule OP on λ_i at output port k {use one WC to convert wavelength λ_l of OP to
                       outgoing wavelength λ_i}
14:                    Set m = m − 1 in set S_{k,l,m} {reduce the number of contending optical packets}
15:                    Exit loop {evaluate the next wavelength i at output port k}
16:                end if
17:            end for
18:        end if
19:    end for
20: end for
21: Drop all optical packets remaining in sets S_{k,i,m} when m > 0
22: END
```

Figure 3.38: Pseudo code of contention resolution scheduling in a core switch with SPC converters

- **CRSA in SPC**: One may use the same CRSA in SPOL for SPC, except that we have $r_{wc} = W$. However, a simpler algorithm can be introduced for SPC as displayed in Fig. 3.38 for limited-range wavelength converters. This algorithm scans every wavelength λ_i at each output link k and schedules an optical packet on this wavelength either without wavelength conversion or with wavelength conversion. If there is an optical packet coming on λ_i, it is directly scheduled on λ_i. However, if there is no optical packet coming on λ_i or the optical packets coming on λ_i have already been scheduled on previous wavelength channels, CRSA searches for an optical packet coming on different wavelengths and schedules it after wavelength conversion to wavelength λ_i. After each scheduling, set $S_{k,i,m}$ or $S_{k,l,m}$ is updated appropriately, where updating means removing the scheduled optical packet index from this set and reducing the number of items m in the set by 1. Finally, those optical packets that cannot be shifted in wavelength are dropped.

Algorithm: Phase 2 of Contention Resolution in SPIL

1: Randomize list U {randomize list U}
2: **for** $z = 0$ to $N - 1$ {for all N input ports} **do**
3:     $C_z = 0$ {initialize the number of used WCs in input port z}
4: **end for**
5: **for** $j = 1$ to $sizeof(U)$ {scan all unused wavelengths in list U} **do**
6:     Get pair (i,k) from $U(j)$ {get the information of an unused output wavelength}
7:     **for** $l = 0$ to $W - 1$ {scan all incoming wavelengths destined to output port k} **do**
8:         Consider list $S_{k,l,m}$
9:         **if** $|l - i| \leq d_c$ **then**
10:            **for** each optical packet OP in $S_{k,l,m}$ **do**
11:                z = input port of OP {incoming on wavelength λ_l}
12:                **if** $C_z < r_{wc}$ **then**
13:                    Remove OP from $S_{k,l,m}$
14:                    Schedule OP on λ_i at output port k {use one WC to convert wavelength λ_l of OP to outgoing wavelength λ_i}
15:                    Set $C_z = C_z + 1$ {update the number of used WCs in input port z converter bank}
16:                    Set $m = m - 1$ in $S_{k,l,m}$ {reduce the number of contending optical packets in set $S_{k,l,m}$}
17:                    Exit both loops {evaluate the next unused wavelength from list U}
18:                **end if**
19:            **end for**
20:        **end if**
21:    **end for**
22: **end for**
23: Drop all optical packets remaining in sets $S_{k,l,m}$ when $m > 0$
24: END

Figure 3.39: Pseudo code of Phase 2 of contention resolution scheduling in a core switch with SPIL converters

- **CRSA in SPIL**: Phase 1 of SPIL is the same as other sharing techniques (see Fig. 3.33). Figure 3.39 shows how the optical packets that cannot be normally switched use the WCs in input ports. For each unused wavelength λ_i at output port k, CRSA makes the set of contending optical packets $S_{k,i,m}$. Then, by scanning every optical packet OP in this set, CRSA saves the input port of OP in parameter z. If there is an available WC in converter bank z dedicated to input port z, then OP is scheduled for wavelength conversion from wavelength λ_l to λ_i. Finally, those optical packets that cannot be wavelength converted are lost.

- **CRSA in SPIW**: When an optical packet arrives on λ_i, the CRSA first checks if wavelength λ_i is free on the corresponding output port. If yes, the optical packet is forwarded based on the algorithm depicted in Fig. 3.33. Otherwise, when an optical packet coming on wavelength λ_i needs wavelength conversion, it is resolved using the algorithm depicted in Fig. 3.40, where there are r_{wc} converters for each wavelength converter bank. First, a randomizing function is applied on list U to randomly change the positions of the elements in U in order to make a fair allocation of WCs to different output ports. Here, C_l keeps the number of used WCs in converter bank l. By scanning all unused wavelengths from list U, the CRSA schedules the incoming optical packets that can be converted to unused outgoing wavelengths. For each unused wavelength λ_i at output port k, CRSA finds a contending optical packet coming on wavelength λ_l and destined to output port k that can be converted to wavelength λ_i through converter bank l

Algorithm: Phase 2 of Contention Resolution in SPIW

```
1: Randomize list U {randomize list U}
2: for l = 0 to W − 1 {for all W output wavelengths } do
3:     C_l = 0 {initialize the number of used WCs for each converter bank}
4: end for
5: for j = 1 to sizeof(U) {scan all unused wavelengths in list U} do
6:     Get pair (i,k) from U(j) {get the information of an unused wavelength}
7:     for l = 0 to W − 1 {scan all incoming wavelengths destined to output port k} do
8:         Consider list S_{k,l,m}
9:         if m > 0 and C_l < r_{wc} then
10:            Remove randomly one optical packet OP from S_{k,l,m} incoming on λ_l
11:            Schedule OP on λ_i at output port k {use one WC to convert wavelength λ_l of OP to outgoing
               wavelength λ_i}
12:            Set C_l = C_l + 1 {update the number of used WCs in converter bank l}
13:            Set m = m − 1 in S_{k,l,m} {reduce the number of contending optical packets}
14:            Exit loop {evaluate the next unused wavelength}
15:         end if
16:     end for
17: end for
18: Drop all optical packets remaining in sets S_{k,i,m} when m > 0
19: END
```

Figure 3.40: Pseudo code of Phase 2 of contention resolution scheduling in a core switch with SPIW converters

with input wavelengths λ_l. An optical packet is lost due to the unavailability of WCs only when all free wavelengths at the destination output port are checked, but no free WC has been found to convert the optical packet.

Algorithm: Phase 2 of Contention Resolution in SPOW

```
1: Randomize list U {randomize list U}
2: for i = 0 to W − 1 {for all W output wavelengths } do
3:     C_i = 0 {initialize the number of used WCs for each converter bank}
4: end for
5: for j = 1 to sizeof(U) {scan all unused wavelengths in list U} do
6:     Get pair (i,k) from U(j) {get the information of an unused output wavelength}
7:     if C_i < r_{wc} then
8:         for l = 0 to W − 1 {scan all incoming wavelengths destined to output port k} do
9:             Consider list S_{k,l,m}
10:            if m > 0 then
11:                Remove randomly one optical packet OP from S_{k,l,m} incoming on λ_l
12:                Schedule OP on λ_i at output port k {use one WC to convert wavelength λ_l of OP to outgoing
                   wavelength λ_i}
13:                Set C_i = C_i + 1 {update the number of used WCs in converter bank i}
14:                Set m = m − 1 in S_{k,l,m} {reduce the number of contending optical packets}
15:                Exit loop {evaluate the next unused wavelength}
16:            end if
17:        end for
18:     end if
19: end for
20: Drop all optical packets remaining in sets S_{k,i,m} when m > 0
21: END
```

Figure 3.41: Pseudo code of Phase 2 of contention resolution scheduling in a core switch with SPOW converters

- **CRSA in SPOW**: When an optical packet arrives on λ_i, the CRSA first checks if wavelength λ_i is free on the corresponding output port. If yes, the optical packet is forwarded as in Fig. 3.33. Otherwise, those optical packets that need conversion are resolved using the code depicted in Fig. 3.41, where there are r_{wc} converters for each wavelength. At the beginning, the positions of the elements in list U are randomly changed in order to make a fair allocation of WCs to different output ports. Here, C_i keeps the number of used WCs in converter bank i. By scanning all unused wavelengths from list U, the CRSA schedules the incoming optical packets that can be converted to unused outgoing wavelengths. For each unused wavelength λ_i at output port k, CRSA tries to find a contending optical packet coming on wavelength λ_l and going to output port k in order to be converted to wavelength λ_i through converter bank i. An optical packet is lost due to WC unavailability only when all free wavelengths at the destination output port are checked, but no unused WC has been found.

3.1.1.2.5 *Issues on Wavelength Converters*   A comparison is made among different WC architectures in the following:

1. The wavelength conversion is the most preferred contention resolution scheme since it does not apply additional delay and jitter on optical packets. It also does not lead to out of ordering of optical packets [7].

2. Increasing the number of WCs can decrease PLR to a certain amount and after that increasing WCs will not change PLR. This is due to utilizing all free wavelengths at output ports.

3. Although the core switches equipped with shared WCs are cheaper than the core switches with the SPC converters, the complexity of their controllers is high. This is because different cases must be evaluated by a controller in order to schedule contending optical packets through wavelength conversion banks.

4. Although the SPN architecture may reduce the number of required WCs and utilize them in an optimum way, it increases the complexity of switching [33]. Note that the SPN architecture can save the number of required WCs by almost 90% compared with the SPC architecture, but this happens at the cost of increasing almost 40% of required number of Semiconductor Optical Amplifier (SOA) gates [49].

5. The SPN-PWC architecture in WB-OPS (see Fig. 3.22a) can provide the same PLR performance as SPN with FRWC (See Fig. 3.19), but with a smaller number of PWCs [51].

6. PLR decreases as the number of PWCs increases using the VRP. The VRP provides higher performance than the NRP [31].

7. Sharing efficiency of WCs is the highest in the SPN architecture compared with other architectures since the SPN architecture can efficiently utilize shared WCs so that no wavelength converter will be wasted [43]. Both SPL architectures use more converters than SPN, but the complexity of controller algorithm in SPIL and SPOL is lower than SPN. In both SPIL and SPOL, wavelength converters may be wasted. This is because not all the optical packets arriving at

an input/output port may need wavelength conversion, thus wasting unused wavelength converters. Whereas at the same time, another input/output port may require more WCs than its available ones [30]. The same problem can be mentioned for SPIW and SPOW, where the number of optical packets arriving/departing on different wavelengths may widely vary. Therefore, some converter banks dedicated to some wavelengths may be less utilized, whereas some other banks may be fully utilized, and even some optical packets may be dropped due to the lack of WCs in the fully utilized banks [67]. For example, assume the same number of WCs is used in the SPN architecture and the SPOL architecture. Consider two architectures for an 8*8 core switch: (1) 16 WCs are used in the SPN architecture, and (2) two WCs are used in each port (i.e., 16 WCs in total) in the SPOL architecture. In this case, the PLR for the SPN architecture would be less than the SPOL architecture because the latter one may not fully utilize the WCs but the former one can fully utilize them.

8. It is better to use LRWC in a network with many wavelengths (such as long-haul networks). On the other hand, FRWC can be used in a network with a small number of wavelengths. This is because in metro networks there is no significant difference between PLR performance between using FRWC or LRWC [7, 68].

9. The SPIL can provide the same PLR as SPOL, but with a smaller number of required WCs. This is because of efficient sharing of WCs in which optical packets destined to a given output link can use all WCs of the core switch. However, the performance of SPIL heavily depends on the contention resolution algorithm which decides the optical packets that must be converted [56, 59].

10. Compared with the SPC architecture, the SPOL architecture can save the required number of WCs by almost 50% (smaller than SPN), but without increasing the complexity of its switching matrix with respect to SPC [49].

11. The SPIW introduces a significant performance improvement to SPOL under realistic traffic scenarios in addition to its architectural benefits; i.e., using less costly FTWCs as opposed to TTWCs used in SPOL [43].

12. SPIW requires more FTWCs than TTWCs in SPN in order to achieve the same performance. However, the implementation complexity and price of FTWCs are lower than those of TTWCs. Hence, the overall cost of a switch with SPIW is lower than that with SPN under the same PLR [63].

13. The SPN and SPOW architectures provide PLR which are significantly lower than SPIW and SPOL, when the conversion ratio is low. This is due to the flexibility provided by SPN and SPOW in exploiting WCs. Note that conversion ratio in an $N * N$ OPS switch with W wavelengths is defined as the ratio of total number of WCs used in the OPS switch to $N \times W$ (i.e., total number of wavelengths). Even, the SPOW performs very close to the SPN, especially for a large W [43].

14. The SPC can result in the lowest PLR because each output port has enough number of full-range WCs to resolve contention.

3.1.1.3 Non-Blocking Receiver (NBR)   In this section, local drop port dimensioning as a hardware-based contention resolution method is studied [26, 66, 67, 69], where an OPS core switch dedicates more than one output links (i.e., drop links) to each edge switch, but may receive optical packets from the edge switch on only one input link. This can lead to almost a Non-Blocking Receiver (NBR), where no optical packet is dropped in the last core switch due to the resource lack to the egress switch. For an example for receiver blocking, consider that a core switch has received four optical packets on wavelength λ_1 to deliver to its local edge switch. Assume that there is no means for contention resolution. Then, only one optical packet can be successfully delivered to the edge switch on λ_1, and three of them should be dropped. In other words, these three optical packets may have traversed a long path and have used some network resources to reach the last core switch, but they are dropped right at the destination.

Define a drop port to be one fiber connecting a core switch to an egress switch for delivering one optical packet on each wavelength from the core to the egress switch. On the other hand, in multi-fiber networks, define a drop link to be f fibers connecting a core switch to an egress switch for delivering f optical packets on each wavelength from the core to the egress switch at the same time. Therefore, a drop link includes f drop ports in a multi-fiber OPS network. The NBR proposed in [26] is for single-fiber OPS, where the number of drop ports in all core switches are equal. However, for multi-fiber OPS, the number of drop links in all core switches could be either fixed in each core switch [66, 69] or different depending on the core switch size and the neighbor core switches [67].

It has been shown in [26, 66] that the NBR technique can resolve contention and reduce PLR because of reducing the blocking probability on egress switches. Since using a number of additional drop links can provide a high delivering capacity for the local optical packets to each egress switch, the optical resources in each core switch can be employed for resolving the contention among transit optical packets, thus improving PLR [26]. For more information on the effectiveness of NBR, see the performance diagrams illustrated in Section 4.2.6.

As stated, the number of drop links in core switches could be different as in [67]. Consider that edge switch m is connected to core switch m, where this core switch has $N_{d,m}$ dedicated drop links to deliver optical packets to edge switch m as the egress switch (see Fig. 3.42). Let the core switch receive N_p optical packets from different routes at the same time to deliver them to edge switch m. If the core switch has enough number of drop ports (i.e., $N_{d,m} \times f \geq N_p$), then no optical packet is dropped due to a collision on the egress switch. Otherwise, $N_p - N_{d,m} \times f$ contending optical packets must be resolved by other contention resolution techniques available at the core switch. Clearly, for this technique, a large switch must be employed. For example, if in an OPS network, a given core switch (at $f = 1$) is connected to an edge switch with $N_{d,m} = 3$ drop links and three other core switches, the given core switch size should be 8*8, whereas, when using a single drop port, a 4*4 core switch

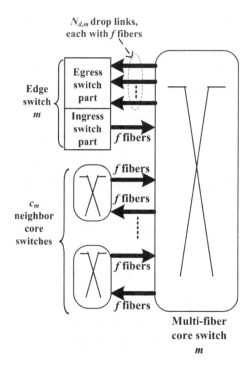

Figure 3.42: Non-blocking receiver architecture

would suffice. Moreover, allocating a number of drop links to the same edge switch, but using only one receiving link from that edge switch wastes the input ports of the core switch. The non-blocking receiver technique used in a core switch is only appropriate for the case that an edge switch is directly connected to the core switch with a number of drop ports and the core switch size allows this.

Let core switch m have c_m neighbor core switches. Note that the structure of core switches usually leads their sizes to be a power of 2 (i.e., 2, 4, ...). Let us choose parameter a as the smallest integer number that satisfies $f \times (c_m + 1) < 2^a$. Hence, a core switch with size $2^a * 2^a$ is used. Then, there are [67]

$$N_{d,m} = \left\lfloor \frac{2^a - f \times c_m}{f} \right\rfloor \tag{3.1}$$

drop links used to connect core switch m to edge switch m in downstream direction, where each drop link includes f fibers and each fiber is connected to an output port (called drop port) of the core switch. By this, whatever port that remains unused in the core switch is connected in downstream direction to its local edge switch to make an almost non-blocking receiver. However, there may be only one upstream connection link from edge switch m to core switch m. By this, the core switch can deliver up to $N_{d,m} \times f$ optical packets arrived on the same wavelength (without using any WCs) to edge switch m at the same time. Using additional drop links, optical packets arriving at the last core switch are not dropped due to lack of drop ports. For example, at $f = 2$ and $c_m = 6$, we should use a 16*16 core switch. Of 16 output ports, 12 output ports connect core switch m to its six neighbor core switches, each neighbor with two fibers, and four ports (i.e., two drop links) connect core switch m to edge switch m.

3.1.2 QoS Differentiation

In this section, hardware-based contention resolution scheme proposed for differentiating optical packets in both asynchronous OPS and slotted OPS networks are studied. Recall that HP, MP, and LP denote to high-priority, mid-priority, and low-priority traffic classes, respectively. The schemes proposed for relative differentiation and absolute differentiation are also discussed in this section.

3.1.2.1 QoS Differentiation with Optical Buffers Most of QoS differentiation algorithms rely on electronic RAM to manage different classes of service. However, as stated, optical RAM does not exist for OPS switches and instead one must use FDLs. However, FDLs are not similar to electronic RAM since

- The size of RAM could be too big, where the size of FDLs are limited.

- Electronic RAMs can be accessed randomly, while data in an FDL must fully circulate an FDL and then it could be reachable. Therefore, conventional preemptive techniques are not easily applicable.

- Holding times of packets in an electronic RAM could be unlimited, while it is limited in FDLs [2].

The FDL architectures studied in 3.1.1.1.1 are not suitable for differentiating class-based optical packets since they are based on the First-In-First-Out (FIFO) scheduling mechanism. Recall that in the FIFO scheduling mechanism, the sequence of output optical packets from an optical buffer must be the same as the incoming sequence. Preemption-based techniques are not applicable in FDL since it is not possible to change the order of optical packets already fed to the FDL. In the following, different techniques and OPS switch architectures proposed for differentiating optical packets are studied.

3.1.2.1.1 Output Circuiting Shared Buffer (OCSB) Figure 3.43 displays the Output Circuiting Shared Buffer (OCSB) architecture in an OPS switch with W switching modules [70]. In each switching module, the dimension of the space switch is $N * (N \times B)$, where each B output ports are connected to a Circuiting Shared Buffer Module (CSBM). Each CSBM includes one 2*2 optical switching unit and $(B - 1)$ 3*2 optical switching units. Each optical switching unit has a circulating FDL. A 3*2 optical switching unit has three input interfaces; one from the space switch, one from the previous optical switching unit, and one from the circulating FDL. On the

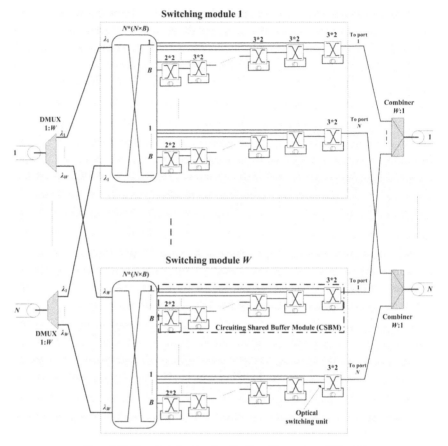

Figure 3.43: The output circuiting shared buffer architecture in an OPS switch

other hand, the 2*2 optical switching unit connected to port B only has two input interfaces: one from the space switch and the other from the circulating FDL. Each CSBM has only one output which is connected to the output combiner.

Scheduling in OCSB is based on this fact that an HP optical packet must be scheduled before an LP optical packet. Assume there are a number of LP optical packets in the optical switching units and an HP optical packet arrives. Then, the LP optical packets must be recirculated in their optical switching units until the HP optical packet leaves the switch through the output port 1 of the space switch. For example, assume $B = 4$ and there are four output ports from the space switch connected to a CSBM. Now consider that there are three LP optical packets OP_1, OP_2, and OP_3 that are in the CSBM. The OP_1 can be sent from output port 1 without experiencing any delay, OP_2 can be scheduled at output port 2 to experience a delay of one time unit, and OP_3 can be scheduled at output port 3 to experience a delay of two time units. Now assume that an HP optical packet OP_4 arrives after exiting OP_1. Then, both OP_2 and OP_3 must be recirculated for an additional time unit until finishing the scheduling OP_4 from output port 1. By this scheduling mechanism, the delay and loss performances of HP optical packets are improved compared with LP optical packets.

3.1.2.1.2 Wavelength Partitioning for FDLs   Here, QoS differentiation is provided in a core switch by partitioning wavelength resources between HP and LP optical packets and then applying a Wavelength Delay Scheduling (WDS) algorithm on each partition to delay the relevant optical packets [71]. A wavelength is called congested when the FDL buffer relevant to it is full and no more optical packets can be injected into it. Two wavelength partitioning strategies can be considered:

- **Fixed-Wavelength Partitioning**: In this partitioning, only K fixed wavelengths are dedicated to HP optical packets; i.e., when wavelengths λ_{K+1} to λ_W are all congested at a given output port, LP optical packets cannot access that output port.

- **Shared Wavelength Partitioning**: Any K wavelengths are dedicated to HP optical packets based on the actual buffer occupancy. In other words, when more than K wavelengths are not congested, both LP and HP optical packets can be transmitted; otherwise only HP optical packets can be transmitted.

Another level of QoS management is to combine the above WDS policy with QoS-based routing in the network level in which the shortest path route is only assigned to HP optical packets, and an alternative route with at most one hop more than the shortest path is assigned to LP optical packets.

3.1.2.1.3 Threshold-Based Technique   In this technique, resource reservation is applied to the FDLs on the time domain [23]. Define delay threshold T_{low} lower than the maximum delay $D_B = D \times B$ in an FDL. The WDS algorithm does not allow an incoming LP optical packet to be accepted on a wavelength if the current buffer occupancy is such that the only available delays on that wavelength are greater than or equal to threshold T_{low}, while HP optical packets can use the whole buffer capacity. For example, let an LP optical packet come on wavelength λ_6 and $T_{low} = 3 \times D$. Now assume that buffers $D_1 = D$ to $D_3 = 3 \times D$ are all occupied on wavelength λ_6. Then, this LP optical packet cannot use the FDL.

Figures 3.44 and 3.45 depict PLR under MINL and MQWS for connectionless and connection-oriented OPS, respectively, for two classes of services at $N = 4$, $W = 16$, and $L_{OP} = 0.8$. The restriction of buffer usage for LP optical packets leads to a high penalty in terms of PLR for the LP traffic, while a good behavior can be achieved for the HP traffic. In addition, adjusting T_{low} tunes the PLR performance of each class of service.

3.1.2.1.4 Wavelength-Based Technique   In this technique, resource reservation is applied to the wavelength domain [23]. HP optical packets can be scheduled to all the wavelengths on a fiber, while LP optical packets are allowed to use only a subset of W_{low} wavelengths. Therefore, LP optical packets are expected to suffer

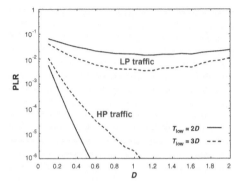

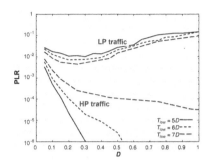

Figure 3.44: PLR under MINL for HP and LP traffic as a function of D for connectionless OPS at $B = 4$. Reprinted from *Computer Networks*, Vol. 44, F. Callegati, W. Cerroni, C. Raffaelli, and P. Zaffoni, Wavelength and time domain exploitation for QoS management in optical packet switches, 569–582, copyright 2004, with permission from Elsevier.

Figure 3.45: PLR under MQWS for HP and LP traffic as a function of D for connection-oriented OPS at $B = 8$. Reprinted from *Computer Networks*, Vol. 44, F. Callegati, W. Cerroni, C. Raffaelli, and P. Zaffoni, Wavelength and time domain exploitation for QoS management in optical packet switches, 569–582, copyright 2004, with permission from Elsevier.

from higher congestion and experience higher losses and delays than HP optical packets.

3.1.2.2 QoS Differentiation with Wavelength Converters
In this section, the contention resolution schemes that use WCs in scheduling of differentiated optical packets are studied. It is common to restrict the access of LP optical packets to the OPS network resources in order to ensure a high performance for HP optical packets. However, this restriction has a negative influence on the performance of LP optical packets where they suffer from higher congestion and delay compared with HP optical packets. Moreover, when reserving resources for a particular traffic class, the resources may be employed less efficiently [72]. This could be the main disadvantage of the access restriction techniques. On the other hand, there are some techniques that do not reserve network resources for a given class, but manage the assignment of the resources to different classes. The following wavelength conversion based approaches can be used to provide QoS for differentiated optical packets in an OPS core switch. Recall that M is the number of traffic classes in the optical domain.

3.1.2.2.1 Reserving Wavelength Converters
Consider a core switch with one of the aforementioned architectures detailed for WCs [58]. The available WCs are divided into three pools. In the first pool, N_1 converters are shared among the optical packets from all traffic classes. In the second pool, N_2 converters are shared between the two HP classes. A smaller subset of N_3 converters is dedicated only to the HP optical packets in the third pool. Therefore, the HP, the MP, and the LP optical packets can access to only $N_1 + N_2 + N_3$, $N_1 + N_2$, and N_1 wavelength converters, respectively. For example, assume that at some time all N_1 wavelength converters are in use and that an LP optical packet arrives. Having insufficient privilege, the LP

optical packet cannot use any converter in other pools and is discarded. However, if a medium-level optical packet arrives, it can use any converter in the second pool, but not the third pool dedicated to the HP traffic. By this assignment, clearly, the HP class has the highest priority to be switched in a core switch, while the LP class may likely be dropped, especially when traffic load is high. The converters may be wasted in this technique because at some time there may be no HP optical packet, whereas there may be many LP optical packets that require wavelength conversion. In this scenario, the reserved WCs for the HP class are wasted, whereas some of the LP packets may be dropped due to the lack of WCs!

3.1.2.2.2 Adaptive Wavelength Converter Allocation Here [73], only two traffic classes are used, and it is attempted to keep the PLR of HP optical packets between two threshold values; i.e., PLR_{MIN} and PLR_{MAX} as the lower and upper bounds for the PLR of HP optical packets, respectively. Unlike the reserving wavelength converters technique in 3.1.2.2.1, this technique adaptively adjusts the number of WCs reserved for the HP class according to the changes in the input traffic pattern because the PLR of HP optical packets varies over time due to variations in the network traffic load. Over time period τ, the PLR is measured for the HP optical packets. If $PLR > PLR_{MAX}$, then the size of the converter pool allocated to HP traffic is increased so that HP traffic has more converters to use. On the other hand, if $PLR < PLR_{MIN}$, then the size of the converter pool dedicated to HP traffic is decreased. In this way, a guaranteed PLR can be provided for HP traffic. The choices of τ, PLR_{MIN}, and PLR_{MAX} are challenging issues. The parameter τ must be chosen large enough to provide reliable results. However, a too large value of τ may lead to a slow reaction to traffic changes. In addition, PLR_{MIN} and PLR_{MAX} must be selected carefully in order to reduce converter under-utilization. The disadvantage mentioned for 3.1.2.2.1 also applies here.

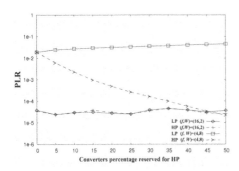

Figure 3.46: PLR as a function of percentage of number of WCs reserved for HP traffic [74]

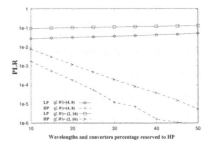

Figure 3.47: PLR as a function of percentage of number of wavelengths and WCs reserved for HP traffic [74]

3.1.2.2.3 Exclusive Reservation of Wavelength Converters for High-Priority Optical Packets Here, among r_{wc} wavelength converters used in a converter bank in a multi-fiber OPS switch output port, N_1 wavelength converters are dynamically reserved for HP optical packets [74]. In dynamic reservation, instead of reserving wavelength converters from index 1 to N_1, the number of available WCs (with any order) is considered. The principle is that when the number of available WCs is less than N_1 (where $0 < N_1 < r_{wc}$), only HP optical packets can be converted, whereas LP optical packets are discarded. Otherwise, both classes can use the available WCs.

A joint reservation of WCs and output port wavelengths in a multi-fiber OPS can be considered for HP optical packets with the threshold mechanism [74]. The reservation percentage for the two domains can be either the same (say 25% of WCs and 25% of wavelengths exclusively reserved for HP optical packets) or not. This reservation can provide better PLR performance for HP optical packets than both only WC reservation and only wavelength reservation.

Figure 3.46 displays PLR in a multi-fiber core switch with SPN architecture as a function of percentage of number of WCs reserved for HP traffic at $(f, W) = \{(4,8),(16,2)\}$, $N_{WC} = 20$ WCs, and 25% of HP traffic. In this case, no wavelength is reserved for HP traffic. By increasing the number of fibers, both HP and LP classes obtain better PLR, but with less differentiation. A good QoS can be reached in configurations with more fibers per input/output link since the multi-fiber switch architecture deals with reduced number of optical packet contentions (i.e., a contention avoidance scheme as detailed in Section 2.2.1). Figure 3.47 illustrates PLR for the joint reservation of WCs and wavelengths for HP traffic (both with the same reservation percentage) at $(f, W) = \{(4,8),(2,16)\}$, $N_{WC} = 20$ WCs, and 25% of HP traffic. Here, a further PLR improvement can be obtained for the HP traffic, while keeping the same performance for the LP traffic [74].

3.1.2.2.4 SPN Wavelength Converters without Reservation Here, SPN wavelength converters are used in a slotted OPS core switch for contention resolution, but without reserving any of them for any class of traffic [75]. In the first phase, required variables are initialized and optical packets destined to each output port are determined. In the second phase of contention resolution for a given output port, optical packets are scheduled without using WCs. When more than one optical packet contend for the same output wavelength channel, the highest priority optical packet is chosen to be directed without wavelength conversion. If more than one highest priority optical packet exists, one of them is randomly scheduled. If there is no highest priority packet, the second HP optical packets are scheduled. The choice of optical packets are at the course of decreasing their priorities. In the third phase of contention resolution, the optical packets needing wavelength conversions are scheduled. The optical packets losing the contention in the second phase are forwarded on free output wavelength channels using the available SPN WCs in the second phase. The optical packets to be scheduled are still selected in order of decreasing their priorities until both wavelength output channels and WCs are available. All the optical packets that cannot be successfully switched are discarded.

3.1.2.2.5 SPIW Wavelength Converters without Reservation The service differentiation is performed in a core switch of a slotted OPS network by giving different priorities to each CoS in accessing to the shared resources (i.e., output wavelength channels and the SPIWs) so that PLR of class h becomes lower than class k, when $1 \leq h < k \leq M$ [64]. In each time slot, the contention resolution scheduler decides whether or not an optical packet of a given class will be directly forwarded, forwarded after wavelength conversion, or dropped. The scheduler should maximize the core switch throughput, minimize the number of conversions required in order to save the required number of SPIWs, and provide an isolation among different classes. The contention resolution scheduler operation is divided into three phases:

- In the first phase, required variables are initialized and optical packets destined to each output port are determined.

- In the second phase, which is run simultaneously on each output port, the optical packets that can be switched without wavelength conversion are evaluated. When a number of optical packets should be scheduled on the same wavelength channel of a given output port, the scheduler will schedule the highest priority optical packet among them. This process is repeated for every wavelength of the given output port. Clearly, if there are more than one optical packet with the highest priority, one of them is chosen randomly for scheduling.

- The third phase is executed in parallel on W different wavelengths. Consider phase 3 for wavelength λ_k that will use the wavelength conversion bank k in which the input of each WC is λ_k. The set of optical packets coming on wavelength λ_k from all N input ports that need wavelength conversion and have the highest priority (called set $S_{k,1}$) is first evaluated. The converters in bank k are used for scheduling of the optical packets in set $S_{k,1}$. However, not all the optical packets in set $S_{k,1}$ could be scheduled because there may be the case in which the desired output wavelength at the desired output port has already been allocated in the second phase. In addition, there may be smaller number of free WCs in the conversion bank than the number of optical packets in set $S_{k,1}$. If there are still unused WCs in bank k, the second set of HP optical packets coming on wavelength λ_k from all N input ports are evaluated (say, set $S_{k,2}$). This process is continued for other priorities, and finally the process of scheduling is stopped when the number of available WCs in bank k becomes zero. All the optical packets that cannot be successfully switched are dropped.

3.1.2.2.6   SPOL Wavelength Converters without Reservation   The service differentiation is provided in a core switch of a slotted OPS network by giving different priorities to each CoS when there is a pool of WCs in each output port (i.e., SPOL wavelength conversion architecture) [76]. The first and second phases of this section are similar to the first and second phases in 3.1.2.2.5 and 3.1.2.2.4. The scheduling of each output port is performed separately. The optical packets that cannot be forwarded directly to their desired output ports and need wavelength conversion are evaluated in the third phase. The unscheduled optical packets destined to output port m are evaluated in the order of decreasing priority. First, the highest-priority optical packets are scheduled by using the WCs in conversion bank m. Then, the second high-priority optical packets are scheduled. This process continues until considering the lowest priority optical packets. The process of scheduling is over when the number of available WCs in bank m becomes zero. All the optical packets that cannot be successfully switched are lost.

3.1.3 Comparison of Hardware-Based Contention Resolution Schemes

Among the hardware-based contention resolution schemes, NBR resolves the contention of local optical packets that have arrived at their destination, whereas the other schemes resolve the contention of transit optical packets. Recall that in NBR, whatever ports that remain unused in a core switch are connected in downstream direction to its local edge switch to make an almost non-blocking receiver. On the other hand, wavelength conversion can provide the best throughout for transit optical packets when increasing network load. Buffering stands at the second rank to reduce PLR and increase network throughput [6].

Access restriction and preemption are two common schemes to provide QoS for HP optical packets. Among the access restriction techniques for the use of WCs, the adaptive wavelength converter allocation could be a reasonable technique because it can alleviate the wastage of network resources, and also can provide a low PLR. When using optical buffering, LP optical packets should arrive at FDL buffers, while HP optical packets should be directly forwarded toward their destinations without buffering. This is because buffering of an HP optical packet that carries video and voice traffic is useless since its destinations will not be able to use an out of order video/voice traffic.

3.2 Software-Based Contention Resolution Schemes

This section studies the algorithmic-based contention resolution schemes that do not need any additional hardware. The algorithms are executed in ingress switches, core switches, or both. It is noted that the hardware-based contention resolution schemes discussed in Section 3.1 can be combined with software-based contention resolution schemes in an OPS network to provide a better performance for OPS networks than using hardware-based or software-based contention resolution schemes solely.

3.2.1 Single-Class Traffic

Here, the software-based contention resolution schemes proposed for single-class traffic in both asynchronous OPS and slotted OPS networks are studied.

3.2.1.1 Routing-Based Schemes   This section reviews different techniques proposed for routing of optical packets in order to resolve contention. Two techniques use the idea of routing to resolve contention:

3.2.1.1.1 Deflection Routing (DR) : In this technique [6, 7, 24, 67, 77–86], some of the contending optical packets are routed along their desired links, while others are routed through non-shortest-path routes. Deflection routing is some sort of buffering where OPS network links behave as FDL to store deflected optical packets.

Consider a multi-fiber network with f fibers on each link and with W wavelengths on each fiber. Suppose a core switch in this network has to send $N_p = \sum_{i=1}^{W} N_{p,i}$ optical packets received from its input ports to a given output link at the same time, where $N_{p,i}$ is the number of optical packets arrived on wavelength λ_i. If there are enough contention resolution techniques (say WCs) at the output link, then the core switch can switch up to $\min(f \times W, N_p)$ optical packets to the output link. Instead of dropping, the core switch can now deflect the remaining $N_p - \min(f \times W, N_p)$ optical packets through other output links toward their egress switches (assuming there are enough resources in other output links). In case there is no WC available at the given output link, up to $y = \sum_{i=1}^{W} \min(N_{p,i}, f)$ optical packets can be switched and $N_p - y$ optical packets should be deflected. Recall that in a multi-fiber OPS network, up to f optical packets on the same wavelength can be switched at an output link.

For example, let core switches X_1 and X_2 be neighbor and core switch X_1 has sent optical packet OP_i to core switch X_2. However, core switch X_2 cannot switch OP_i through the desired output link. The following deflection routing mechanisms can be performed:

- **Deflection Routing without Backward Deflection (DRw/oBD):** This is the most common deflection routing in which backward deflection to the source core switch is not permitted, and core switch X_2 should only deflect OP_i toward other neighbor core switches, except X_1.

- **Deflection Routing with Possible Backward Deflection (DRwBD):** This deflection allows backward deflection to the source core switch as well. This means that core switch X_2 first tries to deflect OP_i to other output links; but if they are all busy, it can deflect OP_i back to X_1 instead of dropping OP_i. This provides an opportunity for OP_i to pass through X_2 later on by sending it back to X_1. Then, X_1 will either be able to switch OP_i or deflect it to X_2 again. This operation resembles the optical link between X_1 and X_2 as an optical buffer that delays the switching of OP_i in X_2. Backward deflection is useful especially when a core switch is connected to a small number of neighbor core switches. Although backward deflection may increase end-to-end delay, it can reduce PLR [67].

- **Reflection Routing:** Reflection routing is a modified version of the deflection routing. In reflection routing, the optical packet to be deflected in core switch X_1 is marked as "reflected" and then sent to neighbor core switch X_2. After receiving the "reflected" optical packet, core switch X_2 acts like a reflector and sends it back to X_1. Actually the link between X_1 and X_2 is considered as an optical buffer. Performance of reflection routing is similar to deflection, but it can be used without any change in the main routes of optical packets [87, 88].

In deflection routing, the desired output link can be chosen by one of the following methods:

- **Uniform Packet Deflection**: in which an optical packet is randomly deflected with equal probability to one of the available output links [79, 80, 85]. Implementation of the uniform deflection is very simple.

- **Shortest Path or Hop Deflection**: in which an available output link with the shortest path or hop that leads to the egress switch of the optical packet is chosen [80–82, 85]. In this deflection, when the desired output link is not available for an optical packet, other output links are checked in the order of shortest path or shortest hop to the relevant egress switch, and therefore the shortest path or hop is selected for deflecting the optical packet. Implementation of the shortest path or hop deflection needs each core switch to know the path lengths or hops to all egress switches from all its output links. Simulation results show that this type of deflection provides similar or better performance than the uniform deflection [85].

- **Shortest Hop and Path Deflection**: in which the available output link according to the shortest hop & path (considering both metrics at the same time) is selected [67]. Here, an $N * N$ core switch keeps a set of output links to each egress switch k; i.e., $P_k = \{OL_0, OL_1, ..., OL_{N-1}\}$. The elements of this set are sorted in an ascending order according to the SHP routing (see 3.2.3.2.3) toward egress switch k. This set is provided in offline and then it is valid until a change happens in the network topology, say due to a node or link failure. For example, OL_0 is the index of the output link with the shortest hop & path to egress switch k, whereas OL_{N-1} is the index of the output link with the longest hop & path to egress switch k.

- **Distributed Deflection Routing**: in which the conflicting optical packets on the same wavelength that cannot be resolved are deflected on different wavelengths [83].

- **Deflection in Hub Core Switches**: in which deflection is only carried out in the core switches that have a high nodal degree [7].

The deflection routing technique has a number of disadvantages:

- The performance of deflection routing highly depends on the topology of the OPS network. It cannot be used in a single-hop OPS network because there is only one path toward a given egress switch.

- Although deflection routing is cheap and simple, it may result in optical packets looping in a multi-hop network for a long time. This looping can be prevented using a Time-To-Live (TTL) field assigned to each optical packet, and dropping the optical packet when its TTL reaches 0. Moreover, a deflection routing algorithm must be intelligent enough to route optical packets so that they go closer to their destinations. In addition, in source routing, one can add a deflection part in an optical packet header for storing deflection routes.

- If a deflected optical packet takes a long path to reach its destination, it would lead to a high end-to-end delay.

- Deflected optical packets are likely to arrive out of sequence at their destinations; however, this is not a considerable issue when using TCP/IP protocol [1, 7].

- Deflected optical packets may experience a higher bit error rate (due to the long paths they travel and additional core switches in which they are switched) than non-deflected optical packets.

3.2.1.1.2 Multi-Path Routing (MPR) Multi-Path Routing (MPR) techniques [24] can be used in OPS networks in order to provide reliability. In particular, when a single link failure happens in an OPS network, optical packets previously routed on that link should be transmitted on alternative paths according to the MPR strategy. Different routing techniques can lead to different PLRs in an OPS network. The MPR uses three routing algorithms:

- **SPR**: A Shortest-Path Routing (SPR) algorithm, using a minimum hop count, is a static routing mechanism that never uses any alternative path. This is called default shortest path.

- **SAP**: A Shortest Alternative Paths (SAP) algorithm uses an alternative path from a set of paths with the same number of hops as the default SPR, but different from it. However, finding such an alternative set may not be possible in some networks depending on the network topology.

- m-**SAP**: An m-Shortest Alternative Paths (m-SAP) algorithm uses m alternative sets of routes where the ith set includes every path with $i-1$ hops more than the SPR one. Note that the first set (set $i = 1$) does not include the default shortest path itself. It is shown that 2-SAP algorithm can be much more effective in reducing PLR than both SPR and SAP [24].

In an OPS core switch, MPR uses SPR as the default technique. However, when a link along the default shortest path becomes congested (say there is no wavelength available for an optical packet), one of the alternative paths from either the SAP set or the m-SAP set is dynamically chosen by the core switch. If there is no alternative path, the packet is dropped. Using m-SAP, MPR provides a lower PLR than SAP and SPR. The MPR tries to reduce PLR by balancing traffic load in an OPS network, thus lowering PLR [24]. However, the computational complexity of the core switch increases under MPR due to the search requirement for the available resources in alternative paths. In addition, the drawbacks mentioned for deflection routing can be stated for MPR. For example, alternative paths may cause optical packets to stay in the OPS network for a long time and the end-to-end delay becomes much higher than SPR.

3.2.1.2 Merit-Based Scheduling (MBS) In an OPS core switch, this technique [89] prioritizes and schedules optical packets based on their merits, and it differentiates the optical packets based on their resource consumptions. The core switch assigns a merit to each optical packet, where the merit could be based on the lifetime of the optical packet in the network, its application real-time demand, the number of resources it has used in the network, and/or the remaining resources along its path in the network. The merit is carried in the header of the optical packet and can be updated in intermediate core switches. OPS core switches resolve the contention among contending optical packets according to their merits. In [89], merit is the number of hops that an optical packet has traveled so that the optical packet that has traveled many hops has a higher merit than the optical packet that has traveled fewer hops. When resolving contention in the core switch, a new arriving optical packet at the core switch with a high-merit can preempt a low-merit optical packet currently in transmission and take over the respective wavelength for its own use. A new arriving optical packet cannot preempt another optical packet in transmission when both have the same merit.

The MBS technique has a lower PLR than the best-effort network because a best-effort network drops all optical packets with the same probability, independent of how many hops an optical packet has traversed. Hence, we often have an unfortunate situation in which an optical packet is dropped even when it is within reach of its destination after traveling many hops, thus wasting the network resources. Therefore, one can see the advantage of MBS. On the other hand, a merit field is required in the header of each optical packet to keep track of resource consumption, thus incurring a small overhead. In addition, the computational complexity of each core switch is slightly high because the merit value of each optical packet must be updated by all intermediate core switches.

3.2.1.3 Retransmission in the Optical Domain Retransmission in the optical domain is another technique to recover the lost traffic in an OPS network in which each ingress switch keeps a copy of the transmitted traffic in its electronic buffer and retransmits whenever required. In other words, instead of retransmitting lost traffic by higher layers, the optical layer manages retransmission. The retransmission technique can be even useful for real-time traffic [90]. There are two retransmission techniques in the optical domain that can be used for OPS networks:

3.2.1.3.1 Random Retransmission (RR) Consider that some traffic is supposed to be sent to a given egress switch. Under the conventional retransmission technique, called Random Retransmission (RR), an ingress switch saves the copy of the traffic in its electronic buffer and then transmits it as an optical packet to the OPS network. Now assume that an intermediate OPS switch should drop the optical packet due to the lack of contention resolution equipments. Then, the OPS switch sends a Negative Acknowledge (NACK) command back to the ingress switch that includes the information of the dropped optical packet. By receiving the NACK command, the ingress switch retransmits the saved traffic to the OPS network as a new optical packet. The new optical packet may also be dropped in an intermediate OPS switch in the network. Then, the ingress switch should resend another optical packet. This continues until a successful transmission [91, 92]. If no NACK command arrives at the ingress switch by some time, the saved traffic is removed from the ingress switch buffer. Since the number of retransmissions is not limited in RR, the multiple retransmissions may lead to further retransmission at high layers such as the TCP layer, which may in turn reduce the TCP throughput in the OPS network. As a drawback, an ingress switch needs a large electronic buffer to save the copy of each transmitted traffic for possible retransmission in future for a long time. In RR, one optical packet may pass through the core switch at its first transmission, and another optical packet may have to be retransmitted for many times.

3.2.1.3.2 Prioritized Retransmission (PR)   The prioritized retransmission technique is proposed and analyzed in [66, 67, 93–98] for slotted OPS networks. In PR, each optical packet transmitted by an ingress switch toward an egress switch carries a priority field δ. The proposed PR is simple since PR only requires a few bits in the header of each optical packet to keep its priority.

Consider that some client traffic is supposed to be sent to a given egress switch in an ingress switch. Under PR, the ingress switch saves the copy of the traffic in its electronic buffer and then transmits it as an optical packet to the OPS network. For the first transmission of the traffic, the ingress switch sets its optical packet priority field to $h - 1$, where h is the number of hops between the ingress switch and the egress switch. This initial priority gives a better chance to the optical packets of long-hop connections to pass through the OPS network.

Assume that the optical packet is supposed to be dropped in an intermediate core switch due to resource lack. Then, the core switch sends a NACK command back to the ingress switch. Next, the ingress switch makes a new optical packet from the client packet traffic kept in its electronic buffer, increases the priority of the optical packet by one, and then retransmits the new optical packet to the network. The new optical packet may also be forced to be dropped in an OPS switch in the network again. Then, the ingress switch should resend another optical packet by an updated priority. This process continues until a successful transmission. For example, for a 4-hop path, the optical packet has an initial priority of $\delta = 3$. If the optical packet is dropped for the first time, we will have $\delta = 4$ in the optical packet of the first retransmission (i.e., the second transmission). For the second drop, we will have $\delta = 5$ for the second retransmission; the priority is increased by one for every retransmission. When no NACK command arrives at the ingress switch after some time, the copy of the traffic is removed from the electronic buffer. However, the time required to keep the copy of the traffic in the electronic buffer is smaller in PR than in RR since the number of retransmissions is not limited in RR, but limited in PR.

On the other hand, each OPS core switch allocates network resources (i.e., wavelengths, optical buffers, and wavelength converters) first to the optical packets with higher δ, and then the remaining resources are given to the optical packets with smaller δ. Thus, the number of retransmissions can be significantly limited. Therefore, even if a newly transmitted optical packet may have a small chance to pass through the core switch in its first transmission during heavy traffic, the higher priority it would acquire in its future retransmission(s) will help it to successfully pass through the core switch, thus cutting down the number of retransmissions. As proved in [94, 97], the number of retransmissions is limited to four times under PR. Therefore, the amount of buffering required in PR is less than in RR. For example, if there are optical packets of up to two retransmissions competing for a tagged output link, then first the available output ports of the link must be given to those optical packets that are retransmitted for the second time, then given to those optical packets retransmitted for the first time, and finally given to the newly transmitted optical packets (for whatever output ports remaining).

Table 3.2: Retransmission probabilities under PR and RR at $L = 0.7$, $N = 100$, and $f = 1, 2, 3, 4$ [94]

	$f = 1$		$f = 2$		$f = 3$		$f = 4$	
	PR	RR	PR	RR	PR	RR	PR	RR
π_0	0.721385	0.721155	0.831128	0.831029	0.881256	0.881296	0.910430	0.910378
π_1	0.253403	0.200887	0.167485	0.138817	0.118678	0.103564	0.089568	0.080587
π_2	0.024995	0.056111	0.001387	0.024522	0.000066	0.013078	0.0000026	0.008019
π_3	0.000218	0.015734	0.0000000	0.004529	0.0000000	0.001766	0.0000000	0.000894
π_4	0.0000001	0.004409	0.0000000	0.000879	0.0000000	0.000252	0.0000000	0.000108
π_5	0.0000000	0.001233	0.0000000	0.000179	0.0000000	0.000037	0.0000000	0.000013
π_6	0.0000000	0.000343	0.0000000	0.000037	0.0000000	0.000006	0.0000000	0.000002

Table 3.2 illustrates the retransmission probabilities under PR and RR in a multi-fiber OPS network at an OPS switch with $N = 100$ at traffic load $L = 0.7$. Here, π_0 shows the transmission probability of new optical packets and π_i is the probability that an optical packet is retransmitted for the ith time. For example, under PR at $f = 1$, $\pi_2 = 0.024995$ shows that 2.4995% of optical packets are successfully switched after the second time retransmission (i.e., the third time transmission). The PR performance results show that by increasing f, the probability of the retransmission at all levels decreases. One can see that for a high number of fibers, most of the dropped optical packets can pass through the core switch even with one retransmission. For example, at $f = 1$ and $f = 4$, an optical packet is retransmitted for the second time with a probability of almost 0.025 and 0.000003, respectively. This is because by increasing the number of fibers, PLR goes down, and therefore we can expect a very small number of retransmissions required in the optical domain. By increasing the number of fibers, the retransmission level under RR decreases as already observed in PR. However, the reduction rate is not as fast as PR, thus proving the superiority of PR. The same trend can be observed in multi-hop slotted OPS networks [94, 97].

In a modified version of PR (called M_PR) [96], an ingress switch sets the initial value for optical packet priority field to $\delta = \alpha \times (h - 1)$, where $\alpha = \{1, 2, 3, ...\}$. Consider two optical packets OP_1 and OP_2 in a given core switch, respectively, with the number of hops h_1 and h_2, where $h_1 > h_2$. Assume that optical packet OP_2 has been dropped κ times in the network, whereas OP_1 is a newly transmitted optical packet. Thus, the priority of optical packets OP_1 and OP_2 are $\delta_1 = \alpha \times (h_1 - 1)$ and $\delta_2 = \alpha \times (h_2 - 1) + \kappa$, respectively. The OP_1 has a higher priority than OP_2 to be switched in the core switch, when we have $\delta_1 > \delta_2$, or $\kappa < \alpha \times (h_1 - h_2)$. Otherwise, when $\delta_1 < \delta_2$, OP_2 finds superiority than OP_1. When $\delta_1 = \delta_2$, the optical packets are comparable and a decision can be made based on the number of client packets they carry. For instance, at $\alpha = 2$, an optical packet transmitted for the first time toward a three-hop egress switch is only comparable with either (a) an optical packet sent toward a two-hop egress switch but has been dropped twice or (b) an optical packet sent toward a one-hop egress switch but has been dropped four times.

Scheduling in a core switch under PR when there is no other means for contention resolution is as follows. Let set $S_{i,l} = \{OP_0, OP_1, ..., OP_{l-1} | 0 < l \leq N\}$ denote the subset of l contending optical packets on wavelength λ_i (where $i = 0, ..., W + W_a - 1$) at an output link of the core switch. Let vector (Src_j, ID_j, δ_j) denote the three parameters for optical packet OP_j carried in an aggregated header, where Src_j is the ingress switch address, ID_j is the optical packet identification, and δ_j is the priority of optical packet j. Recall that δ_j is not used under RR. When optical packet OP_j is dropped, the core switch sends a NACK command that includes ID_j to Src_j under both PR and RR. In PR, the optical packets in set $S_{i,l}$ are sorted in a descending order according to their δ values. Then, the top optical packet is picked for transmission first and the contention of the remaining $l - 1$ optical packets is resolved using PR or RR. As can be observed, under PR, an optical packet with a high priority always finds a high chance to pass through the core switch.

3.2.2 QoT-Aware Scheduling

An Impairment-Aware Scheduling (IAS) mechanism (as a QoT-aware scheduling scheme) has been proposed in [67] that considers Polarization Mode Dispersion (PMD) [99] of optical packets in contention resolution. This mechanism can be used for both asynchronous OPS and slotted OPS networks. Here, optical packet OP_j arriving at a core switch carries a field called PMD_j that includes the amount of fractional pulse broadening of OP_j from its ingress switch until the core switch. Note that PMD_j is set to 0 at the ingress switch.

When OP_j arrives at an intermediate core switch after traversing its kth hop, its PMD_j is updated by

$$PMD_j = PMD_j + B_c^2 \times D_{PMD,k}^2 \times l_k \,, \tag{3.2}$$

where B_c is data rate on each wavelength channel, $D_{PMD,k}$ (in $ps/km^{0.5}$) is the fiber PMD coefficient in the kth hop of the given route, and l_k (in km) is the fiber length of the kth hop. Obviously, the value of $B_c^2 \times D_{PMD,k}^2 \times l_k$ is known for each input link of the core switch and can be computed offline and saved in the core switch.

The scheduler in the core switch should schedule optical packets with acceptable PMD because all optical packets arriving at their egress switches must have acceptable PMD values; otherwise they will be useless. For this, let us define the parameter δ_{max} to be the maximum tolerable fractional pulse broadening for each optical packet, which is usually set to 0.1 in practice, so that PMD for the given optical packet can be tolerated. Then, the core switch can switch OP_j if its PMD_j is acceptable when it is sent toward the next hop core switch; i.e.,

$$PMD_j + B_c^2 \times D_{PMD,k+1}^2 \times l_{k+1} \leq \delta_{max}^2 \,. \tag{3.3}$$

Otherwise, OP_j should be dropped. Note that $D_{PMD,k+1}$ is the fiber PMD coefficient in the $(k+1)$th hop of the given route, and l_{k+1} is the fiber length of the $(k+1)$th hop. Note that the value of $B_c^2 \times D_{PMD,k+1}^2 \times l_{k+1}$ is known in advance for each output link and can be calculated offline and saved in the core switch. In addition, δ_{max}^2 is constant.

3.2.3 QoS Differentiation

In this section, software-based contention resolution schemes proposed for management of differentiated optical packets in both asynchronous OPS and slotted OPS networks are studied. The schemes proposed for relative differentiation and absolute differentiation are also discussed in this section.

The current proposals to provide differentiation in store-and-forward electronic routers/switches use electronic buffers to separate different traffic classes and use queue management algorithms in core switches [100, 101]. Since these proposals rely on buffers, they are not suitable for OPS networks [102–104] due to the lack of optical buffers and the difficulty to have a random access to the present immature optical memory technology. Therefore, new approaches are being sought to provide service differentiation in OPS networks without the use of optical buffers. The following is the list of software-based contention resolution schemes that can support QoS in OPS networks.

3.2.3.1 Drop-Based Schemes This section studies the drop-based schemes in which LP optical packets are dropped in such a way that HP traffic can find a better opportunity to pass through OPS core switches. Two techniques can be classified as follows:

3.2.3.1.1 Preemptive Drop Policy (PDP) In PDP [72, 104, 105], available wavelengths at an output port of a core switch are shared among all classes. When all wavelengths on the output port are busy, a new HP optical packet arriving at the output port can interrupt an LP optical packet currently under transmission and take over the respective wavelength for its own use. This technique is only applicable to asynchronous OPS networks. In slotted OPS, the PDP is not applicable since all optical packets arrive at the same time to the switch fabric.

The preemption could be also performed with a certain probability in which LP optical packets currently in transmission are preempted with probability p [104]. By properly adjusting p, the PLR of HP traffic can be kept below a desirable threshold value [104, 105]. For example, it is easy to see that by setting $p = 1$, the lowest PLR can be achieved for HP optical packets because LP optical packets are always

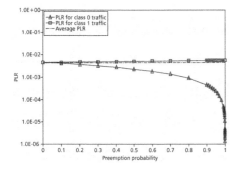
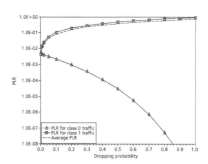

Figure 3.48: PLR as a function of the preemption probability at $W = 16$, $L = 0.5$, 20% class-0 (HP) traffic [2]

Figure 3.49: PLR as a function of the dropping probability at $W = 16$, $L = 0.5$, 20% HP traffic [2]

preempted. By reducing p, the PLR of HP optical packets is increased and a better performance can be achieved for LP optical packets. The major disadvantage of preemption is that the interruption of an LP optical packet under transmission may cause both optical packet fragmentation (i.e., generating worthless optical packets) and extra load in the network [72]. Note that the preemption technique is only applicable to an asynchronous OPS network due to the different operation of a slotted OPS network.

Figure 3.48 displays PLR as a function of the preemption probability for class-0 (HP) and class-1 (LP) traffic. At $p = 0$, both traffic classes experience the same PLR. By increasing p, the HP traffic finds better PLR than the LP traffic [2].

3.2.3.1.2 Intentional Packet Dropping (IPD) This technique [106] intentionally drops LP optical packets with probability p at a core switch before transmitting them to their desired output links. After dropping, the optical packets from all classes will find equal priority to seize the wavelengths of the output link as they are treated equally. This technique can achieve a low PLR for HP optical packets and high PLR for LP optical packets. Clearly, network load decreases by dropping LP optical packets. However, the core switch resources may be under-utilized. For example, consider that there is an available resource that can be used for switching an LP optical packet. However, this LP optical packet is intentionally dropped before reaching its desired output link. This results in wasting the core switch resource. Class isolation can be controlled by adjusting parameter p. With $p = 0$, no intentional dropping occurs and PLR for both LP and HP optical packets could be equal. With $p = 1$, maximum class isolation is obtained and all LP optical packets are dropped. The IDP technique is applicable to both asynchronous and slotted OPS networks.

Figure 3.49 shows PLR as a function of intentional packet dropping probability. By increasing p, the PLR for HP traffic decreases and the PLR for LP traffic increases, as expected [2].

3.2.3.2 Routing-Based Schemes This section reviews different techniques proposed for routing of optical packets in order to resolve their contention. These techniques are applicable to both asynchronous and slotted OPS networks.

3.2.3.2.1 Deflection Routing (DR) Section 3.2.1.1 discussed how the DR technique can be used to reduce the PLR of contending packets in a single-class OPS network. This routing can even be used to provide QoS differentiation for multi-class OPS networks as in [24, 78, 86]. When contention occurs in an OPS switch, LP optical packets are deflected and deliberately routed to longer paths toward their egress switches. On the other hand, HP optical packets are allowed to be routed through their regular shortest paths in order to assure a lower delay for HP traffic than for LP traffic. In other words, the available contention resolution resources at a core switch are first used for HP optical packets.

By deflecting LP optical packets, BER of LP traffic will be increased, whereas no additional BER is incurred for HP traffic. This is because a deflected LP optical packet may pass through additional core switches and traverse a longer path, and therefore it experiences additional physical impairments such as crosstalk, noise, non-linear impairments, and so on (see 1.4.6.2.3).

3.2.3.2.2 Multi-Path Routing (MPR) Quality of service differentiation can be achieved by providing different alternative routes to LP and HP traffic. Define a congested link in a core switch to be the link on which at least N_T wavelengths of W wavelengths are busy. Then, MPR strategy is only used for LP optical packets since optical packets traversing alternative paths may experience high delay, while HP optical packets are always routed according to the Shortest Path Routing (SPR) strategy. In addition, K wavelengths are reserved solely for HP traffic in each output link of any core switch. By increasing K, lower PLR can be obtained for HP optical packets, while the PLR of LP optical packets slightly goes up [24].

Clearly, BER of LP traffic will be increased by routing LP optical packets through alternative routes, whereas no additional BER is incurred for HP traffic. This is due to the reason stated in 3.2.3.2.1.

3.2.3.2.3 Shortest Hop-Path (SHP) For routing of optical packets in an OPS network, the common method is either the shortest path or shortest hop routing. In general, a shortest path routing can lead to the selection of routes with small PMD, Chromatic Dispersion (CD), and Amplified Spontaneous Emission (ASE) noises. On the other hand, a shortest hop routing can result in a path with a small crosstalk because the path will cross a small number of core switches. This is an indirect model to consider signal impairments in an OPS network (i.e., QoT-aware scheduling). To utilize the advantages of both routing types and indirectly considering signal impairments for optical packets, a combined routing technique based on both shortest hop and shortest path can be used. This combined routing is called Shortest Hop-Path (SHP) routing [67].

In SHP, the cost C_k of link k in the network is computed by

$$C_k = 1 + \frac{l_k}{l_{max}^2} ,\qquad(3.4)$$

where l_k is the length of link k (in meters) and l_{max} is the maximum one-hop link length (in meters) in the network. Then, the whole cost of a path with z hops is computed by

$$C = z + \sum_{k=1}^{z} \frac{l_k}{l_{max}^2} .\qquad(3.5)$$

The SHP can be pre-computed in offline mode, where the well-known Dijkstra routing algorithm is used to find a suitable route between each pair of ingress–egress switch pairs based on the link costs C_k. Under SHP, if there are two routes with the same hop count between a pair of ingress–egress switches, then the route with the shortest path is chosen. The SHP concentrates first on the shortest hop aspect of a route and then on the shortest path aspect. This is because crosstalk is an impairment that cannot be easily resolved in an OPS network. However, ASE, PMD, and CD impairments can be resolved to some extent.

Algorithm: Composite Optical Packet Scheduling in Multi-Class OPS

Definitions:
N: number of input links and output links of an OPS switch
l: number of contending packets at the output link, where $0 < l \le N \times f$
OP_j: optical packet j destined to the output link, where $0 \le j \le l - 1$
$S_{i,l}$:   set of l contending optical packets at the output link on wavelength λ_i
$N_{j,c}$:   number of client packets from class c carried in optical packet OP_j

Composite Scheduling:
1: Compute rank $v_j = \sum_{c=1}^{M} N_{j,c} \times V_c$ for each optical packet OP_j
2: Sort the indices of optical packets in Set $S_{i,l}$ in a descending order based on ranks v.
3: Select the top $x = \min(f,l)$ indices from the sorted list.
4: The relevant x optical packets will be transmitted on f wavelengths λ_i of f fibers, where output fibers are randomly chosen in order to provide load balancing on connection links.
5: Resolve the conflicts for the remaining $l - x$ optical packets using an available contention resolution scheme, but with respecting their ranks so that optical packets with high v values are first resolved.
6: Drop the optical packets that cannot be resolved.
7: END

Figure 3.50: Pseudo code of composite optical packet scheduling in multi-class multi-fiber OPS networks. Reprinted from *Computer Networks*, Vol. 53, Akbar Ghaffar Pour Rahbar and Oliver Yang, Distribution-based bandwidth access scheme in slotted all-optical packet-switched networks, 744–758, copyright 2009, with permission from Elsevier.

3.2.3.3 Composite Optical Packet Scheduling (COPS)
Contending optical packets at an output link of an intermediate core switch may have different conditions that must be considered by the contention resolution scheduler such as [67]:

- An optical packet may carry a small amount of traffic, but mostly from the HP class.

- An optical packet may carry a small amount of HP traffic, but a high amount of LP traffic, or vice versa.

- An optical packet may carry a small amount of traffic from the LP class, but has used many resources so far (i.e., it has passed through many core switches and is close to its egress switch).

- An optical packet may carry a high volume of traffic, but has used few network resources so far.

For example, assume that two optical packets OP_1 and OP_2 have collided and one of them must be dropped. Consider that OP_1 carries 15 HP client packets and 3 LP client packets, and OP_2 carries 5 HP client packets and 48 LP client packets. It is really difficult for the contention resolution scheduler to choose one of them for scheduling and drop the other one.

Considering the above conditions, decision making for scheduling significant optical packets and dropping those optical packets that cannot be scheduled becomes a challenging issue. In the following, an intelligent scheduling technique is presented in which each optical packet is given a rank. Then, based on their ranks, significant optical packets find the eligibility to be switched according to available resources in the core switch. In the following, different types of composite optical packet scheduling mechanisms are discussed.

Here, "significance" parameter V_c is assigned to client traffic class c, where $c = \{HP, MP, LP\}$ in general or $c = \{EF, AF, and BE\}$ in differentiated services. By assigning a high "significance" parameter to a class, the traffic of that class is given a high chance to pass through any OPS core switch in the network, and therefore we can expect a high throughput for that class. The V_c values may be specified based on the relative pricing for different classes. For instance, if the price of HP, MP, and LP classes are 30, 15, and 5 units, then we could have $V_{HP} = \frac{30}{50} = 0.6$, $V_{MP} = \frac{15}{50} = 0.3$, and $V_{LP} = \frac{5}{50} = 0.1$, respectively.

3.2.3.3.1 COPS Type 1

This technique [69, 107] can be used when an optical packet is allowed to carry a number of client packets from different classes. The composite packet transmission occurs under CPA (see Section 2.1.2.1.1) and DA (see Section 2.1.2.2). An OPS core switch $N * N$ assigns a rank v value to each one of l contending optical packets at a given output link. Figure 3.50 shows the composite optical packet scheduling in a given output link of a multi-fiber OPS switch with f fibers per connection link and W wavelengths per fiber. If optical packet OP_j carries $N_{j,c}$ client packets from class c and the "significance" parameter for class c is V_c, then the rank v for this optical packet is computed by

$$v_j = \sum_{c=1}^{M} N_{j,c} \times V_c , \tag{3.6}$$

where M is the number of classes of service in the OPS network. Then, the contending optical packets are sorted according to their rank values in a descending order. The optical packets identified from the top of the ranking list would be eligible first to use the available resources in the core switch and pass through the given output link. The remaining optical packets from the ranking list are finally dropped when there are no available resources.

By this technique, a valuable optical packet is given a high opportunity to pass through the OPS network. However, the sorting procedure in this technique imposes an additional computational complexity on the core switch. Using the well-known Quick Sort algorithm [108] can moderate this complexity.

Table 3.3: Example for composite optical packet scheduling

Packet Generator	OP_j	$N_{j,HP}$	$N_{j,MP}$	$N_{j,LP}$	v
CPA, DA	1	10	15	21	**12.6**
	2	12	18	13	**13.9**
	3	18	5	2	**12.5**
	4	7	29	11	**14**
	5	2	23	18	**9.9**
	6	16	1	28	**12.7**
NCPA	1	30	0	0	**18**
	2	0	0	43	**4.3**
	3	0	49	0	**14.7**
	4	24	0	0	**14.4**
	5	0	43	0	**12.9**
	6	0	0	48	**4.8**

For example, consider that there are $M = 3$ classes (i.e., HP, MP, and LP) with "significance" parameters of $V_{HP} = 0.6$, $V_{MP} = 0.3$, and $V_{LP} = 0.1$. Table 3.3 shows $l = 6$ contending optical packets and their rank values in an OPS switch. As can be observed under CPA or DA, OP_4 finds the highest rank, followed by OP_2, OP_6, OP_1, OP_3, and OP_5. Now at $f = 2$, the two optical packets with the highest ranks (i.e., OP_4 and OP_2) can be switched to the output link. The remaining four optical packets should be resolved with the available contention resolution hardware in the OPS switch.

3.2.3.3.2 COPS Type 2 The resolution scheme under the composite optical packet scheduling can even be used for the case that all client packets carried in an optical packet belong to the same class [69]. Recall that this type of optical packet is generated under NCPA (see 2.1.2.1.2). The same algorithm discussed in Fig. 3.50 is also used under this case. However, when computing v, there is just one non-zero item in Eq. (3.6). The second part of Table 3.3 depicts six contending optical packets: two carrying only HP traffic, two carrying only MP traffic, and two carrying only LP traffic. In this situation, OP_1 has the highest ranking followed by OP_3 that carries only MP traffic. In other words, OP_3 with 49 MP client packets is more valuable to be switched before OP_4 that carries 24 HP client packets in the core switch.

3.2.3.3.3 COPS Type 3 The rank of an optical packet can be computed based on different information it carries. In [109], the priority field δ defined in 3.2.1.3.2 is also carried in each optical packet. Therefore, there are four fields in the optical packet OP_j as δ_j, $N_{j,HP}$, $N_{j,MP}$ and $N_{j,LP}$. Then, rank OP_j is computed by $v_j = 1000 \times \delta_j + N_{j,HP} \times V_{HP} + N_{j,MP} \times V_{MP} + N_{j,LP} \times V_{LP}$. By this ranking, if there are two optical packets with different δ values, the optical packet with high δ value finds higher rank, irrespective of the client packets inside the optical packets. However, when they have equal δ values, their rankings will depend on the number of class-based client packets they carry.

3.2.3.3.4 COPS Type 4 In [67], optical packet OP_j arriving at an intermediate core switch carries a number of fields:

- Src_j: ingress switch of OP_j.

- Dst_j: egress switch of OP_j.

- d_j: the number of deflections happened for OP_j from ingress switch Src_j to the core switch. Note that d_j is set to 0 at Src_j. If OP_j is deflected in the core switch, its d_j parameter is incremented by 1.

- H_j: the number of hops obtained by the SHP routing (see 3.2.3.2.3) between ingress switch and egress switch of OP_j.

- h_j: the number of hops that OP_j has traveled from Src_j to the core switch. As soon as arrival of OP_j at the core switch, its h_j is incremented by 1. When an optical packet is deflected, h_j could be higher than H_j. Hence, by increasing $\frac{h_j}{H_j}$, an optical packet finds a high opportunity to be switched in the core switch toward its egress switch.

- $p_{j,HP}$, $p_{j,MP}$, $p_{j,LP}$: percentage of client traffic from HP, MP, and LP classes carried in OP_j (where $p_{j,HP} + p_{j,MP} + p_{j,LP} \leq 100$).

- PMD_j: fractional pulse broadening of OP_j from Src_j till the core switch (see Section 3.2.2 for more information).

Then, rank v_j for OP_j is computed by

$$v_j = 100\alpha_0 \frac{d_j}{D_{max}} + 100\alpha_1 \frac{h_j}{H_j} + \alpha_2 \frac{V_{HP} \times p_{j,HP} + V_{MP} \times p_{j,MP} + V_{LP} \times p_{j,LP}}{V_{HP}},$$

(3.7)

where $V_{LP} \leq V_{MP} \leq V_{HP} \leq 1.0$ are "significance" parameters so that $V_{LP} + V_{MP} + V_{HP} = 1.0$. Equation (3.7) considers three merits for computing rank v_j at the same time: the number of deflections until the core switch, the number of hops OP_j has traversed until the core switch, and the percentage of traffic carried in OP_j. Parameters α_0, α_1, and α_2 are weighting factors that specify the importance of the three merits in Eq. (3.7) with the constraint of $\alpha_0 + \alpha_1 + \alpha_2 = 1$. Parameter D_{max} is a threshold on the number of deflections. Suppose OP_j has already experienced D_{max} deflections. If it still cannot be switched in the core switch, OP_j should be dropped. Although the computation of Eq. (3.7) seems to be high, it can be computed faster by the following considerations. First, the third term can be computed in ingress switch Src_j, and then carried as a separate field in the header of OP_j. Second, the values of $\frac{100 \times \alpha_0}{D_{max}}$ and $100 \times \alpha_1$ can be computed offline.

Table 3.4: Evaluation of rank parameter at $D_{max} = 3$ under composite packet scheduling (Type 4). Reproduced from [67] with permission from De Gruyter.

β_1	β_2	β_3	OP#	d	h	H	$p1\%$	$p2\%$	$p3\%$	Rank at $\alpha_0 = 0.0$, $\alpha_1 = 0.4$, $\alpha_2 = 0.6$	Rank at $\alpha_0 = 0.1$, $\alpha_1 = 0.3$, $\alpha_2 = 0.6$
			1	2	3	8	42	20	45	0.68	0.7
0.4	0.35	0.25	2	0	4	8	60	15	3	0.65	0.60
			3	1	5	6	20	30	23	0.70	0.65
			1	2	3	8	42	20	45	0.46	0.49
0.7	0.25	0.05	2	0	4	8	60	15	3	0.59	0.54
			3	1	5	6	20	30	23	0.53	0.48

Table 3.4 shows the rank parameter at $D_{max} = 3$ under composite packet scheduling (Type 4) for three optical packets OP_1 to OP_3 in a core switch. The "significance" parameters β_1 to β_3 in the first set are close to each other, whereas there is a big difference among the "significance" parameters in the second set. Consider $\alpha_0 = 0.0$, $\alpha_1 = 0.4$, and $\alpha_2 = 0.6$. Under the first set, OP_3 appears at the top of the sorted list, followed by OP_1 and OP_2. Although OP_3 includes traffic with a smaller significance compared to OP_1 and OP_2, it is located at the first rank for switching because it is too close to its egress switch. Under the second set, OP_2 is located at the first position in the sorted list followed by OP_3 and OP_1. This is because of the high value of β_1 that pushes OP_2 to the top of the list. In the last column, the rank parameter has been evaluated under $\alpha_0 = 0.1$, $\alpha_1 = 0.3$, and $\alpha_2 = 0.6$. Considering the deflection number (i.e., when $\alpha_0 > 0$), the order of optical packets have changed in some cases compared with the case $\alpha_0 = 0$. Hence, deflected optical packets find more chance to be switched in the core switch.

3.2.3.4 Wavelength Access Restriction (WAR) Schemes   Here, the access of LP optical packets to the OPS network resources is restricted in order to ensure a higher performance for HP optical packets than LP ones. This restriction has a negative influence on the performance of LP optical packets as they suffer from higher congestion and delay compared with HP ones. Moreover, when reserving resources for a particular traffic class, the resources are employed less efficiently [72]. This could be the main disadvantage of the access restriction techniques. The WAR schemes are applicable to both asynchronous and slotted OPS networks. The following approaches manage the wavelengths in an OPS core switch in order to provide QoS differentiation for optical packets.

3.2.3.4.1 Exclusive Wavelength Reservation for HP Optical Packets Here, a set of W_1 wavelengths (i.e., wavelengths λ_1 to λ_{W_1}) are exclusively reserved for HP optical packets at each core switch in a static manner [110]. Then, the remaining $W - W_1$ wavelengths from a pool of W wavelengths are shared among other classes. When traffic load of HP optical packets is relatively high, then all the W_1 wavelengths can be fully utilized. Otherwise, some of them may be wasted.

A dynamic issue can also be considered for reserving wavelengths, where W_1 wavelengths (say $0 < W_1 < f \times W$ in a multi-fiber OPS) are dynamically reserved for HP optical packets [74, 111]. In dynamic reservation, instead of static wavelength numbering from λ_1 to λ_{W_1}, the number of available wavelengths is considered. The principle is that when the number of available wavelength channels at an output port is less than W_1, only HP optical packets can be forwarded, whereas LP optical packets are discarded. Otherwise, both classes can use the available wavelengths. Assume that $k = N_{HP} + N_{LP}$ optical packets have arrived at an output port of a core switch at $f = 1$, where N_{HP} and N_{HP} are the number of HP and LP optical packets, respectively. In this case, four contention scenarios may happen as following [111]:

- Case $k \leq W$: In this case, no optical packets are lost.

- Case $k > W$ and $N_{HP} \leq W_1$: In this case, $k - W$ LP optical packets are only lost.

- Case $k > W$ and $N_{HP} > W_1$ and $N_{LP} \leq W - W_1$: In this case, $k - W$ HP optical packets are only lost, where no LP optical packets is lost.

- Case $k > W$ and $N_{HP} > W_1$ and $N_{LP} > W - W_1$: In this case, $N_{HP} - W_1$ HP optical packets and $k - W - (N_{HP} - W_1)$ LP optical packets are lost.

3.2.3.4.2 Free Wavelength Usage for HP Optical Packets The W wavelengths at an output fiber are divided into two pools [2, 106, 112]: an HP pool with W_1 wavelengths and an LP pool with W_2 wavelengths (i.e., $W = W_1 + W_2$). An incoming optical packet can use a wavelength from these two pools only if its priority level is high enough. In other words, LP optical packets can only access to the wavelengths in the LP pool (only W_2 wavelengths), while HP optical packets can access to the wavelengths from both pools (with a total of $W_1 + W_2$ wavelengths). Clearly, by decreasing W_2, the parameter PLR for HP optical packets decreases because more resources are reserved exclusively for HP optical packets. By properly adjusting the wavelengths assigned to HP optical packets, a desired PLR can be achieved for them [106].

Since the reserved wavelengths for HP optical packets may not always be used, wavelength wastage is an important disadvantage for this technique, whereas at the same time a number of LP optical packets may be dropped due to the lack of wavelengths.

3.2.3.4.3 Adaptive Access Restriction Mechanism This technique [102, 113] is a software-based contention resolution technique proposed for slotted OPS networks in which a desired PLR is considered for HP traffic (i.e., PLR_{req}) so that the loss rate of HP optical packets (i.e., PLR_{HP}) should never exceed this threshold value.

Define l_{HP} to be the maximum number of HP optical packets per output link of an OPS switch that is guaranteed to be switched in the case of contention. By this, the service classes at a core switch are isolated in such a way that a certain number of wavelengths, l_{HP}, are allocated to HP optical packets. This technique ensures that l_{HP} HP optical packets can be transmitted at a given output link in a time slot when contention happens.

Define the measuring interval as the number of time slots in which PLR_{HP} is measured. The interval of 10,000 time slots is required when PLR_{req} is larger than a certain threshold. However, a larger interval should be used if $PLR_{req} < 10^{-4}$ because more reliable measurements are necessary. However, in this case, adaptation to an environment change will be slow.

Each OPS switch collects the optical packets for each output link and makes the dropping decisions according to the value of l_{HP}. At the end of a measuring interval, the OPS network measures PLR_{HP} in any OPS switch from the received and lost optical packets in the interval, and the maximum of PLR_{HP} among all OPS switches is considered as the network PLR_{HP} because all OPS switches operate independently from each other. Then, the measured PLR_{HP} is used to adjust l_{HP} in the next time period in order to keep PLR_{HP} less than PLR_{req}. By properly adjusting l_{HP} based on the PLR of HP optical packets, changes in the network load, and relative share of HP traffic, the PLR of HP optical packets can be kept below the desired value PLR_{req}.

The measured PLR_{HP} must be kept inside a tolerated region between two bounds K_{min} and K_{max}. When the experienced PLR_{HP} violates $PLR_{req} \times K_{max}$ at two consecutive intervals, l_{HP} should be increased. Consequently, HP optical packets would find more opportunity to pass through core switches, thus reducing PLR of HP optical packets. On the other hand, when we have $PLR_{HP} < PLR_{req} \times K_{min}$ at two consecutive intervals, l_{HP} should be decreased, thus giving a better chance for LP optical packets to use the network resources. This range is necessary for changing l_{HP} before the occurrence of the loss rate violation in order to maintain the absolute QoS bounds all the time. The best results for K_{max} is obtained when $0.8 < K_{max} < 0.95$, and for K_{min} when $0.2 < K_{min} < 0.5$. Note that parameter K_{max} should be small enough such that a small increase in system load would be tolerated without violating PLR_{req}. In addition, the distance between K_{min} and K_{max} should be large enough in such a way that for every system load, a suitable l_{HP} could be found that results in a PLR_{HP} inside the limits. Otherwise, the value of l_{HP} will oscillate, causing the experienced PLR_{HP} to oscillate as well.

Performance evaluation results at measuring interval size of 10,000 time slots, $K_{min} = 0.2$, $K_{max} = 0.85$, and $PLR_{req} = 0.0015$ show that PLR_{HP} goes up by increasing traffic load. When PLR_{HP} exceeds $PLR_{req} \times K_{max}$, parameter l_{HP} is increased and then PLR_{HP} drops. This is because there are more wavelengths available for HP optical packets by increasing l_{HP}. This is even true when the share of HP traffic increases, and the network can easily keep PLR_{HP} below PLR_{req}. On the other hand, when PLR_{HP} goes below $PLR_{req} \times K_{min}$, parameter l_{HP} is decreased and fewer wavelength channels are provided for HP optical packets, thus increasing PLR_{HP}. However, this technique may under-utilize allocated wavelengths for HP optical packets, when there are not enough number of HP optical packets to use the allocated wavelengths.

3.2.4 Comparison of Software-Based Contention Resolution Schemes

This section compares the efficiency of different software-based techniques proposed for contention resolution.

- Both the merit-based scheduling and the prioritized retransmission techniques assign a merit value to each optical packet, where this value is updated within an OPS network. In a core switch, an optical packet having a high merit value has a high priority to use the core switch resources, thus having a high chance to pass through the core switch. On the other hand, a routing-based technique redirects the route of an optical packet with the hope that the optical packet finds a less congested route. However, a routing-based technique never considers the network resources used by an optical packet. Hence, an optical packet that has used many network resources may be deflected, whereas another optical packet that has recently arrived at the network may be switched at a core switch. This results in some sort of unfairness and even wasting network resources. It is suggested that we combine the routing-based and merit-based techniques in order to avoid wasting network resources.

- Deflection routing is highly dependent on the network topology and can provide good results at high degree of connectivity between core switches and low-load networks. The DR could not be used independently and should be combined with other contention resolution methods. To exploit the benefit of fiber delay lines, deflection with backward feature can be employed [6]. The deflection routing and composite packet scheduling techniques never waste network resources because they never reserve network resources solely for HP traffic. These two techniques can be combined with each other where the "significance" parameter of an optical packet would be the criterion that could be used by the deflection routing technique. Here, the optical packets with high "significance" values are routed through the regular shortest path, while other optical packets with low "significance" values are deflected.

- The prioritized retransmission scheme is shown to be a very simple but efficient protocol for slotted OPS networks. Note that the PR technique is some sort of merit-based scheduling that assigns a merit value to each optical packet, where this merit value is updated within an OPS network by each retransmission. Then, an optical packet having a high merit value has a great opportunity to use each core switch resources, thus passing through the core switch with a high chance. The PR mechanism can also be used to increase the fairness among different traffic streams as opposed to the RR scheme where one optical packet may pass through the core switch at the first transmission and another one may be (theoretically) retransmitted forever. The PR can provide better throughput than RR for OPS networks [93, 94, 97]. However, a core switch has a high computational complexity for handling optical packets with different priorities δ. The priority field also allocates a small amount of the network bandwidth. Finally, an ingress switch needs electronic buffer to save the copy of each transmitted traffic for possible retransmission in future, but less than RR. By limiting the number of retransmissions, the extra load injected into the network due to the higher layer retransmissions can be reduced, which in turn helps to increase the network throughput. The PR is much more effective than RR whenever PLR is high. We expect such PLR whenever (1) traffic load is high and (2) a small number of contention resolution/avoidance hardware has been used in a core switch. Therefore, it is recommended to use PR in networks with a medium or high loss rate. In an OPS network with a low PLR (this low loss rate network can be achieved when using so many expensive contention resolution hardware in core switches), there may be no significant difference between the performance of PR and RR, although PR is always better than RR. Thus, RR is recommended for such a network because of its low computational complexity.

- Two techniques have been used to intelligently consider physical layer impairments when scheduling of optical packets. First, an optical packet experienced an unacceptable PMD along its path is dropped under IAS. Second, a combined routing indirectly takes into account optical signal impairments under SHP.

- To improve packet loss and delay performances, a number of merits can be intelligently combined to provide criteria to switch optical packets in core switches based on the number of deflections an optical packet has experienced, the number of hops an optical packet has traversed, and the percentage of class-based user packets carried in an optical packet.

- Access restriction in a buffer-less OPS core switch can be used mainly for accessing wavelengths on fibers, drop ports on the switch, and WCs for contention resolution. However, in OPS with optical buffers, access restriction can even be applied for accessing to FDLs with respect to packets priority. Among the access restriction techniques for the use of wavelengths, the technique in which HP optical packets are allowed to use all wavelengths could result in a lower PLR for HP optical packets than the other two techniques that allow HP packets to use only a certain number of reserved wavelengths.

- The PDP has the least reduction and IPD has the most reduction in the throughput performance. The reason is that in IPD, all LP optical packets are dropped with probability p even if all the wavelengths at the tagged output port are idle, thus significantly increasing PLR of LP traffic [2]. However, the PDP technique at least provides an opportunity for LP packets to be switched. The PDP has the most implementation complexity, followed by the WAR and IPD. In IPD, no additional hardware and scheduling are required, as LP optical packets are randomly dropped as soon as arriving. In WAR, no additional hardware is required. In PDP, a scheduler needs to know which wavelength has been occupied to which class at any time. The PDP scheduler also needs to know the last arrived LP optical packets in order to preempt them when required. In addition, the PDP scheduler needs additional hardware to erase parts of the preempted LP optical packets. It has been shown that the PDP provides the best performance, followed by the WAR and IPD. The difference is much clearer at high traffic loads. Note that a network often needs QoS differentiation at high traffic loads, where PLR increases for all classes of optical packets [2].

- To provide the lowest PLR for HP traffic, both the access restriction and drop-based schemes may sacrifice network resources, which is the major disadvantage for these techniques.

- Dynamic management of both the probability p in IPD and the number of reserved resources in the WAR schemes not only can provide better performance results than the case with static values for them but also can alleviate their drawbacks.

3.3 Summary

This chapter has presented hardware-based and software-based solutions for resolving the contention of optical packets. For hardware-based schemes, optical buffering, wavelength conversion, and non-blocking receivers have been introduced. For software-based schemes, this chapter has introduced deflection routing, retransmission in the optical domain, routing, and so on.

In the studied schemes, QoS differentiation and QoT-aware scheduling have also been studied. Under QoT-aware scheduling, some techniques directly take into account physical impairments and some other indirectly consider the impairments when routing of optical packets. The QoS differentiation and QoT-aware scheduling are too important considerations to be considered in developing future contention resolution and avoidance schemes.

REFERENCES

1. L. Xu, H. Perros, and G. Rouskas. Techniques for optical packet switching and optical burst switching. *IEEE Communications Magazine*, 39:136–142, Jan. 2001.

2. H. Øverby, N. Stol, and M. Nord. Evaluation of QoS differentiation mechanisms in asynchronous bufferless optical packet-switched networks. *IEEE Communications Magazine*, 44(8):52–57, 2006.

3. J. Bowers, E. Burmeister, and D. Blumenthal. Optical buffering and switching for optical packet switching. In *Photonics in Switching Conference*, Heraklion, Crete, 2006.

4. J. Yang, A. O. Karalar, S. S. Djordjevic, N. K. Fontaine, C. Yang, W. Chen, S. Chu, B. E. Little, and S. J. Yoo. Variable slowlight buffers in all-optical packet switching routers. In *Optical Fiber communication/National Fiber Optic Engineers Conference*, San Diego, CA, May 2008.

5. I. Chlamtac, A. Fumagalli, L. G. Kazovsky, P. Melman, W. Nelson, P. Poggiolini, M. Cerisola, A. N. M. Masum Choudhury, T. K. Fong, R. T. Hofmeister, C. L. Lu, A. Mekkittikul, D. J. M. Sabido, C. J. Suh, and E. W. M. Wong. CORD: Contention resolution by delay lines. *IEEE Journal on Selected Areas in Communications*, 14(5):1014–1029, 1996.

6. S. Yao, B. Mukherjee, and S. Dixit. Advances in photonic packet switching: an overview. *IEEE Communications Magazine*, 38(2):84–94, Feb. 2000.

7. S. Yao, B. Mukherjee, S. J. B. Yoo, and S. Dixit. A unified study of contention-resolution schemes in optical packet-switched networks. *IEEE Journal of Lightwave Technology*, 21(3):672, 2003.

Quality of Service in Optical Packet Switched Networks, First Edition.
By Akbar Ghaffarpour Rahbar Copyright © 2015 IEEE. Published by John Wiley & Sons, Inc.

8. J. Liu, T. T. Lee, X. Jiang, and S. Horiguchi. Blocking and delay analysis of single wavelength optical buffer with general packet size distribution. *IEEE Journal of Lightwave Technology*, 27(8):955–966, Apr. 2009.

9. I. Chlamtac, A. Fumagalli, and C. J. Suh. Multibuffer delay line architectures for efficient contention resolution in optical switching nodes. *IEEE Transactions on Communications*, 48(12):2089–2098, Dec. 2000.

10. D. K. Hunter and I. Andonovic. Approaches to optical Internet packet switching. *IEEE Communications Magazine*, 38(9):116–122, 2000.

11. C. H. Chang, S. K. Shao, M. R. Perati, and J. Wu. Performance study of various packet scheduling algorithms for variable-packet-length feedback type WDM optical packet switches. In *IEEE High Performance Switching and Routing*, Poznan, 2006.

12. T. Ismaila, H. S. Hamzab, K. Elsayedc, and J. Elazab. An enhanced OPS architecture with optical buffers. In *6th International Symposium on High-Capacity Optical Networks and Enabling Technologies (HONET)*, pages 165–171, Alexandria, 2009.

13. M. Maier. *Optical Switching Networks*. Cambridge University Press, 2008.

14. M. C. Chia, D. K. Hunter, I. Andonovic, P. Ball, I. Wright, S. P. Ferguson, K. M. Guild, and M. J. OMahony. Packet loss and delay performance of feedback and feed-forward arrayed-waveguide gratingsbased optical packet switches with WDM inputs–outputs. *IEEE Journal of Lightwave Technology*, 19(9):1241–1254, Sept. 2001.

15. S. L. Danielsen, C. Joergensen, B. Mikkelsen, and K. E. Stubkjaer. Analysis of a WDM packet switch with improved performance under bursty traffic conditions due to tuneable wavelength converters. *IEEE Journal of Lightwave Technology*, 16(5):729–735, May 1998.

16. H. S. Hamza, T. Ismail, and K. E. Sayed. On the design of asynchronous optical packet switch architectures with shared delay lines and converters. *Photonic Network Communications*, 22:191–208, 2011.

17. A. G. Reza, T. S. Chin, and F. M. Abbou. Hybrid buffering architecture using feedforward and feedback share fiber delay lines. In *International Conference on Photonics (ICP)*, Langkawi, Kedah, 2010.

18. S. Y. Liew, G. Hu, and H. J. Chao. Scheduling algorithms for shared fiber-delay-line optical packet switchespart i: The single-stage case. *IEEE Journal of Lightwave Technology*, 23:1586–1600, 2005.

19. R. Srivastava and Y. N. Singh. Feedback fiber delay lines and AWG based optical packet switch architecture. *Optical Switching and Networking*, 7:75–84, 2010.

20. L. Li, S. D. Scott, and J. S. Deogun. A novel fiber delay line buffering architecture for optical packet switching. In *IEEE GlobeCom*, pages 2809–2813, 2003.

21. C. Guillemot, M. Renaud, P. Gambini, C. Janz, I. Andonovic, R. Bauknecht, B. Bostica, M. Burzio, F. Callegati, M. Casoni, D. Chiaroni, F. Clerot, S. L. Danielsen, F. Dorgeuille, A. Dupas, A. Franzen, P. B. Hansen, D. K. Hunter, A. Kloch, R. Krahenbuhl, B. Lavigne, A. LeCorre, C. Raffaelli, M. Schilling, J. C. Simon, and L. Zucchelli. Transparent optical packet switching approach: European ACTS KEOPS project approach. *Journal of Lightwave Technology*, 16(12):2117–2134, 1998.

22. C. Papazoglou, G. Papadimitriou, and A. Pomportsis. Design alternatives for optical-packet-interconnection network architectures. *OSA Journal of Optical Networking*, 3(11):810–825, 2004.

23. F. Callegati, W. Cerroni, C. Raffaelli, and P. Zaffoni. Wavelength and time domain exploitation for QoS management in optical packet switches. *Computer Networks*, 44:569–582, 2004.

24. F. Callegati, W. Cerroni, and C. Raffaelli. Routing techniques in optical packet-switched networks. In *IEEE International Conference on Transparent Optical Networks (ICTON)*, Barcellona, Luglio, July 2005.

25. G. Muretto and C. Raffaelli. Combining contention resolution schemes in WDM optical packet switches with multifiber interfaces. *Journal of Optical Networking*, 6(1):74–89, Jan. 2007.

26. F. Xue, Z. Pan, Y. Bansal, and J. Cao. End-to-end contention resolution schemes for an optical packet switching network with enhanced edge routers. *IEEE Journal of Lightwave Technology*, 21(11):2595–2604, Nov. 2003.

27. S. J. B. Yoo. Wavelength conversion technologies for WDM network applications. *IEEE Journal of Lightwave Technology*, 1996(6):955–966, June 14.

28. B. Chen, J. Bracken, F. Verleyen, Y. Wang, and C. Xu. Analysis of all-optical intra-cavity wavelength conversions in fiber ring resonators. *Elsevier Optics Communications*, 247(1-3):57–64, Mar. 2005.

29. N. Akar, E. Karasan, and C. Raffaelli. Fixed point analysis of limited range share per node wavelength conversion in asynchronous optical packet switching systems. *Photonic Network Communications*, 18(2):255–263, Feb. 2009.

30. A. G. Rahbar and O. Yang. Contention avoidance and resolution schemes in bufferless all-optical packet-switched networks: a survey. *IEEE Communications Surveys and Tutorials*, 10(4):94–107, Dec. 2008.

31. N. Kitsuwan, R. Rojas-Cessa, M. Matsuura, and E. Oki. Performance of optical packet switches based on parametric wavelength converters. *IEEE/OSA Journal of Optical Communications and Networking*, 2(8):558–569, 2010.

32. N. Kitsuwan and E. Oki. Optical packet switch based on dynamic pump wavelength selection. *Journal of Optical Communications and Networking*, 3(2):162–171, 2011.

33. V. Eramo, M. Listanti, and P. Pacifici. A comparison study on the number of wavelength converters needed in synchronous and asynchronous all-optical switching architectures. *IEEE Journal of Lightwave Technology*, 21(2):340–355, Feb. 2003.

34. H. Øverby. Performance modeling of synchronous bufferless OPS networks. In *International Conference on Transparent Optical Networks*, Wroclaw, Poland, July 2004.

35. V. Eramo. Performance of scheduling algorithms in optical packet switches equipped with limited-range wavelength converters. *Journal of Optical Networking*, 4(12):856–869, 2005.

36. H. L. Liu, B. Zhang, and S. L. Shi. A novel contention resolution scheme of hybrid shared wavelength conversion for optical packet switching. *IEEE Journal of Lightwave Technology*, 30(2):222–228, 2012.

37. H. Liu, Y. Chen, and B. Zhang. An optical packet contention resolution supporting QoS differentiation with LRWC wavelength converters for OPS. *Photonic Network Communications*, 24:71–76, 2012.

38. V. Eramo, M. Listanti, and M. D. Donato. Performance evaluation of a bufferless optical packet switch with limited-range wavelength converters. *IEEE Photonic Technology Letters*, 16(2):644–645, Feb. 2004.

39. V. Eramo, M. Listanti, and A. Germoni. Cost evaluation of optical packet switches equipped with limited-range and full-range wavelength converters for contention resolution. In *IEEE GLOBECOM*, 2007.

40. R. C. Almeida, J. F. Martins-Filho, and H. Waldman. Limited-range wavelength conversion modeling for asynchronous optical packet-switched networks. In *SBMO/IEEE MTT-S International Conference on Microwave and Optoelectronics*, Kharkov, Ukraine, 2005.

41. K. Dogana, Y. Gunalayb, and N. Akar. Comparative study of limited range wavelength conversion policies for asynchronous optical packet switching. *Journal of Optical Networking*, 6(2):134–145, Feb. 2007.

42. M. Savi C. Raffaelli and E. Karasan N. Akar. Packet loss analysis of synchronous buffer-less optical switch with shared limited range wavelength converters. In *IEEE High Performance Switching and Routing (HPSR)*, Brooklyn, NY, 2007.

43. N. Akar, C. Raffaelli, M. Savi, and E. Karasan. Shared-per-wavelength asynchronous optical packet switching: A comparative analysis. *Computer Networks*, 54(13):2166–2181, Sept. 2010.

44. S. A. Y. Fukushima, H. Harai, and M. Murata. Design of wavelengthconvertible edge nodes in wavelength-routed networks. *Journal of Optical Networking*, 5(3):196–209, 2006.

45. M. Savi, H. Øverby, N. Stol, and C. Raffaelli. Contention resolution using parametric wavelength converters: Performance and cost analysis. In *IEEE Globecom*, Anaheim, CA, Dec. 2012.

46. K. C. Lee and V. O. K. Li. Optimization of a WDM optical packet switch with wavelength converters. In *IEEE INFOCOM*, Boston, MA, Apr. 1995.

47. M. Nord. Performance analysis of a low-complexity and efficient QoS differentiation algorithm for bufferless optical packet switches with shared wavelength converters in asynchronous operation. In *The First International Conference on Broadband Networks (BROADNETS04)*, 2004.

48. V. Eramo M. Listanti, C. Nuzman, and P. Whiting. Optical switch dimensioning and the classical occupancy problem. *International Journal of Communication Systems*, 15:127–141, 2002.

49. V. Eramo and M. Listanti. Comparison of unicast/multicast optical packet switching architectures using wavelength conversion. *Optical Network Magazine*, 3(3):63–75, March-April 2002.

50. V. Eramo, M. Listanti, and M. Spaziani. Dimensioning models in optical packet switches equipped with shared limited-range wavelength converters. In *IEEE Globecom*, Dallas, TX, 2004.

51. C. Okonkwo, R. C. Almeida, R. E. Martin, and K. Guild. Performance analysis of an optical packet switch with shared parametric wavelength converters. *IEEE Communications Letters*, 12:596–598, Aug. 2008.

52. N. Kitsuwan, H. N. Tan, M. Matsuura, N. Kishi, and E. Oki. Recursive parametric wavelength conversion scheme for optical packet switch. *IEEE Journal of Lightwave Technology*, 29(11):1659–1670, 2011.

53. H. N. Tan, N. Kitsuwan, M. Matsuura, N. Kishia, and E. Okia. Demonstration of bufferless optical packet switch with recursive stages of parametric wavelength converter. *Optical Switching and Networking*, 9(4):336–342, 2012.

54. K. Dogan and N. Akar. A performance study of limited-range partial wavelength conversion for asynchronous optical packet/burst switching. In *IEEE ICC*, Istanbul, Turkey, June 2006.

55. R. E. Gordon and L. R. Chen. New control algorithms in an optical packet switch with limited-range wavelength converters. *IEEE Communications Letters*, 10(6):495–497, 2006.

56. V. Eramo, M. Listanti, and M. Tarola. Advantages of input wavelength conversion in optical packet switches. In *IEEE Globecom*, San Francisco, CA, Dec. 2003.

57. S. L. Danielsen, P. B. Hansen, and K. E. Stubkjaer. Wavelength conversion in optical packet switching. *IEEE Journal of Lightwave Technology*, 16(12):2095–2108, Dec. 1998.

58. F. Callegati, G. Corazza, and C. Raffaelli. Exploitation of DWDM for optical packet switching with quality of service guarantees. *IEEE Journal on Selected Areas in Communications*, 20(1):190–201, Jan. 2002.

59. V. Eramo, M. Listanti, and A. Valletta. Scheduling algorithms in optical packet switches with input wavelength conversion. *Computer Communications*, 28:1456–1467, 2005.

60. A. Cianfrani, V. Eramo, A. Germoni, C. Raffaelli, and M. Savi. Loss analysis of multiple service classes in shared-per-wavelength optical packet switches. *Journal of Optical Communications and Networking*, 1(2):A69–A80, July 2009.

61. V. Eramo, A. Germoni, C. Raffaelli, and M. Savi. Multi-fiber shared-per-wavelength all-optical switching: architectures, control and performance. *IEEE Journal of Lightwave Technology*, 26(5):537–551, Mar. 2008.

62. V. Eramo, A. Germoni, C. Raffaelli, and M. Savi. Packet loss analysis of shared-per-wavelength multi-fiber all-optical switch with parallel scheduling. *Computer Networks*, 53(2):202–216, Feb. 2009.

63. M. Savi, H. Øverby, N. Stol, and C. Raffaelli. Loss, complexity and cost analysis of an optical switching fabric using fixed converters. In *16th International Conference on Optical Network Design and Modeling (ONDM)*, 2012.

64. V. Eramo, A. Germoni, A. Cianfrani, C. Raffaelli, and M. Savi. Evaluation of QoS differentiation mechanism in shared-per-wavelength optical packet switches. In *International Conference on Optical Network Design and Modeling (ONDM)*, Braunschweig, Feb. 2009.

65. C. Raffaelli and M. Savi. Sharing simple wavelength converters in optical packet switching: Performance and power consumption evaluation. In *IEEE Globecom*, Miami, FL, Dec. 2010.

66. A. G. Rahbar. Cost-effective combination of contention resolution/avoidance schemes in bufferless slotted OPS networks. *Elsevier Optics Communications*, 282(5):798–808, Mar. 2009.

67. A. G. Rahbar. Impairment-aware merit-based scheduling in QoS-capable multi-fiber OPS networks. *Journal of Optical Communications*, 33(2):103–121, June 2012.

68. J. Li. Performance of an optical packet switch with limited wavelength converter. *Springer Intelligence Computation and Evolutionary Computation*, pages 223–229, 2013.

69. A. G. Rahbar and O. Yang. Distribution-based bandwidth access scheme in slotted all-optical packet-switched networks. *Elsevier Computer Networks*, 53(5):744–758, Apr. 2009.

70. W. Dapeng, W. Ruyan, H. Sheng, Z. Jie, and L. Keping. Service differentiating supporting output circuiting shared optical buffer architecture. *Photonic Network Communications*, 22:73–78, 2011.

71. F. Callegati, W. Cerroni, C. Raffaelli, and M. Savi. QoS differentiation in optical packet-switched networks. *Computer Communications*, 29:855–864, 2006.

72. S. Bjrnstad and H. Øverby. Quality of service differentiation in optical packet/burst switching: a performance and reliability perspective. In *International Conference on Transparent Optical Networks (ICTON)*, Barcelona, Spain, July 2005.

73. H. Øverby. An adaptive service differentiation algorithm for optical packet switched networks. In *International Conference on Transparent Optical Networks (ICTON)*, Warsaw, Poland, July 2003.

74. D. Careglio, G. Muretto, C. Raffaelli, J. Sole-Pareta, and E. Vines. Quality of service in a multi-fiber optical packet switch. In *Photonics in Switching Conference*, Heraklion, Crete, Oct. 2006.

75. V. Eramo, A. Germoni, A. Cianfrani, and F. L. Buono. Performance evaluation of a QoS technique for bufferless optical packet switches. In *International Conference on Transparent Optical Networks*, Portugal, July 2009.

76. V. Eramo. Performance evaluation of bufferless optical packet switches supporting quality of service. *IET Communications*, 3(3):428–440, 2009.

77. T. Chich, J. Cohen, and P. Fraingniaud. Unslotted deflection routing: a practical and efficient protocol for multihop optical networks. *IEEE/ACM Transactions on Networking*, 9(1):47–59, Feb. 2001.

78. T. S. El-Bawab and J. Shin. Optical packet switching in core networks: between vision and reality. *IEEE Communications Magazine*, 40(9):60–65, 2002.

79. S. Bregni and A. Pattavina. Performance evaluation of deflection routing in optical IP packet-switched networks. *Cluster Computing*, 7:239–244, 2004.

80. A. Pattavina. Performance of deflection routing algorithms in IP optical transport networks. *Computer Networks*, 50(2):207–218, 2006.

81. S. Yamano, F. Xue, and S. J. B. Yoo. Load-sensitive deflection routing for contention resolution in optical packet switched networks. In *IEEE International Conference on Computer Communications and Networks*, 2003.

82. E. W. M. Wong, J. Baliga, M. Zukerman, A. Zalesky, and G. Raskutti. A new method for blocking probability evaluation in OBS/OPS networks with deflection routing. *IEEE Journal of Lightwave Technology*, 27(23):5335–5347, 2009.

83. T. Eido, F. Pekergin, M. Marot, and T. Atmaca. Multiservice optical packet switched networks: modeling, performance evaluation and QoS mechanisms in a mesh slotted architecture. In *IEEE International Conference on Networking and Services*, Valencia, Spain, 2009.

84. A. G. Fayoumi, A. Jayasumana, and J. Sauer. Performance of multihop networks using optical buffering and deflection routing. In *IEEE Conference on Local Computer Networks (LCN2000)*, Tampa, FL, Nov. 2000.

85. G. Maier and A. Pattavina. Deflection routing in IP optical networks. *Telecommunication Systems*, 52(1):51–60, Jan. 2013.

86. R. Nejabati, G. Zervas, D. Simeonidou, M. J. OMahony, and D. Klonidis. The OPORON project: Demonstration of a fully functional end-to-end asynchronous optical packet-switched network. *IEEE Journal of Lightwave Technology*, 25(11):3495–3510, Nov. 2007.

87. H. Yokoyama, H. Nakamura, and S. Nomoto. Reflection routing: a simple approach for contention resolution in all-optical packet switched networks. *IEICE Transaction on Communications*, E87-B(6):1561–1568, June 2004.

88. H. Yokoyama and H. Nakamura. Mechanisms and performance of reflection routing for optical packet switched networks. In *IEEE Optical Fiber Communication (OFC)*, 2002.

89. T. K. Moseng, H. Øverby, and N. Stol. Merit based scheduling in asynchronous bufferless optical packet switched networks. In *Norsk Informatikk Konferanse (NIK)*, Stavanger, Norway, Anov. 2004.

90. S. Pejhan, M. Schwartz, and D. Anastassiou. Error control using retransmission techniques in multicast transport protocols for real-time media. *IEEE/ACM Transactions on Networking*, 4(3):413–427, June 1996.

91. E. Modiano. Random algorithms for scheduling multicast traffic in WDM broadcast-and-select networks. *IEEE/ACM Transactions on Networking*, 7(3):425–434, June 1999.

92. Q. Zhang, V. M. Vokkarane, Y. Wang, and J. P. Jue. Evaluation of burst retransmission in optical burst-switched networks. In *IEEE Broadnets*, Boston, MA, Oct. 2005.

93. A. G. Rahbar and O. Yang. Retransmission in slotted optical networks. In *IEEE High Performance Switching and Routing (HPSR)*, pages 21–26, Poznan, Poland, June 2006.

94. A. G. Rahbar and O. Yang. Prioritized retransmission in slotted all-optical packet-switched networks. *Journal of Optical Networking*, 5(12):1056–1070, Dec. 2006.

95. A. G. P. Rahbar and O. Yang. Contention resolution by retransmission in single-hop OPS metro networks. *Journal of Networks (JNW)*, 2(4):20–27, Aug. 2007.

96. A. G. Rahbar. Improving throughput of long-hop TCP connections in IP over OPS networks. *Springer Photonic Network Communications*, 17(3):226–237, June 2009.

97. A. G. Rahbar and O. Yang. *OPS Networks: Bandwidth Management & QoS*. VDM Verlag, Germany, 2009.

98. A. G. Rahbar. An approach to improve TCP throughput of long-hop connections in bufferless OPS networks. In *IEEE International Symposium on High Capacity Optical Networks and Enabling Technologies (HONET07)*, UAE, Dubai, Nov. 2007.

99. S. V. Kartalopoulos. *DWDM: Networks, Devices, and Technology*. Wiley-IEEE Press, 2003.

100. N. Christin and J. Liebherr. A QoS architecture for quantitative service differentiation. *IEEE Communications Magazine*, 41(6):38–45, 2003.

101. B. Wydrowski and M. Zukerman. QoS in best-effort networks. *IEEE Communications Magazine*, 40(12):44–49, 2002.

102. A. Undheim, H. Øverby, and N. Stol. Absolute QoS in synchronous optical packet switched networks. In *Norsk Informatikk Konferanse (NIK)*, Stavanger, Norway, 2004.

103. A. Kaheel, T. Khattab, A. Mohamed, and H. Alnuweiri. Quality of Service mechanisms in IP-over WDM networks. *IEEE Communications Magazine*, 40(12):38–43, Dec. 2002.

104. H. Øverby and N. Stol. Quality of service in asynchronous bufferless optical packet switched networks. *Springer Telecommuication Systems*, 27(2-4):151–179, Oct. 2004.

105. H. Øverby and N. Stol. Providing absolute QoS in asynchronous bufferless optical packet/burst switched networks with the adaptive preemptive drop policy. *Computer Communications*, 28(9):1038–1049, 2005.

106. H. Øverby and N. Stol. QoS differentiation in asynchronous bufferless optical packet switched networks. *Wireless Networks*, 12(3):383–394, June 2006.

107. A. G. Rahbar and O. Yang. Slot contention resolution for distributed TDM scheduling in an optical star network. In *IEEE Canadian Conference on Electrical and Computer Engineering*, pages 286–289, Saskatoon, Canada, May 2005.

108. T. H. Cormen, C. E. Leiserson, R. L. Rivest, and C. Stein. *Introduction to Algorithms*. MIT Press and McGraw-Hill, 2001.

109. A. G. Rahbar and O. Yang. Agile bandwidth management techniques in slotted all-optical packet interconnection networks. *Elsevier Computer Networks*, 54(3):387–403, Feb. 2010.

110. H. Elbiaze and T. Atmaca. Traffic management in multi-service optical network. In *IEEE International Conference on Networking (ICN)*, Colmar, France, July 2001.

111. H. Øverby. QoS in slotted bufferless optical packet switched networks. In *International Conference on Transparent Optical Networks*, pages 334–337, July 2004.

112. F. Callegati, W. Cerroni, C. Raffaelli, and P. Zaffoni. DWDM for QoS management in optical packet switches. In *Lecture Notes in Computer Science*, volume 2601, pages 447–459. Springer, 2003.

113. H. Øverby. Packet loss rate differentiation in slotted optical packet switched networks. *IEEE Photonics Technology Letters*, 17(11):2469–2471, Nov. 2005.

CHAPTER 4

HYBRID CONTENTION AVOIDANCE/RESOLUTION IN OPS NETWORKS

Many studies have evaluated the performance of OPS networks using an individual technique such as either using only WCs, using only optical buffers, or using only a multi-fiber architecture. In these studies, when a very low loss rate is required, the required number of WCs [1], the required number of optical buffers [2], the required number of fibers in a multi-fiber architecture [3], and the required number of additional drop ports [4] drastically increases. In other words, the role of using more and more hardware such as WCs to improve network throughput decreases in practice. In addition, the increase of hardware usage is not cost-effective. In other words, a single contention resolution scheme cannot solely provide significant improvement in reducing PLR.

Since there are some limitations in terms of technical possibilities and costs that prevent us from using a high density of contention resolution mechanisms (such as limited number of FDLs in each core switch, limited number of WCs in each core switch, limited number of wavelengths on each fiber, and marginal effectiveness of deflection routing due to the network topology), one should combine these mechanisms for providing the best performance and lowest PLR in OPS networks. Instead of using the same technique to achieve a very low PLR, the idea of hybrid contention

Quality of Service in Optical Packet Switched Networks, First Edition.
By Akbar Ghaffarpour Rahbar Copyright © 2015 IEEE. Published by John Wiley & Sons, Inc.

resolution is to combine different contention resolution schemes at the same time, but using a small amount of each scheme.

Three categories can be stated for hybrid contention avoidance/resolution in OPS networks:

- **Hybrid Contention Avoidance Schemes**: Here, a number of hardware-based and a number of software-based contention avoidance schemes are combined in an OPS network. Since this combination cannot reduce PLR in an efficient manner in a mesh network, it could be suitable for single-hop OPS networks (see Section 6.1).

- **Hybrid Contention Resolution Schemes**: Here, a number of hardware-based and a number of software-based contention resolution schemes are combined in an OPS network (see Section 4.1).

- **Hybrid Contention Avoidance and Resolution Schemes**: Here, a number of hardware-based/software-based contention resolution schemes and a number of hardware-based/software-based contention avoidance schemes are combined in an OPS network (see Section 4.2).

4.1 Hybrid Contention Resolution Schemes

In this section, the hybrid contention resolution schemes proposed for OPS networks are studied and evaluated.

4.1.1 Hybrid WC Techniques

There are a number of architectures proposed to combine different WC types in OPS switches detailed in the following.

4.1.1.1 Hybrid SPN-LRWC + SPOF-PWC Architecture
The Hybrid Shared Wavelength Conversion (HSWC) architecture proposed in [5] combines SPN-LRWCs with SPOF-PWCs (see Fig. 4.1 for an $N * N$ core switch). Under HSWC, C LRWCs are shared per OPS switch and r_{wc} PWCs are shared at each output port. Each PWC can simultaneously accept up to x (where $x \leq W$) wavelengths through a $x : 1$ coupler. Hence, the switch size is $(N \times W + C) \times (N \times x \times r_{wc} + N \times W + C)$.

Scheduling of WCs in HSWC could be based on two mechanisms:

- **HSWC with PWC First Available (HSWC-PFA)**: Here, PWCs are first used to resolve contention and then LRWCs are used to resolve contention of optical packets. Finally, the combination of remaining PWCs and LRWCs are employed to resolve the contention of remaining optical packets. The HSWC-PFA tries to first utilize entire PWCs conversion capabilities. Then, LRWCs are used to resolve those optical packets that are either at the guard band of PWCs or cannot be converted due to the lack of conversion pairs. The HSWC-PFA can maximize the utilization of WCs.

$(N{\times}W + C) * (N{\times}x{\times}r_{wc} + N{\times}W + C)$

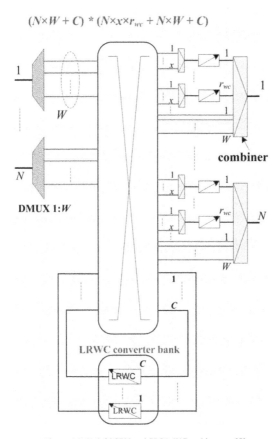

Figure 4.1: Hybrid SPN and SPOL WC architecture [5]

- **HSWC with LRWC First Available (HSWC-LFA)**: Here, LRWCs are first used, and then PWCs are used to resolve contention of optical packets. Finally, the combination of remaining PWCs and LRWCs are employed to resolve the contention of remaining optical packets. The HSWC-LFA could not be as efficient as HSWC-PFA since it is more likely that almost all LRWCs are consumed at first and then there will be no LRWC available for the combination of LRWC and PWC at the end.

The Phase 1 of both HSWC scheduling mechanisms is similar to the Phase 1 depicted in Fig. 3.33. For Phase 2 of both HSWC-LFA and HSWC-PFA, the Phase 2 algorithm depicted in Fig. 3.35 for SPN with PWC, and the Phase 2 algorithm displayed in Fig. 3.34 for SPN with LRWC can be used. However, note that the Phase 2 algorithm for SPN with PWC must be individually used for each output port with r_{wc} PWCs in both HSWC-LFA and HSWC-PFA.

Simulation results show that with equal number of WCs, the HSWC has the lowest PLR followed by SPN architecture with LRWC converters and SPN architecture with PWC converters. The PLR performance under HSWC-PFA is always smaller than that under HSWC-LFA.

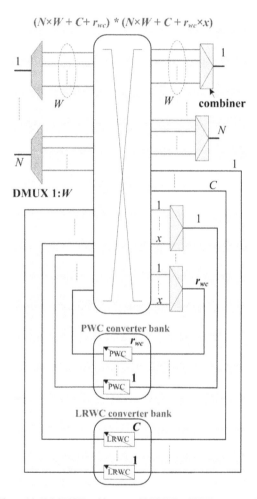

Figure 4.2: Hybrid SPN architecture with LRWC and PWC converters [6]

4.1.1.2 Hybrid SPN-LRWC + SPN-PWC Architecture

The hybrid SPN architecture proposed in [7, 8] combines C LRWCs and r_{wc} PWCs as shared-per-node (see Fig. 4.2). This is an extension of sharing of PWCs in [6, 9]. Each PWC can simultaneously accept up to x (where $x \leq W$) wavelengths. Therefore, the switch size is $(N \times W + C + r_{wc}) \times (N \times W + C + r_{wc} \times x)$. This architecture can reduce PLR, improve utilization of WCs, and reduce switch cost. This hybrid SPN architecture needs more ports on switch fabric than HSWC, while it requires less PWCs than HSWC.

Consider that there are M priorities for optical packets, where Class 1 optical packets have the highest priority and Class M optical packets have the lowest priority. Scheduling of optical packets in this hybrid SPN architecture has two major phases. The first phase schedules the optical packets that do not need wavelength conversion, but HP optical packets are scheduled before LP ones. The Phase 1 of this scheduling is similar to the Phase 1 depicted in Fig. 3.33, except that when creating set $S_{k,i,m}$ (i.e., the optical packets coming on wavelength λ_i and destined to output port k), it is filled first with Class 1 optical packets, then with Class 2 optical packets, and finally with Class M optical packets. The second phase first uses PWCs and then LRWCs for resolving contention of optical packets. The Phase 2 algorithm depicted in Fig. 3.35 for SPN with PWC and the Phase 2 algorithm depicted in Fig. 3.34 for SPN with LRWC can be used here, except that optical packets must be removed in sequence from set $S_{k,i,m}$ when scheduling. By this, the contention of HP optical packets are resolved before LP ones; therefore, PLR of HP optical packets always becomes lower than LP optical packets.

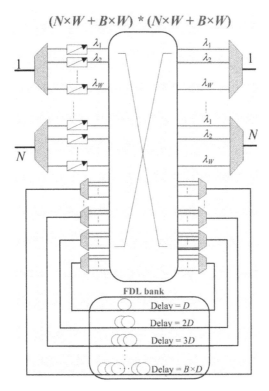

Figure 4.3: Hybrid OPS switch architecture with FDL and LRWC [10]

4.1.1.3 Two-Layer Wavelength Conversion (TLWC)

In TLWC proposed for a slotted OPS core switch, there are two layers of wavelength conversion, one before the switching module and one after the switching module. The first layer is equipped with LRWCs and the second layer is equipped with FRWCs. The first layer performs low-range wavelength conversions, thus decreasing the need for more expensive FRWCs performing such functions. The second layer provides large-range wavelength conversions and is used for the case that the LRWC layer cannot convert optical packets. Two different architectures can be stated for TLWC:

- **TLWC-SPOF**: In this architecture, one LRWC is used per each input wavelength channel in the first layer and one pool of r_{wc} FRWCs at each output port as the SPOL architecture at the second layer. In total, there are $N \times W$ LRWC wavelength converters at the first layer and $N \times r_{wc}$ wavelength converters at the second layer. Any incoming optical packet enters the switching module either directly or after limited-range wavelength conversion. After switching, the optical packet can be sent either directly or after using FRWC to the relevant output port [11, 12].

- **TLWC-SPN**: This is similar to TLWC-SPOF, but with the exception that the second layer uses shared-per-node architecture with N_{WC} FRWCs. Switched optical packets may enter the shared-per-node FRWC bank at the second layer for further wavelength conversion [11].

Performance evaluation results prove that the proposed architectures provide better PLR results than one layer conversion architectures (i.e., SPN and SPOL). In addition, by increasing conversion degree of LRWCs, the need for FRWC conversions reduces. Therefore, the proposed architectures are suitable for the case that LRWCs have small conversion degree.

4.1.2 Hybrid FDL + WC Techniques

Different architectures have been proposed to combine FDLs and WCs in OPS core switches detailed in the following. Using wavelength converters with optical buffers can provide much better PLR performance than using buffers solely because the chance of finding free buffers increases [13].

4.1.2.1 Hybrid SPN FDL + LRWC Architecture
The OPS switch architecture proposed for slotted OPS in [10, 14] uses LRWCs before switching fabric and recirculation FDLs as depicted in Fig. 4.3. The length of each FDL is an integer number of time-slot sizes (with size D) from 1 to B. An FDL can buffer up to W optical packets with different wavelengths at the same time. Clearly, this architecture can reduce PLR compared with the switch with either LRWCs or FDLs solely.

When an optical packet destined to output port k arrives at an OPS switch on wavelength λ_i, the switch first checks whether λ_i is idle at output port k in order to directly switch the optical packet. If no, the switch checks whether there is an idle wavelength λ_l at output port k to switch the optical packet through converting its wavelength from λ_i to λ_l. If no, the switch will then try to temporarily route it to the FDL bank. For this, all FDLs from FDL 1 to FDL B are linearly searched for idle wavelength λ_i. If wavelength λ_i is idle on FDL m, then the optical packet is routed through FDL m. When the optical packet recirculates FDL m and returns back to the switch fabric, it is treated as a new optical packet, and then the switch tries to send it on output port k or the FDL bank again. If no FDL can be found to accommodate the optical packet on wavelength λ_i, the FDL 1 to FDL B are linearly searched for an idle wavelength λ_l. Let FDL m include idle wavelength λ_l. Then, after wavelength conversion from λ_i to λ_l, the optical packet is injected in FDL m on wavelength λ_l. Finally, if no idle wavelength can be found, it is dropped.

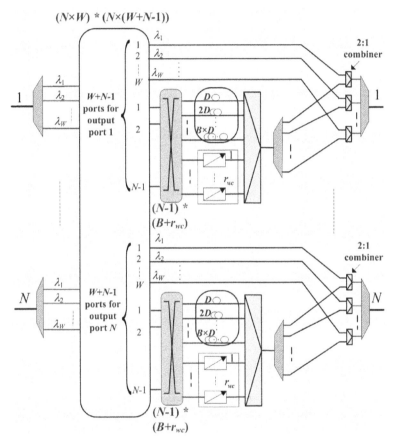

Figure 4.4: Shared FDL and wavelength conversion per output port in an OPS switch

4.1.2.2 Hybrid Shared FDL + WC per Output Port Architecture

Figure 4.4 combines B FDLs and r_{wc} WCs per output port [15]. The second small switch space with size $(N-1)*(B+r_{wc})$ directs a contenting optical packet in a given output port to either the WC bank or the FDL bank. Optical packets leaving the banks are aggregated by a combiner and then demultiplexed to wavelengths λ_1 to λ_W and then sent to the 2:1 combiners. There is at most one optical packet entering a 2:1 combiner: either the optical packet that does not need contention resolution or the optical packet that needs contention resolution (through conversion or delaying). Clearly, in some cases, no optical packet may arrive at the 2:1 combiner.

In this architecture, the contention resolution scheduler tries first to send a contending optical packet through one of the r_{wc} WCs. If there is no free WC, one of the FDLs is chosen to delay the contending optical packet. However, the scheduler attempts to first send it through short FDLs (less delaying) and then through long FDLs. If there is no FDL available, the contending optical packet is discarded.

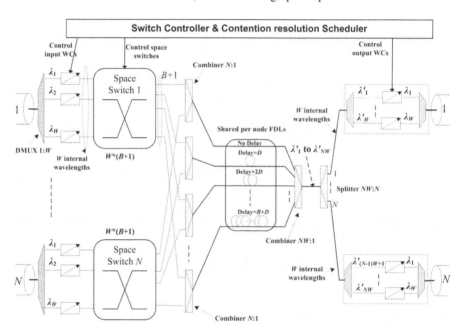

Figure 4.5: Hybrid contention resolution with WC and shared FDLs [16]

4.1.2.3 Hybrid WC + Shared per Node FDL Architecture   Wavelength converters can be combined with shared-per-node FDLs in an OPS core switch as proposed in [16, 17]. Figure 4.5 depicts this combined architecture, where there are W wavelengths (λ_1 to λ_W) out of the core switch, but $N \times W$ wavelengths (i.e., λ'_1 to $\lambda'_{N \times W}$) inside the core switch, called internal wavelengths. The internal wavelengths $\lambda'_{W(k-1)+1}$ to $\lambda'_{W \times k}$ are used for sending W optical packets to output port k, where $1 \leq k \leq N$. There are B FDLs shared among all output ports. For each optical packet arriving at an input port, the contention resolution scheduler first converts its wavelength to an appropriate unused internal wavelength considering its desired output port, assigns a suitable delay (ranging from 0 to $B \times D$ time units where D is a constant time value) that it must tolerate in order to avoid contention, and then routes it toward the suitable FDL. Each FDL can provide some delay for a number of optical packets (at most N) on different wavelengths at the same time. At the end of the

buffering stage, wavelengths λ_1' to λ_W' are sent toward output port 1, wavelengths λ_{W+1}' to $\lambda_{2\times W}'$ are sent toward output port 2, and in the same manner to other output ports, and finally wavelengths $\lambda_{W(N-1)+1}'$ to $\lambda_{W\times N}'$ are sent toward output port N. For a given output port, all W optical packets directed to that output port on internal wavelengths are converted to wavelengths λ_1 to λ_W, then multiplexed, and finally sent to the output port. Note that each input/output port uses W FRWCs, totally $2 \times N \times W$ converters.

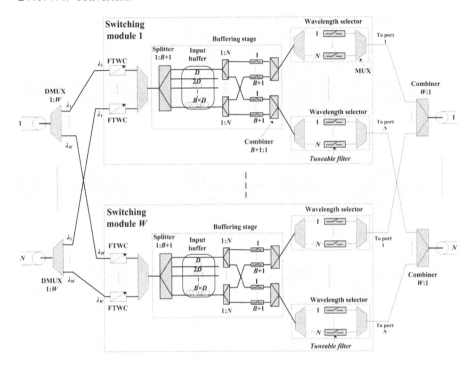

Figure 4.6: A feed-forward input buffering FDL architecture combined with wavelength conversion in an OPS switch [18]

4.1.2.4 Hybrid WC + Feed-Forward Input Buffering Architect ure
The Keys to Optical Switching (KEOPS) architecture for feed-forward input buffering for a slotted OPS switch has been depicted in Fig. 4.6 [18]. A demultiplexing module separates the optical packets from each incoming fiber on W wavelengths. The operation of this architecture is managed by a control unit with the following three steps:

- All optical packets coming on wavelength λ_i from different input ports are routed to switching module i. In any switching module, there are N FTWCs to convert the optical packets coming on the same wavelength to different wavelengths. There are W switching modules in the OPS switch, where the figure shows only switching modules 1 and W.

- The buffering stage in a switching module includes a splitter, B FDLs, and a space-switching stage implemented by means of splitters, optical gates, and combiners. The outputs of the N wavelength converters are combined and then distributed through a splitter into B FDLs and a 0 delay line (totally $B+1$ FDLs). Each FDL has a different delay which is an integer number of time-slots. Note that by this distribution, all N optical packets are stored simultaneously into all $B+1$ FDLs in the input buffer. Hence, at the beginning of the next time-slot, a maximum of $(B+1) \times N$ optical packets exit from the buffering unit, and among them, up to N optical packets must be directed to their destination output ports without any collisions. The output signal from each FDL (i.e., N multiplexed optical packets) goes through a splitter which distributes the output signal over N outputs. The signal from output j of each splitter is directed to output port j. Since there are $B+1$ such splitters, there are $B+1$ such output signals, of which only one should be chosen and directed to output port j.

- The selected output signal in the previous step is routed into a wavelength selector, which demultiplexes it into N wavelengths. As can be seen, a wavelength selector block consists of N wavelength channel selectors implemented by demultiplexers, optical gates, and multiplexers. Having only one gate connected in a wavelength selector, only one optical packet is sent toward the output port. At the output side, each group of W wavelengths are combined together through a combiner onto the same output port.

4.1.2.5 Hybrid WC + Feed-Forward Output Buffering Architecture
The OPS switch illustrated in Fig. 4.7 works in slotted mode in which the amount of delays of FDLs vary from 0 to B time slots. The OPS switch consists of three sections as the optical packet encoder, the space switch, and optical buffers [18].

First, the spectrum of each fiber is demultiplexed into W separate wavelengths, and then each of these wavelengths is converted using Tunable WCs (of type FTWC) to a suitable wavelength which is chosen by the control unit for contention resolution purposes. Then, in the switching space, each wavelength is divided into N equal signals and each of them again is divided into $B+1$ equal signals. Next, each signal is inserted to an on–off switch controlled by the control unit. In the on–off switch part, only one of those $N \times (B+1)$ signals will be switched to pass to the relevant FDL bank. At the last step, forwarded optical packets with different wavelengths to each FDL are combined using a combiner and fed to an FDL. Finally, delayed optical packets will be multiplexed together in each output interface and sent to the output port. The control unit always selects the FDL with the lowest delay. In this architecture, the wavelengths of contending optical packets are converted to idle wavelengths in FDLs and then transmitted to the FDL bank. This clearly increases the utilization of FDLs and decreases the number of required FDLs.

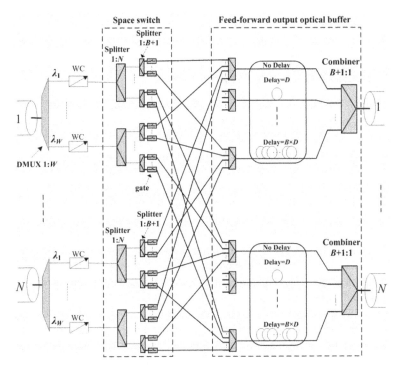

Figure 4.7: A feed-forward shared-per-output buffer architecture in an OPS switch [18]

4.1.3 Hybrid WDS + MPR Architecture

In this architecture, contention at a given output port of a core switch can be resolved by means of load balancing that exploits both the wavelength and time dimensions. Here, to resolve contention, the controller of the core switch chooses a suitable wavelength and delay for a contending optical packet; i.e., choice of the best (wavelength, delay) according to resource availability. This problem is referred to as the Wavelength and Delay Selection (WDS) problem as detailed in 3.1.1.1.3 [19].

The basic idea here is to combine the flexibility of MPR with the efficiency of packet multiplexing over a large set of wavelengths by means of an effective WDS policy. For load balancing, the WDS policy can be applied on an entire set of links, not only on a single link (either default or alternative). Two methods can be stated for this policy:

- **Shared Shortest Paths (SSP)**: The WDS is performed over all the wavelengths on any shortest path link, including the default Shortest Path Routing (SPR) one.

- **m-Shared Shortest Paths (m-SSP)**: The WDS is performed over all the wavelengths on any link belonging to paths with up to $m - 1$ hops more than the default SPR.

It is shown that 2-SSP can significantly reduce PLR compared with SSP in OPS networks.

4.1.4 Hybrid WC + FDL + DR Architecture

The work in [20] has combined three contention resolution domains: wavelength conversion, optical buffering, and deflection routing. Deflection routing has low impact on throughput and PLR due to network topology since it is more effective on high-connectivity topologies (such as ShuffleNet). When the optical packet OP contends with other optical packets, the core switch first checks the wavelength conversion domain. If there is any idle wavelength at the desired output port of OP and also there is an available WC which can convert the optical packet wavelength to the idle wavelength, OP is scheduled through wavelength conversion. If there is no such a wavelength or an available WC, the core switch checks idle FDLs in order to delay OP. If there is no available FDL, the core switch searches for an idle wavelength at a deflection output. When all three domains fail, OP is dropped. This order of wavelength conversion, optical buffering, and deflection routing for contention resolution can provide the best performance of PLR and delay [20, 21].

As another architecture, a slotted OPS switch based on AWG, FDLs, and WCs has been proposed in [22]. This switch architecture includes three units in sequence: a synchronization unit for each input wavelength, an FDL unit for each input wavelength, and a switching unit. The FDL units are located before the switching unit. Each FDL unit for an input wavelength includes one WC and a bank of 0 to $B \times D$ FDls. An FDL unit stores optical packets to accomplish optical buffering and scheduling for resolving contention at output ports. The WC in an FDL unit is used to route optical packets to the chosen delay line. By this, the optical scheduling algorithm sets variable delays for optical packets entering the switching unit. The switching unit includes an AWG switching matrix and two stages of WCs (one stage before the switching matrix and one stage after the switching matrix), where the first stage of WCs is used to route optical packets to the desired output and the second WC stage is responsible to convert the signal to a suitable wavelength in order to avoid two optical packets to be transmitted on the same wavelength. When there is no way to resolve the contention of an optical packet to its desired output port, it is deflected into other output ports with equal probability, but with a hop limit H. This hop limit is used for discarding the optical packets pinging too long inside the network.

4.2 Hybrid Contention Resolution and Avoidance Schemes

Neither a contention avoidance scheme nor a contention resolution scheme alone can effectively decrease PLR in an OPS network to a very small value. To obtain a better performance, however, a number of contention resolution schemes and a number of contention avoidance schemes should be used at the same time in the OPS network in order to design even a loss-free OPS network. As categorized in Chapters 2 and

3, there are four major schemes that can deal with the contention problem in an OPS network: software-based contention avoidance schemes, hardware-based contention avoidance schemes, software-based contention resolution schemes, and hardware-based contention resolution schemes.

The combination can be carried out at two levels. At the first level, any combination of the four major schemes can coexist at the same time in an OPS network. At the second level of combination, different techniques within the same major scheme can be combined with each other. For an example, in the contention avoidance phase, any possible combination of the inexpensive software-based techniques can be used together. By using the multi-fiber architecture and/or using additional wavelengths per fiber, the hardware-based schemes can be involved in handling contention. In the contention resolution phase, shared-per-node WCs and FDLs can reduce PLR to a desired value. Finally, prioritized retransmission in the optical domain and deflection routing techniques both can be used to reduce PLR to almost zero.

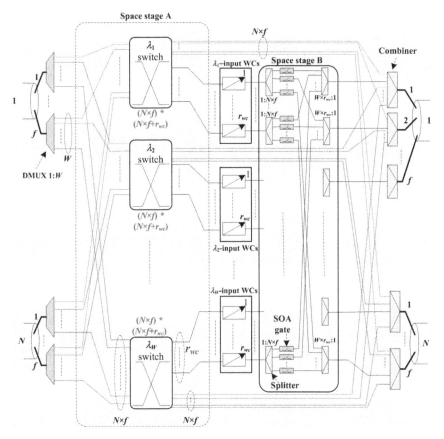

Figure 4.8: Multi-fiber OPS switch with SPIW WC architecture [23]

Algorithm: Phase 1 of Contention Resolution in Multi-fiber SPIW Architecture

1: **for** $k = 0$ to $N - 1$ {scan all N output links} **do**
2: **for** $i = 0$ to $W - 1$ {scan all wavelengths on a fiber} **do**
3:   Make set $S_{k,i,m}$ {make the set of optical packets coming on wavelength λ_i from all f fibers and destined to output link k}
4:   **for** $v = 0$ to $f - 1$ {scan all fibers on output link k for λ_i} **do**
5:    **if** $m = 0$ {no optical packet contending for λ_i on link k or all contending ones scheduled} **then**
6:     Add triple (v,i,k) to list U {save unused wavelength i in list U}
7:    **else**
8:     {schedule one of the m optical packets contending for wavelength i on output link k}
9:     Remove randomly one optical packet OP from $S_{k,i,m}$
10:    Schedule OP on λ_i of fiber v at output link k
11:    Set $m = m - 1$ in $S_{k,i,m}$ {update m in list $S_{k,i,m}$}
12:   **end if**
13: **end for**
14: **end for**
15: **end for**

Algorithm: Phase 2 of Contention Resolution in Multi-fiber SPIW Architecture

1: Randomize list U {randomize list U}
2: **for** $l = 0$ to $W - 1$ {for all W output wavelengths} **do**
3: $C_l = 0$ {initialize the number of used WCs for each converter bank}
4: **end for**
5: **for** $j = 1$ to $sizeof(U)$ {scan all unused wavelengths in list U} **do**
6: Get triple (v,i,k) from $U(j)$ {get the information of an unused wavelength}
7: **for** $l = 0$ to $W - 1$ {scan all incoming wavelengths destined to output link k} **do**
8:   Consider list $S_{k,l,m}$
9:   **if** $m > 0$ and $C_l < r_{wc}$ **then**
10:    Remove randomly one optical packet OP from $S_{k,l,m}$ incoming on λ_l
11:    Schedule OP on λ_i of fiber v at output link k {use one WC to convert wavelength λ_l of OP to outgoing wavelength λ_i of fiber v}
12:    Set $C_l = C_l + 1$ {update the number of used WCs in converter bank l}
13:    Set $m = m - 1$ in $S_{k,l,m}$ {reduce the number of contending optical packets}
14:    Exit loop {evaluate the next unused wavelength}
15:   **end if**
16: **end for**
17: **end for**
18: Drop all optical packets remaining in sets $S_{k,i,m}$ when $m > 0$
19: END

Figure 4.9: Pseudo code of contention resolution scheduling in a multi-fiber core switch with SPIW converters

4.2.1 Hybrid MF + SPIW Architecture

Figure 4.8 shows the architecture of a multi-fiber OPS switch with SPIW architecture, where there are r_{wc} SPIW converters in each one of the W converter banks. Here, there are $W \times N \times f \times (N \times f + r_{wc})$ SOA gates in switching space A and $W \times N \times f \times r_{wc}$ SOA gates in switching space B, totally $W \times (N \times f)^2 + 2 \times W \times N \times f \times r_{wc}$ in the whole switch. Consider that $f \times W$ is constant. It is proved that by increasing f (i.e., decreasing W) and keeping r_{wc} constant, PLR reduces [23–25].

Figure 4.9 illustrates CRSA in multi-fiber OPS with SPIW architecture as follows:

- In Phase 1, all output links and all wavelengths on each output link are scanned to schedule the optical packets that do not need wavelength conversion. First, for each output link k and wavelength λ_i, set $S_{k,i,m}$ is created that includes the set of m contending optical packets coming on wavelength λ_i of all f input fibers and destined to output link k. If we have $m \leq f$, then all optical packets in set

$S_{k,i,m}$ are scheduled on λ_i at output link k, and $f - m$ unused wavelengths are added to list U. Otherwise, all f wavelengths λ_i at output link k are scheduled, and $m - f$ optical packets remain in $S_{k,i,m}$ as contending ones to be resolved in Phase 2. The scheduling issue happens in the third loop that scans each one of f fibers. The triple (v, k, i) as the information of an unused wavelength is appended to list U if $m = 0$. Otherwise, one of the contending optical packets in $S_{k,i,m}$ is scheduled on λ_i of fiber v. Finally, set $S_{k,i,m}$ is updated and m is reduced by 1.

- Phase 2 is similar to the phase 2 of single-fiber architecture depicted in Fig. 3.40, except that triple (v, k, i) is extracted from list U, and then a contending optical packet from set $S_{k,i,m}$ is scheduled on wavelength λ_i of fiber v at output link k.

The SPIW architecture is a sharing strategy to save the number of required WCs in an OPS switch. Hence, the higher the sharing degree, the higher the WC saving. In a SPIW architecture, the sharing degree is high when the number of optical packets carried on the same wavelength is high. In a single-fiber OPS switch (i.e., $f = 1$), this occurs when N is high, and therefore a high number of optical packets are carried on the same wavelength, thus improving the performance of SPIW. In a multi-fiber OPS, the sharing degree depends on $N \times f$; i.e., high number of WC saving can be achieved when this number is high (either N is high or f is high) [24].

4.2.2   Hybrid MF + SPN WCs Architecture

Figure 4.10 depicts a general multi-fiber OPS core switch architecture with f fibers per input/output link and $N_{WC} = r_{wc}$ shared-per-node wavelength converters [25–30]. Space switch A is a $(N \times f \times W) * (N \times f \times W + r_{wc})$ all-optical switching module that forwards optical packets either directly or through the SPN wavelength conversion bank to the $(W + r_{wc}) : 1$ combiners. A $1 : (N \times f)$ splitter is also used at the output of each WC to split a wavelength converted optical packet to $N \times f$ output signals. Switching stage B switches each optical packet to one of the $N \times f$ output fibers according to its destination. The output signal from a $(W + r_{wc}) : 1$ combiner is fed into a demultiplexer for separating wavelength channels. Then, the optical packet on each wavelength channel is sent to the synchronizer unit so that all optical packets are synchronized. Then, the optical packets on different wavelengths are multiplexed and sent to the next core switch.

Figure 4.11 displays the scheduling algorithm for a multi-fiber OPS switch in which each wavelength converter is a LRWC with conversion degree d_c. Here, Phase 1 is the same as Phase 1 in Fig. 4.9 detailed in Section 4.2.1. Phase 2 is similar to Phase 2 in Fig. 3.34, but with this exception that each item in list U carries a triple (v, i, k); where this triple shows that wavelength λ_i of fiber v at output link k is idle. Therefore, a contending optical packet extracted from set $S_{k,i,m}$ should be scheduled after wavelength conversion on wavelength λ_i of fiber v at output link k.

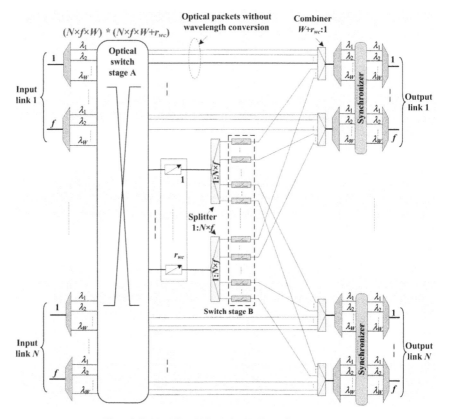

Figure 4.10: Multi-fiber OPS switch with SPN WC architecture [26]

4.2.3   Hybrid MF + SPOL WC Architecture

The multi-fiber architecture can be combined with SPOL WCs (see Fig. 4.12) to efficiently reduce PLR in a slotted OPS network [31]. The Phase 1 of scheduling in this architecture is the same as Phase 1 in Fig. 4.9 detailed in Section 4.2.1. Phase 2 is similar to Phase 2 in Fig. 3.37, but with the exception that each item in list U carries a triple (v, i, k) as stated in Section 4.2.2.

Now, an analysis is adapted from [31] to compute PLR in a buffer-less multi-fiber slotted OPS core switch with SPOL wavelength conversion, but where traffic distribution to output links of the OPS core switch is asymmetric. The OPS switch has $N \times f$ input fibers and $N \times f$ output fibers, where each fiber has W wavelength channels (λ_0 to λ_{W-1}) and r_{wc} WCs are shared-per-output link.

Consider a given time slot, where there are $N \times f$ time slots on wavelength λ_w at the input ports of the core switch. Let the set of traffic distribution probabilities to output links of the core switch be $P = \{p_1, p_2, \ldots, p_N\}$, where $\sum_{d=1}^{N} p_d = 1$, and p_d denotes the probability of forwarding optical packets to output link d. Clearly, $p_d = \frac{1}{N}$ leads to a symmetric traffic distribution (i.e., equal traffic distribution from an

Algorithm: Phase 1 of Contention Resolution in Multi-fiber SPN Architecture

1: **for** $k = 0$ to $N - 1$ {scan all N output links} **do**
2:    **for** $i = 0$ to $W - 1$ {scan all wavelengths on a fiber} **do**
3:       Make set $S_{k,i,m}$ {make the set of optical packets coming on wavelength λ_i from all f fibers and destined to output link k}
4:       **for** $v = 0$ to $f - 1$ {scan all fibers on output link k for λ_i} **do**
5:          **if** $m = 0$ {no optical packet contending for λ_i on link k or all contending ones scheduled} **then**
6:             Add triple (v, i, k) to list U {save unused wavelength i in list U}
7:          **else**
8:             {schedule one of the m optical packets contending for wavelength i on output link k}
9:             Remove randomly one optical packet OP from $S_{k,i,m}$
10:            Schedule OP on λ_i of fiber v at output link k
11:            Set $m = m - 1$ in $S_{k,i,m}$ {update m in list $S_{k,i,m}$}
12:         **end if**
13:      **end for**
14:   **end for**
15: **end for**

Algorithm: Phase 2 of Contention Resolution in Multi-Fiber SPN

1: $C = 0$ {initialize the number of used WCs in core switch}
2: Randomize list U {randomize list U}
3: **for** $j = 1$ to $sizeof(U)$ {scan all unused wavelengths in list U} **do**
4:    **if** $C = N_{WC}$ **then**
5:       Exit loop {end of phase 2 if all WCs are used}
6:    **end if**
7:    Get triple (v, i, k) from $U(j)$ {get the information of an unused output wavelength}
8:    **for** $l = 0$ to $W - 1$ {scan all incoming wavelengths destined to output port k} **do**
9:       Consider list $S_{k,l,m}$
10:      **if** $m > 0$ AND $|l - i| \leq d_c$ **then**
11:         Remove randomly one optical packet OP from $S_{k,l,m}$ incoming on λ_l
12:         Schedule OP on λ_i of fiber v at output port k {use one WC to convert OP from incoming wavelength λ_l to outgoing wavelength λ_i}
13:         Set $C = C + 1$ {update the number of used WCs}
14:         Set $m = m - 1$ in $S_{k,l,m}$ {reduce the number of contending optical packets}
15:         Exit loop {evaluate the next unused wavelength}
16:      **end if**
17:   **end for**
18: **end for**
19: Drop all optical packets remaining in sets $S_{k,i,m}$ when $m > 0$
20: END

Figure 4.11: Pseudo code of contention resolution scheduling in a multi-fiber core switch with SPN converters

input wavelength to all output links of the core switch), where its analysis has been performed in [3, 26, 33]. However, traffic distribution is not symmetric in practice, and therefore the symmetric traffic assumption is not realistic.

Define L to be the probability of the arrival of optical packets on each wavelength channel at the input ports of the core switch. Let $q_d = L \times p_d$ denote the probability that an optical packet arriving at an input wavelength channel is destined to output link d (where $d = 1, 2, \ldots, N$). In the following, PLR is first computed for a given output link d with f output ports. Then, PLR is calculated for the core switch.

A set of f the same wavelength channels at output link d is referred to as the same-wavelength group, as depicted in Fig. 4.13, for the same-wavelength groups λ_2 and λ_4. The W the same-wavelength groups can be divided into two categories:

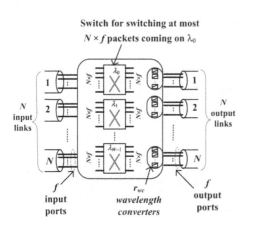

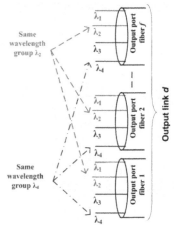

Figure 4.12: Core switch model for a MF + SPOL WC core switch [32]

Figure 4.13: An example for the same-wavelength groups [32]

- j the same-wavelength groups on which no optical packets have arrived on any of the wavelengths of each group; and

- $u = W - j$ the same-wavelength groups on which at least one optical packet has arrived on one wavelength channel of each group. For simplicity, the channels in the u groups are assigned alias names from λ_1' to λ_u'.

Let X be the number of the same-wavelength groups on which no optical packets have arrived on any of the wavelengths of each group, and let $q_{0,d}$ be the probability of having no optical packets arriving on the wavelength channels of a given same-wavelength group at output link d. Considering the fact that there are $N \times f$ same-wavelength channels at the input ports of the core switch, one can compute $q_{0,d} = (1 - q_d)^{N \times f}$. Thus, the probability of having $X = j$ at output link d is computed by

$$\text{Prob.}\{X = j\} = \left(\begin{array}{c} W \\ j \end{array} \right) (q_{0,d})^j (1 - q_{0,d})^{W-j}, \qquad \text{for } j \in \{0, \ldots, W-1\}. \quad (4.1)$$

Define Y_i to be the number of optical packets (where $Y_i \geq 1$) arriving on the wavelength channels of the same-wavelength group i (i.e., the group of f wavelengths λ_i'). The probability of having $y > 0$ optical packets on f wavelengths λ_w on output link d is computed by

$$\text{Prob.}\{Y_i = y\}_d = \frac{\left(\begin{array}{c} N \times f \\ y \end{array} \right) (q_d)^y (1 - q_d)^{N \times f - y}}{1 - q_{0,d}}, \qquad \text{for } y > 0. \quad (4.2)$$

Then, the probability of having $y_1 > 0$ optical packets arriving on f wavelengths of the same-wavelength group 1, $y_2 > 0$ optical packets on f wavelengths of the same-wavelength group 2, till $y_u > 0$ optical packets on f wavelengths of the same-wavelength group u can be computed by

$$\prod_{k=1}^{u} \text{Prob.}\{Y_k = y_k\}_d .$$ (4.3)

Among y_k optical packets arriving on f wavelengths of the same-wavelength group k, $\min(f, y_k)$ of them can be easily switched on f wavelengths λ'_k. In other words, of $\sum_{k=1}^{u} y_k$ optical packets arriving at output link d, $\sum_{k=1}^{u} \min(f, y_k)$ optical packets can be switched. In addition, some of the remaining optical packets can then be switched after using r_{wc} shared-per-link WCs located at output link d. First, recall that there are $f \times j$ wavelength channels on which no optical packets have arrived at output link d. Moreover, among the u same-wavelength groups, there are η wavelength channels on which no optical packets have arrived:

$$\eta = \sum_{k=1}^{u} \max(0, f - y_k) .$$ (4.4)

Therefore, there are totally $n_s = \min(r_{wc}, f \times j + \eta)$ optical packets that can be switched using r_{wc} shared-per-link WCs. Thus, for a given state $(j, y_1, y_2, ..., y_u)$, the number of lost optical packets is computed by

$$\alpha_{(j,y_1,y_2,...,y_u)} = \max\left(\left(\sum_{k=1}^{u} (y_k - \min(f, y_k)) \right) - n_s, 0 \right) .$$ (4.5)

The average number of lost optical packets at output link d for a given j can then be computed by

$$Z_{j,d} = \sum_{\substack{1 \le y_1 \le N \times f \\ 1 \le y_u \le N \times f}} \left(\prod_{k=1}^{u} \text{Prob.}\{Y_k = y_k\}_d \times \alpha_{(j,y_1,y_2,...,y_u)} \right),$$ (4.6)

where the upper limit $N \times f$ for y_k denotes the total number of optical packets that could be destined to output link d on f wavelengths λ'_k in the worst-case. Hence, at output link d, the average number of lost optical packets on all the same-wavelength groups can be computed by

$$n_{loss}(d) = \sum_{j=0}^{W} \left(\text{Prob.}\{X = j\} \times Z_{j,d} \right),$$ (4.7)

where $\text{Prob.}\{X = j\}_d$ is computed from Eq. (4.1), and $\text{Prob.}\{Y_k = y_k\}_d$ is given by Eq. (4.2). Of $N \times f \times W \times L$ optical packets arriving at N input links of the core switch, on average $n_{dlv}(d) = N \times f \times W \times q_d$ optical packets are delivered to output link d. Hence, the average PLR at output link d is given by

$$Loss(d,N,f,W,q_d) = \frac{n_{loss}(d)}{n_{dlv}(d)} \, . \tag{4.8}$$

Finally, average PLR in the core switch is computed by total number of lost optical packets on all output links divided by total number of optical packets forwarded to all output links of the core switch as

$$PLR(N,f,W) = \frac{\sum\limits_{d=1}^{N} (N \times f \times W \times q_d \times Loss(d,N,f,W,q_d))}{\sum\limits_{d=1}^{N} (N \times f \times W \times q_d)}$$

$$= \frac{\sum\limits_{d=1}^{N} (q_d \times Loss(d,N,f,W,q_d))}{L} \, . \tag{4.9}$$

Consider a core switch at $N = 8$ and asymmetric traffic distribution to its output links with $P = \{0.10, 0.03, 0.29, 0.01, 0.19, 0.22, 0.11, 0.05\}$. Assume that $r_{wc} \in \{0,2,4,6\}, f \in \{1,2,4\}$, and $f \times W = 8$. Client packets arrive at each edge switch according to a Poisson process and saved in electronic buffers, and then sent as optical packets at time-slot boundaries to the core switch.

Figures 4.14a to 4.14c display PLR under asymmetric traffic. According to these figures, analysis results match well with simulation results. By increasing f or r_{wc} or both, PLR decreases. However, there is no significant reduction in PLR at $r_{wc} = 6$. At $L = 0.9$, the core switch experiences the lowest PLR of 0.296 at $f = 2$ and $r_{wc} = 6$. This high PLR is due to the asymmetric traffic distribution in which many optical packets are lost, especially at output links 3, 5, and 6, respectively, with traffic probability distributions of 0.29, 0.19, and 0.22.

Figure 4.14d depicts PLR under symmetric traffic distribution at $f = 2$, $N = 8$, $r_{wc} \in \{0,2,4\}$, where the analysis results provided in this section (denoted by Ana1) well match with the analysis results in [34] (displayed with Ana2). The analysis results also follow well the simulation results. As expected, we have the lowest PLR under symmetric traffic distribution when $p_k = \frac{1}{8}$ compared to the asymmetric traffic distribution. For example, according to the simulation and analysis results at $L = 0.9$, the smallest PLR of 0.092 is obtained under symmetric traffic at $f = 2$ and $r_{wc} = 6$ (comparable to PLR of 0.296 under asymmetric traffic). The worst-case analysis (i.e., the whole traffic is destined only to one output link of the core switch) shows a PLR of 0.821 at $f = 2$ and $r_{wc} = 6$.

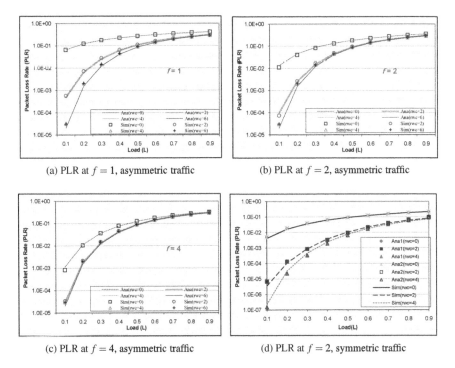

(a) PLR at $f = 1$, asymmetric traffic

(b) PLR at $f = 2$, asymmetric traffic

(c) PLR at $f = 4$, asymmetric traffic

(d) PLR at $f = 2$, symmetric traffic

Figure 4.14: PLR in a core switch with asymmetric and symmetric traffic distribution. Reprinted from *Optics Communications*, Vol. 124, Akbar Ghaffarpour Rahbar, Analysis of optical packet loss rate under asymmetric traffic distribution in multi-fiber synchronous OPS switches, 769–772, copyright 2013, with permission from Elsevier.

As proved, the provided analysis can accurately obtain PLR under both asymmetric and symmetric traffic distributions. The performance evaluation results show that the symmetric traffic can significantly reduce PLR compared with the asymmetric traffic case.

4.2.4   Hybrid MF + SPOL WCs + FDL Architecture

The combination of multi-fiber architecture, shared-per-output link WCs, and FDLs in an OPS core switch is presented in Fig. 4.15 [35], where r_{wc} converters are used in each conversion bank dedicated to each output link. The optical switch space stage 1 is a $(N \times f \times W) * (N \times f \times W)$ switch that chooses the proper output link for switching an incoming optical packet. The optical packet may be sent either directly or through wavelength conversion (if it needs conversion) to the relevant optical switch space stage 2. The second space switch picks the right output fiber within the relevant output link for a given optical packet. The second stage switch is provided with a small buffer to provide additional contention resolution in time domain. By this, the optical packet can be delayed before being multiplexed with other optical packets on other wavelengths.

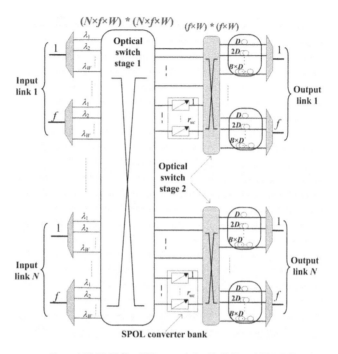

Figure 4.15: Multi-fiber OPS core switch with SPOL and FDL [35]

This architecture not only resolves contention in both wavelength and time domains, but also allows to reuse the same wavelength on different output ports of the same output link due to the multi-fiber architecture. Two different contention resolution schemes can be used for this architecture in a given output link:

- **Wavelength Before Time (WT):** Under WT, contention resolution is first performed in the wavelength domain and then the time domain. Here, for any arriving optical packet, the control unit of the core switch first checks whether it can be switched without wavelength conversion. If there are some fibers in the output link that can accept this optical packet, it is simply switched. Otherwise, the unallocated wavelengths in the output link are considered for conversion of contending optical packets. After wavelength conversion, some contending optical packets may need some delay to be successfully switched in the core switch.

- **Time Before Wavelength (TW):** Under TW, contention resolution is first applied in the time domain to the whole set of f fibers of the output link and then the proper wavelength and fiber are chosen. A Wavelength and Delay Selection (WDS) algorithm [36] is used. When an optical packet arrives at the core switch, the control unit checks among all the wavelengths of the output link and chooses the one that minimizes the gap in the corresponding FDL (the MING algorithm).

This architecture can be used to support differentiation of optical packets by converter reservation. Assume two optical packet classes with different priorities have to be served in OPS network as High-Priority (HP) and Low-Priority (LP). Let C_1 wavelength converters be reserved for HP optical packets. Upon the arrival of an optical packet, the control unit first checks whether it belongs to the HP or LP class. Assume there are S free WCs and $S \leq C_1$. In this case, an HP optical packet can be converted, but an LP optical packet cannot be converted. Otherwise, when $S > C_1$, there is no restriction for the LP packets, and they can be converted by any WC.

Besides reserving WCs, the output wavelengths with their associated FDLs can be reserved for HP optical packets. Define C_2 to be the number of wavelengths reserved for HP optical packets. Let the control unit find out that U wavelengths are free in the corresponding FDL when processing an optical packet. If $U \leq C_2$, only HP optical packets are served whereas the LP optical packets are dropped. Clearly, we should have $C_2 < f \times W$. Hence, a joint reservation of WCs and output wavelengths can optimize the design of the quality of service for the two classes.

Performance evaluation results under the same buffer depth show that for a small number of WCs, the WT contention resolution results in lower PLR than the TW contention resolution. However, when using more converters, TW outperforms WT in PLR performance.

4.2.5   Hybrid DA + M_PR + SPN WC Architecture

In [37], ingress switches use the DA scheme for traffic transmission in a single-fiber slotted OPS. On the other hand, each core switch uses the Modified PR (M_PR) (see Section 3.2.1.3.2) and $N_{WC} = r_{wc}$ shared-per-node wavelength converters in total for contention resolution purposes. Figure 4.16 illustrates the contention resolution algorithm in an OPS core switch that includes three phases:

- Phase 1: It simply resolves all contentions occurred on each wavelength channel at each output port. The optical packets in set $S_{k,i,m}$ are first sorted using the Quick Sort algorithm in a descending order according to their priority values δ (see 3.2.1.3.2). Then, the algorithm selects one optical packet from the top of the sorted list, schedules it for transmission on wavelengths λ_i at output port k, and updates the number of items in the list (i.e., m). When there is an unused wavelength on a fiber of an output port, the information of this wavelength including output port number k and wavelength number i are added to list U. At the end of Phase 1, the optical packets remaining in different sets $S_{k,i,m}$ are all unresolved optical packets and the wavelengths saved in list U are all unused.

- Phase 2 uses N_{WC} SPN WCs to resolve the contention among the unresolved collisions at Phase 1. First, a randomizing function is applied on list U to randomly change the position of the elements in this list. Then, the elements in list U are evaluated sequentially where for each element from this list a pair (k,i) is obtained. Then, list $S_{k,l,m}$ is found in a way that $m > 0$. In this case, a WC is allocated to output port k in order to schedule an optical packet from the top of

Algorithm: Contention Resolution in DA+MP+SPN
Definitions:

$S_{k,i,m}$:   set of m contending optical packets on wavelength λ_i at output port k

Phase 1 of Contention Resolution:
1: **for** $k = 0$ to $N - 1$ {scan all N output ports} **do**
2:     **for** $i = 0$ to $W - 1$ {scan all wavelengths on port k} **do**
3:         Make set $S_{k,i,m}$
4:         **if** $m = 0$ {no optical packet contending for wavelength i at output port k} **then**
5:             Add pair (i,k) to list U {save unused wavelength i in list U}
6:         **else**
7:             {m optical packets contending for wavelength i on port k}
8:             Sort optical packets in $S_{k,i,m}$ {according to the priority field}
9:             Select one optical packet from top of $S_{k,i,m}$ {for scheduling on wavelength i}
10:            Set $m = m - 1$ {update m in list $S_{k,i,m}$}
11:         **end if**
12:     **end for**
13: **end for**

Phase 2 of Contention Resolution:
1: $N_u = 0$ {initial the number of used WCs}
2: Randomize list U {randomize list U}
3: **for** $j = 1$ to $sizeof(U)$ {scan all unused wavelengths in list U} **do**
4:     **if** $N_u = N_{WC}$ **then**
5:         Exit loop {end of phase 2 if all converters are used}
6:     **end if**
7:     Get pair (i,k) from $U(j)$ {get information of an unused wavelength}
8:     **for** $l = 0$ to $W - 1$ {scan all wavelengths on port k} **do**
9:         Consider list $S_{k,l,m}$
10:        **if** $m = 0$ **then**
11:            continue {evaluate the next wavelength}
12:        **end if**
13:        Schedule the optical packet from top of $S_{k,l,m}$ for sending on wavelength i {after using one WC}
14:        Set $N_u = N_u + 1$ {update the number of used WCs}
15:        Set $m = m - 1$ {reduce the number of contending optical packets}
16:        Exit loop {evaluate the next unused wavelength}
17:     **end for**
18: **end for**

Phase 3 of Contention Resolution:
1: **for** $k = 0$ to $N - 1$ {scan all N ports} **do**
2:     **for** $i = 0$ to $W - 1$ {scan all wavelengths on port k} **do**
3:         Consider $S_{k,i,m}$ {consider contending optical packets for wavelength i on port k}
4:         **for** $r = 1$ to m {scan all contending optical packets from set $S_{k,i,m}$} **do**
5:             $SrcNode$ = Source of optical packet OP_r {get ingress switch address of OP_r from $S_{k,i,m}$}
6:             Send a NACK command to $SrcNode$ {inform $SrcNode$ for retransmission}
7:         **end for**
8:     **end for**
9: **end for**
10: END

Figure 4.16: Pseudo code of contention resolution in a core switch under DA + M_PR + SPN. Reprinted from Akbar Ghaffar Pour Rahbar, Improving throughput of long-hop TCP connections in IP over OPS networks, *Photonic Network Communications* 2009, 17:226–237, Figure 2, with kind permission from Springer Science and Business Media.

set $S_{k,l,m}$ for transmission on wavelength λ_i after wavelength conversion from wavelength λ_l. Phase 2 is over whenever all WCs are used (i.e., $N_u = N_{WC}$).

- Phase 3 announces each source ingress switch to retransmit its dropped optical packets. For dropped optical packet OP_r, the core switch informs its ingress switch address with a NACK command including the dropped packet identification code (under both PR and RR). If there are more than one dropped optical packet for a particular ingress switch at a time slot, the information of all dropped optical packets are encapsulated in only one NACK command and then sent back to the ingress switch, thus reducing the volume of the NACK commands in the OPS network.

An eight-node network with $W = 4$ wavelengths per fiber and diameter of 3 is first considered [37]. Considering $S_T = 9\mu s$ and $S_O = 1\mu s$, an integer number of IP packets up to 22,500 bits can be aggregated in an optical packet. The TCP sources and sinks connected to each edge switch use TCP Reno with an average IP packet size of 576 bytes. The window size of each TCP sender/receiver is at most 1000 IP packets. A set of 700 TCP source nodes and a 700 TCP sink nodes are connected to each edge switch so that there are 100 TCP connections between any ingress and egress switch pair in the OPS network.

In the following, three types of performance evaluations are performed under DA+M_PR+SPN, DA+RR+SPN (that uses simple random retransmission technique), and DA+NR+SPN (that uses No Retransmission (NR)). The contention resolution under DA+NR+SPN and DA+RR+SPN are similar to Fig. 4.16, but with some slight changes. For DA+NR+SPN, there is no retransmission management phase (i.e., Phase 3). Under both DA+NR+SPN and DA+RR+SPN, the sorting operation in Phase 1 is replaced with a randomizing function.

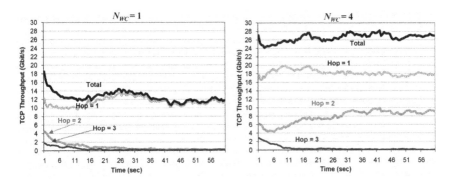

Figure 4.17: TCP throughput under DA + NR + SPN combination at $N_{WC} = 1$ and $N_{WC} = 4$. Reprinted from Akbar Ghaffar Pour Rahbar, Improving throughput of long-hop TCP connections in IP over OPS networks, *Photonic Network Communications* 2009, 17:226–237, Figures 3 and 5, with kind permission from Springer Science and Business Media.

For DA+NR+SPN, Fig. 4.17 shows TCP throughput of 1-hop connections, 2-hop connections, 3-hop connections, and the total network throughput under $N_{WC} = 1$ and $N_{WC} = 4$. Define an i-hop TCP connection as the connection in which there are i hops between the ingress–egress switch pair of the TCP connection. Note that the average TCP throughput for all i-hop TCP connections (where $i = 1, 2, 3$) is displayed with Hop = i in the figures. As can be observed, TCP throughput goes up by increasing the number of SPN WCs. However, 3-hop connections experience the lowest throughput due to the high PLR for long-hop connections, whereas 1-hop connections greatly utilize the network bandwidth. As the number of hops increases for a connection between an ingress–egress switch pair, PLR increases for those optical packets that belong to this long-hop connection, and therefore TCP throughput of long-hop connections reduces compared with short-hop connections. This shows some sort of unfairness for long-hop TCP connections.

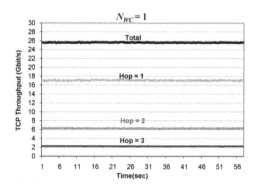

Figure 4.18: TCP throughput under DA + RR + SPN combination at $N_{WC} = 1$. Reprinted from Akbar Ghaffar Pour Rahbar, Improving throughput of long-hop TCP connections in IP over OPS networks, *Photonic Network Communications* 2009, 17:226–237, Figure 6, with kind permission from Springer Science and Business Media.

For DA+RR+SPN, Fig. 4.18 shows TCP throughput of 1-hop connections, 2-hop connections, 3-hop connections, and the total network throughput under $N_{WC} = 1$. Using only one SPN WC, the TCP throughput of DA+RR+SPN becomes comparable with DA+NR+SPN at $N_{WC} = 4$. In other words, using the RR technique in DA+RR+SPN can significantly reduce PLR and increase TCP throughput. Further performance evaluations show that the number of TCP timeout events is almost zero after elapsing few seconds from the beginning of the simulation. This is because the RR in the optical domain handles most of the retransmissions, and TCP sources are less aware of traffic loss in the OPS network. The DA+RR+SPN has noticeably increased the throughput of long-hop connections compared with DA+NR+SPN.

For DA+M_PR+SPN, Fig. 4.19 depicts TCP throughput of 1-hop connections, 2-hop connections, 3-hop connections, and the total network throughput under $N_{WC} = 1$ and $N_{WC} = 4$ for $\alpha = \{1, 2, 3\}$, where α is a coefficient in the prioritized retransmission technique which is used to set an initial priority of an optical packet. For the role of parameter α in M_PR, refer to Section 3.2.1.3.2. Unlike RR, throughputs of different TCP connections have become closer to each other under M_PR so that TCP throughputs of 3-hop connections and 2-hop connections have increased, and TCP throughput of 1-hop connections have decreased. In addition, total TCP throughput is a little higher in M_PR than in RR. At $\alpha = 2$, TCP throughputs of two-hop and three-hop connections are very close to each other compared with $\alpha = 1$. At $\alpha = 3$, TCP throughputs of two-hop and three-hop connections become very close to each other so that TCP throughputs of three-hop connections is even better than two-hop connections. Using $\alpha = 3$, the total TCP throughput has also increased by 1 Gbits/s compared with $\alpha = 1$.

Additional performance evaluations under the PacNet topology with diameter 5 shows the same behavior as discussed for the network with diameter 3. By using

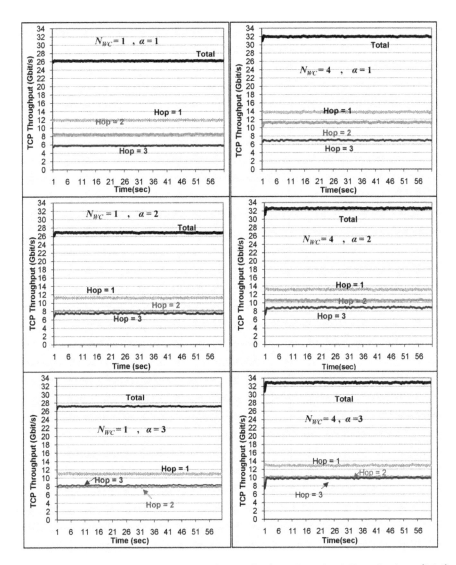

Figure 4.19: TCP throughput under DA + M_PR + SPN combination at $N_{WC} = 1$ and $N_{WC} = 4$ and $\alpha = \{1,2,3\}$. Reprinted from Akbar Ghaffar Pour Rahbar, Improving throughput of long-hop TCP connections in IP over OPS networks, *Photonic Network Communications* 2009, 17:226–237, Figures 7 to 12, with kind permission from Springer Science and Business Media.

M_PR, the throughputs of 5-hop, 4-hop, 3-hop, 2-hop, and 1-hop connections become closer to each other, especially at $\alpha = 3$. However, total TCP throughput has reduced by 0.93 Gbits/s at $\alpha = 3$ compared with $\alpha = 1$. Therefore, $\alpha = 2$ could be a good choice to approach reasonable TCP throughput performance for both small and large networks.

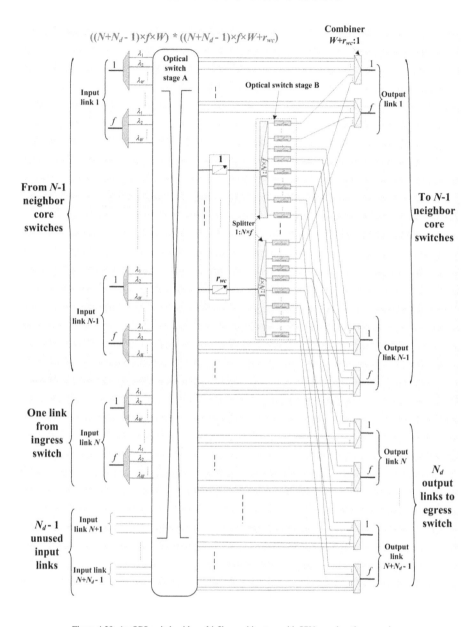

Figure 4.20: An OPS switch with multi-fiber architecture with SPN wavelength conversion

4.2.6 Hybrid DA + MF + PR + NBR + SPN WC Architecture

In [29], two contention avoidance schemes (DA + multi-fiber architecture) are combined with three contention resolution schemes (non-blocking receiver, prioritized retransmission, and shared-per-node wavelength conversion) in slotted OPS networks. Consider that core switch z is connected to $N-1$ neighbor core switches. It is also connected to its local edge switch with one add link and $N_{d,z}$ drop links. For simplicity, we show $N_{d,z}$ with N_d in the following discussions. Since each link has f fibers, there are $f \times N_d$ drop ports in total from the core switch to its local edge switch. Figure 4.20 shows the multi-fiber OPS core switch with $N_{WC} = r_{wc}$ SPN wavelength converters. Space switch A is a $((N+N_d-1) \times f \times W) * ((N+N_d-1) \times f \times W + r_{wc})$ that forwards optical packets either directly or through wavelength conversion to the $(W+r_{wc}):1$ combiners. A $1:(N \times f)$ splitter used at the output of each WC splits an optical packet to $(N+N_d-1) \times f$ signals. Switching stage B switches the optical packet to one of the $(N+N_d-1) \times f$ output ports, which is the destination of the optical packet. One additional wavelength is used on each fiber for transmitting header of optical packets (see Aggregated Header in Section 1.4.4).

4.2.6.1   Scheduling in DA + MF + PR + NBR + SPN WC Architecture   Let the

vector $(Src_j, Dst_j, ID_j, \delta_j)$ denote four parameters carried in the header sent for optical packet OP_j, where Src_j is the ingress switch address, Dst_j is the egress switch address, ID_j is the optical packet identification, and δ_j is the priority of OP_j. Based on the egress switch addresses of the optical packets arriving at its input ports, the core switch groups the optical packets to send on its output links. Now, let list $S_{k,i,m} = \{OP_0, OP_1, ..., OP_{m-1} | 1 \leq m \leq N \times f, 0 \leq k \leq N-1\}$ denote the list of m contending optical packets on wavelength λ_i (where $i = 0, ..., W-1$) at the given output link k. Note that all N_d drop links are considered as one output link in the scheduling procedure as they are all connected to the same egress switch.

Figure 4.21 depicts the pseudo code of contention resolution in a core switch under the DA + MF + PR + NBR + SPN mechanism. This algorithm has three phases:

- Phase 1 (shown with Contention Resolution) is performed for any wavelength on each output link, where the optical packets in list $S_{k,i,m}$ are first sorted either using the Quick Sort algorithm or the Heap Sort algorithm in a descending order according to their priority values. Then, the algorithm picks $m_1 = \min(f, m)$ optical packets from the top of the sorted list, schedules them for transmission on wavelengths λ_i of output link k (note that there are f wavelengths λ_i on link k), and updates the number of items m in the list accordingly. Then, if output link k is not connected to the local edge switch, the contention resolution for wavelength λ_i is over. Otherwise, the algorithm selects $m_2 = \min(f \times (N_d - 1), m - m_1)$ optical packets from the remaining unresolved optical packets in the sorted list, schedules them for transmission on $N_d - 1$ additional output links, and updates parameter m of $S_{k,i,m}$ accordingly. In any case, when there is an unused wavelength on a fiber of an output link, the information of this wavelength including the output link number (k), the fiber number (b), and the wavelength number (i) is added to list U. At the end of Phase 1, the optical packets remaining in all lists $S_{v,i,m}$ (where $0 \leq v \leq N-1$) are all unresolved optical packets and the wavelengths in list U are all unused. List U is required in Phase 2 for contention resolution using WCs.

Algorithm: Contention Resolution in DA + MF + PR + NBR + SPN
Phase 1 of Contention Resolution:

1: **for** $k = 0$ to $N - 1$ {scan all N output links} **do**
2:    **for** $i = 0$ to $W - 1$ {scan all wavelengths on link k} **do**
3:       Provide set $S_{k,i,m}$
4:       **if** $m = 0$ {no optical packet contending for wavelength λ_i at output link k} **then**
5:          For all unused fibers b, add (k,b,i) to list U {save unused wavelength λ_i in list U}
6:       **else**
7:          {m optical packets contending for wavelength λ_i on link k}
8:          Sort optical packets in $S_{k,i,m}$ {according to the priority field}
9:          Schedule $m_1 = \min(f,m)$ optical packets from top of $S_{k,i,m}$ for scheduling on λ_i at link k
10:          **if** $m_1 < f$ **then**
11:             For all unused fibers b, add (k,b,i) to list U {save unused wavelength λ_i in list U}
12:          **else**
13:             Set $m = m - m_1$ for list $S_{k,i,m}$ {update m in list $S_{k,i,m}$}
14:             **if** link k is connected to local edge switch **then**
15:                Schedule $m_2 = \min(f \times (N_d - 1),m)$ optical packets from top of $S_{k,i,m}$ for scheduling on $N_d - 1$ additional output links connected to local edge switch
16:                Set $m = m - m_2$ for list $S_{k,i,m}$ {update m in list $S_{k,i,m}$}
17:             **end if**
18:          **end if**
19:       **end if**
20:    **end for**
21: **end for**

Phase 2 of Contention Resolution (Resolution using
WCs):

1: $N_u = 0$ {initial the number of used WCs}
2: Randomize list U {randomize list U}
3: **for** $j = 1$ to $sizeof(U)$ {scan all unused wavelengths in list U} **do**
4:    **if** $N_u = N_{WC}$ **then**
5:       Exit loop {end of phase 2 if all converters are used}
6:    **end if**
7:    Get triple (k,b,i) from $U(j)$ {get the information of an unused wavelength}
8:    **for** $l = 0$ to $W - 1$ {scan all wavelengths on link k} **do**
9:       Consider list $S_{k,l,m}$
10:       **if** $m > 0$ **then**
11:          Schedule the top optical packet of $S_{k,l,m}$ on link k, fiber b, wavelength λ_i after using a WC
12:          Set $N_u = N_u + 1$ {update the number of used WCs}
13:          Set $m = m - 1$ for list $S_{k,l,m}$ {reduce the number of contending optical packets}
14:          Exit loop {evaluate the next unused wavelength}
15:       **end if**
16:    **end for**
17: **end for**
18: Make new header for each output link

Phase 3 of Contention Resolution (Retransmission Management):

1: **for** $k = 0$ to $N - 1$ {scan all N links} **do**
2:    **for** $i = 0$ to $W - 1$ {scan all wavelengths on link k} **do**
3:       Consider list $S_{k,i,m}$ {consider contending optical packets for wavelength λ_i on link k}
4:       **for** $j = 1$ to m {scan all contending optical packets from set $S_{k,i,m}$} **do**
5:          $SrcNode$ = Source of optical packet OP_j {get ingress switch address OP_j from $S_{k,i,m}$}
6:          Send a NACK command to $SrcNode$ {inform $SrcNode$ for retransmission}
7:       **end for**
8:    **end for**
9: **end for**
10: END

Figure 4.21: Pseudo code of contention resolution under the DA + MF + PR + NBR + SPN mechanism. Reproduced from *Optics Communications*, Vol. 282, Akbar Ghaffarpour Rahbar, Cost-effective combination of contention resolution/avoidance schemes in bufferless slotted OPS networks, 798–808, copyright 2009, with permission from Elsevier.

- Phase 2 (displayed with Resolution using WCs) is executed for all output links because the WCs are shared-per-node. First, a randomizing function is applied on list U to randomly change the positions of the elements in this list. Then, the elements in list U are evaluated sequentially where for each element from this list a triple (k, b, i) is obtained. Then, list $S_{k,l,m}$ is found randomly in such a way that $m > 0$. In this case, one WC is allocated to fiber b of link k in order to schedule an optical packet from the top of list $S_{k,l,m}$ for transmission on wavelength λ_i after wavelength conversion from wavelength λ_l. Phase 2 ends whenever N_u (i.e., the number of used WCs) equals N_{WC}. At the end of this phase, a new header is made for the optical packets departing from each output link in order to be transmitted on the control channel of the output link toward either the next hop core switch or the next hop egress switch. Under both Phase 1 and Phase 2, an optical packet with a high priority always finds a higher opportunity to pass through each core switch.

- In Phase 3 (i.e., Retransmission Management), the information of dropped optical packets are sent back to their source ingress switches. Note that when optical packet OP_j is dropped, the core switch informs Src_j with a NACK command (under PR). In the NACK command, ID_j is sent back to Src_j to identify the dropped optical packet OP_j. Note that if there are more than one dropped optical packet belonging to a particular ingress switch, the information of all dropped optical packets are encapsulated in only one NACK command and then sent back to Src_j, thus reducing the volume of the NACK commands in the OPS network.

The above algorithm can be slightly changed to be used under NR (i.e., no retransmission in the optical domain) so that instead of the sorting procedure used for PR in Phase 1, a randomizing function is applied on list $S_{k,i,m}$ to randomly change the position of optical packets in list $S_{k,i,m}$. Then, optical packets are randomly chosen for transmission from the randomized list $S_{k,i,m}$. In addition, Phase 3 is not necessary under NR.

4.2.6.2   Performance Evaluation in DA + MF + PR + NBR + SPN WC Arc hitecture   The network under simulation is the PacNet with $n = 15$ core switches and edge switches under slotted OPS operation. There are f fibers on each direction of a connection link, where each fiber carries W wavelength channels (each with bandwidth $B_c = 2.5$ Gbits/s) so that total number of available wavelength channels required for each ingress switch is $\omega = f \times W = 4$. There are 350 TCP source nodes and 350 TCP sink nodes connected to each edge switch. In each ingress switch, all the TCP streams are equally divided among fourteen other egress switches in the network. Hence, there are 350/14 = 25 TCP connections (referred to as a group connection) between any ingress and egress switch pair. The TCP group connections can be grouped based on their hops as single-hop connections, two-hop connections, three-hop connections, and so on. An average IP packet size of 576 bytes is used in the TCP experiments performed in this section. The maximum window size of each TCP sender/receiver is set to 576 Kbytes. In an ingress switch, each buffer can

save at most 5.76 MBytes for each egress switch. Setting $S_T = 900$ ns and $S_O = 100$ ns, an integer number of IP packets up to 22.5 kbits can be aggregated in an optical packet and transmitted to the network.

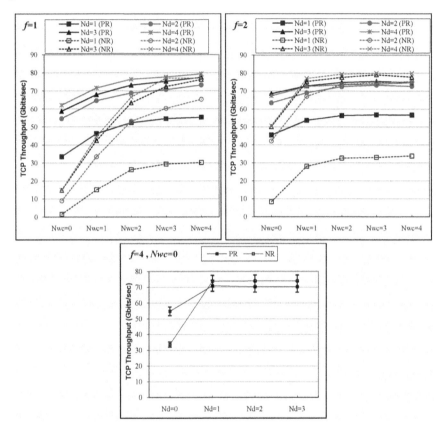

Figure 4.22: Network-wide TCP throughput under the DA + MF + PR + NBR + SPN mechanism. Reprinted from *Optics Communications*, Vol. 282, Akbar Ghaffarpour Rahbar, Cost-effective combination of contention resolution/avoidance schemes in bufferless slotted OPS networks, 798–808, copyright 2009, with permission from Elsevier.

Figure 4.22 displays the average network-wide TCP throughput under different values of f fibers, N_d drop links, N_{WC} SPN wavelength converters, and retransmission types of PR and NR. At $f = 1$, network TCP throughput goes up by increasing N_{WC}. However, at $N_{WC} \geq 3$, there is no significant increase for the throughput. Similarly, network TCP throughput increases when using additional drop links in each core switch. However, the rate of this increase reduces after $N_d \geq 3$. In the $f = 1$ diagram, it can be observed that PR can always provide significantly higher throughput than NR. At $f = 2$, the throughput goes up similar to $f = 1$. However, when using both $N_d \geq 3$ and $N_{WC} \geq 1$, NR slightly outperforms PR because some part of the network bandwidth is allocated for retransmitting dropped optical packets under PR. At $f = 4$, the throughput increases similar to $f = 2$ so that NR slightly outperforms PR at $N_d \geq 2$.

Figure 4.22 also shows cost-effective combinations of various scenarios. Consider the original OPS network at $f = 1$, $N_d = 1$, and $N_{WC} = 0$ under NR. Here, the network-wide TCP throughput is around 1.59 Gbits/s. In this case, using $N_{WC} = 4$ and no additional drop links at each core switch, the throughput becomes around 30.3 Gbits/s. On the other hand, when using no WCs and $N_d = 4$ drop links at each core switch, the throughput is around 14.9 Gbits/s. The throughput is around 33.5 Gbits/s under $(PR, N_d = 1, N_{WC} = 0)$. This shows that PR by itself can improve TCP throughput more than using any other hardware-based scheme. The TCP throughput goes up around 79.4 Gbits/s when using $(PR, N_d = 4, N_{WC} = 4)$. Since this combination may not be so cost-effective, one can choose the combination of say $(PR, N_d = 3, N_{WC} = 2)$ to obtain throughput of around 73.2 Gbits/s. At the multi-fiber OPS architecture with $f = 2$, the cost-effective combination could be under either using $(PR, N_d = 3, N_{WC} = 1)$, where throughput is 72.9 Gbits/s, or $(NR, N_d = 3, N_{WC} = 1)$, where throughput is 75.4 Gbits/s. Finally, the cost-effective combination at the multi-fiber OPS architecture with $f = 4$ could be under $(NR, N_d = 2, N_{WC} = 0)$, where throughput reaches up to 73.8 Gbits/s.

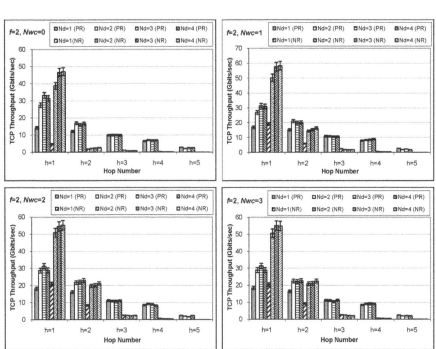

Figure 4.23: Per-hop TCP throughput under the DA + MF + PR + NBR + SPN mechanism. Reprinted from *Optics Communications*, Vol. 282, Akbar Ghaffarpour Rahbar, Cost-effective combination of contention resolution/avoidance schemes in bufferless slotted OPS networks, 798–808, copyright 2009, with permission from Elsevier.

Figure 4.23 illustrates per-hop TCP throughput under the DA + MF + PR + NBR + SPN mechanism. Under NR, TCP throughput significantly reduces by increasing the number of hops (i.e., h) so that 5-hop connections can only obtain a TCP throughput of almost 0.07 Gbits/s in the best case, whereas 1-hop connections obtain a TCP throughput of almost 58 Gbits/s in the best case. On the other hand, PR can always provide a better distribution (i.e., much fairer) of TCP throughput than NR among h-hop ($1 \le h \le 5$) connections.

4.2.6.3 Network Cost Evaluation in DA + MF + PR + NBR + SPN WC Architecture

Now, the cost of OPS network is evaluated to compare the cost of different network scenarios with each other. Since the aim is to achieve a desired TCP throughput but at the expense of a reasonable network cost, one should find out the combination that would give the best performance while minimizing the hardware cost. Assume that the base OPS network to be designed requires ω wavelength channels on each connection link between any edge-to-core and core-to-core switch to carry the full traffic load on each wavelength channel. Let the configuration parameters of the base network scenario be $f = 1$, $W = \omega$, $N_{WC} = 0$, and $N_d = 1$.

First, the cost of a core switch is computed in general. Consider that a given core switch within the network is connected to k neighbor core switches. Hence, the number of input ports and the number of output ports required in the core switch is $N = k \times f + f \times N_d$. The costs of different components relevant to the core switch are as follows:

- Wavelength Conversion Cost (C_{WC}): Since there are N_{WC} shared-per-node WCs at each core switch, the relevant cost is computed by $C_{WC}(N_{WC}) = c_{WC} \times N_{WC}$, where c_{WC} is the cost of one wavelength converter.

- Multiplexing/Demultiplexing Cost (C_{MUX}): The core switch has N units of 1 * ($\frac{\omega}{f}$) wavelength demultiplexing modules and N units of ($\frac{\omega}{f}$) * 1 wavelength multiplexing modules. Wavelength demultiplexers and multiplexers can be constructed using cascaded Multilayer Interference (MI) filters [38]. By assuming the same cost for multiplexing and demultiplexing modules, the total number of the multiplexing and demultiplexing modules required in the core switch is $2 \times N$. Since ($\frac{\omega}{f} - 1$) MI filters are required to construct one multiplexing/demultiplexing module with $\frac{\omega}{f}$ input wavelength channels [38], the cost of multiplexing/demultiplexing modules inside the core switch is given by $C_{MUX}(f, N_d) = 2 \times N \times c_{MI} \times (\frac{\omega}{f} - 1)$, where c_{MI} is the cost of one MI filter.

- Switching Cost (C_{SW}): There are $\frac{\omega}{f}$ wavelength switches inside the core switch each with size $N * N$. Let each wavelength switch be a non-blocking, where $G(N)$ is the number of 2*2 switching elements used inside a wavelength switch. Then, the cost of switching elements inside the core switch is $C_{SW}(f, N_d) = (\frac{\omega}{f}) \times c_{2*2} \times G(N)$, where c_{2*2} is the cost of one 2*2 switching element. For instance, the number of 2*2 switching elements used in a non-blocking Batcher–Banyan switch of size $N * N$ is $G(N) = N \times \frac{(3 + \log_2(N)) \times (\log_2(N))}{4}$ [39].

- Core Switch Cost (C_{OXC}): the cost of core switch is

$$C_{OXC}(f,N_d,N_{WC}) = C_{SW}(f,N_d) + C_{MUX}(f,N_d) + C_{WC}(N_{WC}). \qquad (4.10)$$

The network cost is calculated by summing the cost of n core switches and fibers in the network:

$$C_{net}(f,N_d,N_{WC}) = \sum_{i=1}^{n} C_{OXC,i}(f,N_d,N_{WC}) + 2 \times f \times c_f \times \sum_{j=1}^{N_e} e_j. \qquad (4.11)$$

where $C_{OXC,i}(f,N_d,N_{WC})$ is the cost of core switch i obtained from Eq. (4.10), N_e is the number of links in the whole network, and e_j is the length of link j in the network in kilometers. Note that $2 \times f$ for the fiber cost accounts for f bi-directional links on each edge of the network topology under study. The parameter c_f is the cost of providing fibers per kilometer by telecommunication companies. Define $\delta(f,N_d,N_{WC})$ to be the rate of cost increase of the network scenario with respect to the base network $(C_{net}(1,1,0))$:

$$\delta(f,N_d,N_{WC}) = \frac{C_{net}(f,N_d,N_{WC}) - C_{net}(1,1,0)}{C_{net}(1,1,0)}. \qquad (4.12)$$

Define $U_{net}(R,f,N_d,N_{WC})$ to be the network-wide TCP throughput under the retransmission type of R (where the parameter R could be either NR or PR). Then, the network-wide TCP throughput of the base network is given by $U_{net}(NR,1,1,0)$. Then, the rate of TCP throughput increase of the network scenario with respect to the base network is computed by

$$\Delta(R,f,N_d,N_{WC}) = \frac{U_{net}(R,f,N_d,N_{WC}) - U_{net}(NR,1,1,0)}{U_{net}(NR,1,1,0)}. \qquad (4.13)$$

Define $\theta_T \geq 1$ to be the minimum (threshold) criterion for investing on the network hardware. Setting $\theta_T = 10$, for example, the network is cost-effective if the increase rate of TCP throughput is at least 10 times of the increase rate of the hardware cost. Now, the hardware used for increasing the network-wide TCP throughput is cost-effective if

$$\theta(R,f,N_d,N_{WC}) = \frac{\Delta(R,f,N_d,N_{WC})}{\delta(f,N_d,N_{WC})} \geq \theta_T. \qquad (4.14)$$

Let us set relative cost parameters to $c_{MI} = 1$, $c_f = 1$, $c_{2*2} = 10$, and $c_{WC} = 500$. Then, considering that each wavelength switch inside any core switch uses the Batcher–Banyan architecture, the tradeoff between network cost and TCP throughput is studied. Tables 4.1 and 4.2 show $\theta(NR,f,N_d,N_{WC})$ and $\theta(PR,f,N_d,N_{WC})$, respectively. These tables highlight the combinations which result in an acceptable θ at $\theta_T = 15$. One can observe that by increasing the number of hardware components toward higher values, θ reduces because not much TCP throughput gain can be achieved. Therefore, using a large number of hardware components may not be so cost-effective. At the last row of each table, the most cost-effective combinations can be observed.

Table 4.1: Study parameter $\theta(NR, f, N_d, N_{WC})$ under NR. Reprinted from *Optics Communications*, Vol. 282, Akbar Ghaffarpour Rahbar, Cost-effective combination of contention resolution/avoidance schemes in bufferless slotted OPS networks, 798–808, copyright 2009, with permission from Elsevier.

N_{WC}	$f = 1$ N_d				$f = 2$ N_d				$f = 4$ N_d			
	1	2	3	4	1	2	3	4	1	2	3	4
0	∞	∞	9.62	6.84	5.52	33.31	15.27	12.40	10.10	**22.77**	12.48	10.63
1	19.35	45.69	19.70	16.29	13.85	34.19	19.04	16.20	—	—	—	—
2	17.73	36.94	22.33	19.44	11.88	27.61	**16.62**	14.55	—	—	—	—
3	13.30	28.09	20.43	18.69	9.51	22.08	14.70	12.95	—	—	—	—
4	10.30	22.88	17.95	**15.88**	8.06	18.37	12.76	11.34	—	—	—	—
best-case	$N_{WC} = 4$, $N_d = 4$, Throughput = 77.11 Gbits/s, $\theta = 15.88$				$N_{WC} = 2$, $N_d = 3$, Throughput = 77.58 Gbits/s, $\theta = 16.62$				$N_{WC} = 0$, $N_d = 2$, Throughput = 73.84 Gbits/s, $\theta = 22.77$			

Table 4.2: Study parameter $\theta(PR, f, N_d, N_{WC})$ under PR. Reprinted from *Optics Communications*, Vol. 282, Akbar Ghaffarpour Rahbar, Cost-effective combination of contention resolution/avoidance schemes in bufferless slotted OPS networks, 798–808, copyright 2009, with permission from Elsevier.

N_{WC}	$f = 1$ N_d				$f = 2$ N_d				$f = 4$ N_d			
	1	2	3	4	1	2	3	4	1	2	3	4
0	∞	∞	41.57	31.05	36.00	50.78	21.08	16.63	16.77	**21.88**	11.85	10.09
1	64.11	90.46	32.00	26.46	27.22	35.27	18.40	15.26	—	—	—	—
2	36.40	48.26	25.86	22.41	20.98	27.12	**16.00**	13.43	—	—	—	—
3	25.36	33.08	21.27	18.85	16.69	21.67	14.01	11.95	—	—	—	—
4	19.28	25.71	18.24	**16.44**	13.74	17.72	12.25	10.68	—	—	—	—
best-case	$N_{WC} = 4$, $N_d = 4$, Throughput = 79.45 Gbits/s, $\theta = 16.44$				$N_{WC} = 2$, $N_d = 3$, Throughput = 74.75 Gbits/s, $\theta = 16.00$				$N_{WC} = 0$, $N_d = 2$, Throughput = 70.97 Gbits/s, $\theta = 21.88$			

4.2.7 Hybrid PA + MF + NBR + DRwBD + IAS + SHP + SPN WC Architecture

An intelligent impairment-aware scheduling in slotted OPS networks has been proposed in [40] with the following features. To reduce PLR, a number of contention resolution and avoidance schemes are combined in each OPS core switch; two contention avoidance schemes (i.e., Packet Aggregation (PA) and Multi-Fiber (MF) architecture) and five contention resolution schemes (i.e., Impairment-Aware Scheduling (IAS), Deflection Routing with Backward Deflection (DRwBD), Shortest Hop-Path routing (SHP), Non-Blocking Receiver (NBR), and shared-per-node WCs).

The scheduling in core switches is based on the method discussed in 3.2.3.3.4. The scheduler in a core switch schedules optical packets only when they have acceptable Polarization Mode Dispersion (PMD). Furthermore, the combined routing scheme based on both shortest hop and SHP is used to indirectly consider signal impairments in the network. The PA includes CPA, NCPA, and even DA bandwidth access schemes.

In this core switch architecture, an optical packet is dropped in three cases:

- when the optical packet finds an unacceptable PMD along the path toward its destination

- when there are no available resources to switch the optical packet in an OPS switch (either by wavelength conversion or by deflection routing)

- when the optical packet experiences a threshold on the number of deflections along its path.

There are N_{WC} Shared-Per-Node (SPN) WCs used at each core switch. The SPN is preferred over other sharing schemes due to its efficiency in reducing packet loss rate and not wasting WCs. Each WC performs limited-range conversion with conversion degree d_c. In multi-fiber OPS core switch m with size $(N \times f) * (N \times f)$, define set $S_{w,b,u} = \{OP_0, OP_1, ..., OP_{b-1} | 0 < b \leq N \times f\}$ to denote the set of indexes of b contending optical packets destined to output link u that will be arriving on wavelength λ_w (where $w = 0, ..., W - 1$). One additional wavelength is used on each fiber for transmitting header of optical packets (see Aggregated Header in Section 1.4.4). Core switch m has $N_{d,m}$ drop links (see Section 3.1.1.3) to edge switch m.

Using the scheduling algorithm depicted in Fig. 4.24, core switch m resolves the contention on its output links, but among those contending optical packets that have acceptable PMD (see Eq. (3.3) in Section 3.2.2). The rank parameter for an optical packet is the criterion in scheduling, where the optical packets with high ranks are routed through their regular shortest paths and other optical packets with low ranks are deflected.

The rank of optical packets are computed based on the method discussed in 3.2.3.3.4. For any given wavelength on each output link, the contention resolution procedure first switches the transit optical packets with high ranks through their desired paths toward their destinations on f fibers in a balanced way. If there are optical packets destined to local edge switch m, at most $N_{d,m} \times f$ of them can be delivered to the local edge switch on the given wavelength. The optical packets that cannot be switched are saved in set Y. Then, the index of optical packets in set Y is sorted based on their ranks in a descending order to resolve contention of optical packets in set Y using WCs. During conversion, the optical packets with the highest ranks are first switched through their normal paths toward their destinations. The optical packets that cannot be resolved are saved in set Z.

Note that set Z includes the index of already sorted optical packets that cannot be switched through their desired output links. Therefore, the scheduler can first deflect the highest rank optical packets among all output links of the core switch. Considering two optical packets with the same traffic and distance traveled, a previously deflected optical packet is routed toward its destination with a high probability compared with the optical packet without previous deflection. When deflecting OP_j

Algorithm: Contention resolution in core switch m

Step 1:   Drop optical packets with unacceptable PMD

Step 2:   Resolve contention at output link u

1: **for** each wavelength λ_w at output link u **do**
2:     Compute rank for each optical packet OP_j in set $S_{w,b,u}$
3:     Choose $\min(f,b)$ optical packets with the highest rank.
4:     Schedule the transmission of selected optical packets on wavelengths λ_w of $\min(f,b)$ output fibers, where output fibers are randomly chosen to provide load balancing on connection links.
5:     Append the index of unused wavelengths to set E_u
6:     Set $x = b - \min(f,b)$
7:     **if** output link u is connected to edge switch m **then**
8:         Switch $\min(x,(N_{d,m}-1) \times f)$ remaining optical packets on the additional drop links to edge switch m
9:     **end if**
10:     Append the index of the remaining optical packets that cannot be switched to set Y.
11: **end for**

Step 3:   Use SPN conversion for contention resolution

1: Sort the index of optical packets in set Y based on their ranks in a descending order
2: Resolve the conflicts for the optical packets in set Y using N_{WC} converters, where the optical packets with high ranks are first resolved. An unused wavelength for the optical packet supposed to be scheduled at output port u is searched from set E_u, while considering conversion degree.
3: The set of optical packets that cannot use WCs is called Z.

Step 4:   Use deflection routing for contention resolution

1: **for** $j = 1$ to Size (Z) **do**
2:     **if** d_j in $OP_j \geq D_{max}$ **then**
3:         Drop OP_j
4:     **else**
5:         $k = Dst_j$ in OP_j
6:         **for** each output link v in set $P_{m,k}$ **do**
7:             **if** λ_w on output link v is free **then**
8:                 **if** $PMD_j + PMD_v$ is acceptable **then**
9:                     $d_j = d_j + 1$
10:                     Schedule OP_j on λ_w at output link v
11:                     break loop
12:                 **end if**
13:             **else**
14:                 **if** wavelength $\lambda_i \neq \lambda_w$ available at output link v and WC available **then**
15:                     **if** $PMD_j + PMD_v$ is acceptable **then**
16:                         $d_j = d_j + 1$
17:                         Schedule OP_j at output link v on λ_i after conversion from λ_w
18:                         break loop
19:                     **end if**
20:                 **end if**
21:             **end if**
22:         **end for**
23:         **if** no schedule found for OP_j **then**
24:             Drop OP_j
25:         **end if**
26:     **end if**
27: **end for**

Step 5:   Update header in output link u

1: Make new header H_u for the optical packets departing from output link u considering the updated hop number, deflection number, and PMD.
2: Transmit H_u on the control channel of output link u to the next hop core or egress switch.
3: END

Figure 4.24: Pseudo code of optical packet scheduling in a core switch under the PA + MF + NBR + DRwBD + IAS + SHP + SPN mechanism. Reproduced from [40] with permission from De Gruyter.

at output link v, the PMD of OP_j should be acceptable when it arrives at the next hop core switch; i.e., the current PMD until core switch m (shown with PMD_j) plus additional PMD that will be experienced at output link v (shown with PMD_v) must be acceptable (see Eq. (3.3)). Otherwise, the deflection of OP_j will be useless. This is why the deflection algorithm at core switch m computes in advance the PMD parameter that OP_j will experience along the next hop (i.e., PMD_v). If the PMD of OP_j will be acceptable at the next hop core switch, it is deflected. Otherwise, another output link is picked or it is dropped. The DRwBD deflection mechanism used here allows backward deflection to the source core switch (see 3.2.1.1.1).

At the end of scheduling, an aggregated header is made for each output link u that includes the information of all optical packets departing from this output link. In the new aggregated header, the updated hop number, deflection number, and PMD of the optical packets should be inserted. Then, the aggregated header is sent over the control channel of output link u to the next hop core switch or egress switch.

To evaluate the complexity of the scheduling in core switch m, the worst-case scenario is considered that happens when there is an optical packet arriving from each input wavelength of the core switch and all the optical packets from N input links are destined to the same output link (i.e., totally $N \times f \times W$ optical packets). The complexities of five steps in Fig. 4.24 are as follows:

- **Step 1**: The PMD of at most $N \times f \times W$ optical packets are evaluated with the complexity of $O(N \times f \times W)$.

- **Step 2**: The loop is repeated W times. In each time, the complexity of computing the rank parameters is $O(N \times f)$ in the worst-case. The complexity of scheduling of at most f selected optical packets is $O(f)$. The complexity of switching optical packets to edge switch m is at most $O(f \times N_{d,m}) \leq O(N \times f)$ since we always have $N > N_{d,m}$. The complexity of adding the indexes of non-resolved optical packets to set Y is $O(N \times f - f) \leq O(N \times f)$. Having $O(1)$ complexity for other lines, the Step 2 complexity is $O(N \times f \times W)$.

- **Step 3**: Considering the $f \times W$ optical packets scheduled in the output link, the sorting should be executed on list Y with at most $(N - 1) \times f \times W$ items in the worst-case. In the worst-case, no WCs will be used since there would be no unused wavelength in set E_u. Therefore, using quick sort, the complexity of this step is $O((N - 1) \times f \times W \times \log_2((N - 1) \times f \times W)) \leq O(N \times f \times W \times \log_2(N \times f \times W))$.

- **Step 4**: The outer loop runs for $(N - 1) \times f \times W$ times since $Z = Y$. The inner loop runs at most for $N - 1$ times. Thus, the complexity of this step is $O((N - 1)^2 \times f \times W) \leq O(N^2 \times f \times W)$.

- **Step 5**: The complexity of this step is $O(f \times W)$ since at most $f \times W$ optical packets can be scheduled at the output link.

Therefore, the worst-case complexity of scheduling is $O(N \times f \times W \times (N + \log_2(N \times f \times W)))$, where the worst-case scenario might rarely happen in practice. In slotted OPS, according to this worst-case complexity, time-slot gap S_O should be chosen large enough for performing the scheduling process. Clearly, the processing time returns to the speed of processing elements used inside each output link of a core switch and the switch controller.

Now the performance of the PA + MF + NBR + DRwBD + IAS + SHP + SPN mechanism is evaluated in slotted OPS under mesh topologies using the OPNET simulator. Common network parameters for both scenarios are: $S_T = 9$ μs, $S_O = 1$ μs, and $D_{PMD} = 0.08$ on any link. In CPA and NCPA, the aggregated packet transmission buffer (APTB) size is taken large enough to avoid dropping of aggregated user packets in APTB buffers at ingress switches. Inside DA, the Loan-Grant-based Round Robin (LGRR) scheduling algorithm [41] is used with the following parameters: $C_1 = 1.0$, $C_2 = 0.6$, and $C_3 = 0.5$ (see [41] for definition of these parameters). The frame size for DA is set to 500 μs. The traffic in each ingress switch to any egress switch consists of 25%, 35%, and 40% of HP, MP, and LP traffic, respectively. The sizes of client packets are as follows: 18% of size 1500 bytes, 18% of size 576 bytes, 18% of size 552 bytes, and 46% of size 40 bytes. Two network scenarios are evaluated in the following:

- **Scenario 1**: The performances of CPA, NCPA, and DA are evaluated in an eight-node mesh network under the case that no deflection routing is performed in core switches and at $\alpha_0 = \alpha_1 = 0$ and $\alpha_2 = 1.0$ (see 3.2.3.3.4 for description of these parameters). The network parameters are $f = 2$, $W = 4$, $B_c = 10$ Gbits/s, and $N_{WC} = 8$ SPN WCs. Two different "significance" sets are evaluated as (see 3.2.3.3.4): Significance Set1 with ($V_{HP} = 0.6$, $V_{MP} = 0.3$, $V_{LP} = 0.1$) and Significance Set2 with ($V_{HP} = 0.5$, $V_{MP} = 0.3$, $V_{LP} = 0.2$). Under Significance Set1 and Significance Set2, NCPA, CPA, and DA results are represented with (NCPA1 and NCPA2), (CPA1, CPA2), and (DA1, DA2), respectively. Two different timeout values are also evaluated for NCPA and CPA: (1) Tout1 with $\{T_{HP} = 20$ μs, $T_{MP} = 40$ μs, $T_{LP} = 100$ μs $\}$ for $\{HP,MP,LP\}$ traffic, respectively; and (2) Tout2 with $\{T_{HP} = 15$ μs, $T_{MP} = 25$ μs, $T_{LP} = 50$ μs $\}$ for $\{HP,MP,LP\}$ traffic, respectively. Note that DA uses packet scheduling to aggregate user packets from different classes in an optical packet. Hence, performance results for DA are independent of timeout values.

Figure 4.25 depicts average packet loss rate for HP, MP, and LP client packets among all source–destination pairs within the network. Here, no optical packet is lost due to PMD impairment because of small wavelength channel bandwidth of $B_C = 10$ Gbits/s and network size. All diagrams show that the loss of HP traffic has increased from Significance Set1 to Significance Set2, whereas the loss of LP traffic has decreased from Significance Set1 to Significance Set2. This is because V_{LP} is smaller in Significance Set1 than in Significance Set2, and V_{HP} is larger in Significance Set1 than in Significance Set2.

Since L_{OP} (i.e., the optical packet generation rate over optical packet service rate in an ingress switch) is higher under NCPA compared with CPA (see Fig. 2.21), NCPA has a higher collision rate than CPA. Therefore, CPA can significantly reduce loss rate for all classes compared with NCPA. One can observe that by reducing timeout values, loss rate under NCPA has significantly increased due to the huge number of optical packets generated due to small timeouts. How-

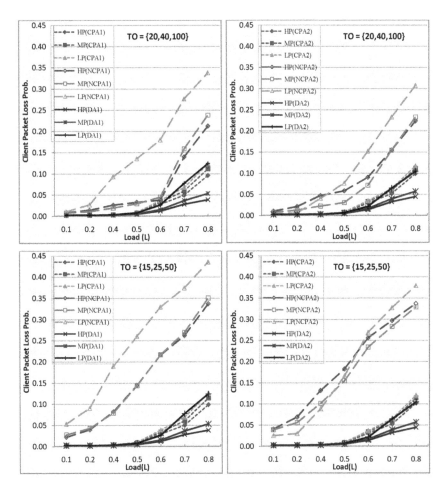

Figure 4.25: Client traffic loss rate under the PA + MF + NBR + DRwBD + IAS + SHP + SPN mechanism without deflection at $f = 2$, $W = 4$, and $N_{WC} = 8$. Reproduced from [40] with permission from De Gruyter.

ever, loss rate under CPA has slightly increased under Tout2. This is because of the aggregation of user packets from different classes in the same optical packet. The DA provides a lower loss rate for both HP and MP traffic compared with CPA, especially at heavy traffic loads. However, LP traffic loss rate under DA is sometimes (at high loads) higher than CPA. In short, CPA provides less packet differentiation compared with DA.

Tables 4.3, 4.4, and 4.5 show average network-wide end-to-end delay, where there are some traffic loads (mostly in NCPA) in which end-to-end delay is infinity. This infinity delay is due to buffering delay in edge switches that occurs when $L_{OP} \geq 1.0$ (at $L > 0.5$ under Tout1, and at $L > 0.2$ under Tout2). When $L_{OP} \geq 1.0$, an edge switch cannot service the optical packets waiting in its APTB (i.e., more than $f \times W$ optical packets are generated per time slot).

Table 4.3: End-to-end delay (ms) in Scenario 1 under Tout1. Reproduced from [40] with permission from De Gruyter.

	L = 0.1	L = 0.2	L = 0.4	L = 0.5	L = 0.6	L = 0.7	L = 0.8
HP(CPA1)	3.053	3.037	3.040	3.057	3.026	3.019	∞
MP(CPA1)	3.060	3.048	3.037	3.042	3.009	3.018	∞
LP(CPA1)	3.049	3.053	3.047	3.052	3.024	3.006	∞
HP(CPA2)	3.053	3.037	3.040	3.057	3.026	3.017	∞
MP(CPA2)	3.060	3.048	3.037	3.042	3.010	3.019	∞
LP(CPA2)	3.049	3.053	3.048	3.052	3.026	3.009	∞
HP(NCPA1)	3.050	3.029	3.027	3.044	∞	∞	∞
MP(NCPA1)	3.064	3.052	3.038	3.040	∞	∞	∞
LP(NCPA1)	3.081	3.065	2.991	2.959	∞	∞	∞
HP(NCPA2)	3.048	3.024	3.011	3.024	∞	∞	∞
MP(NCPA2)	3.064	3.051	3.036	3.038	∞	∞	∞
LP(NCPA2)	3.086	3.081	3.037	3.014	∞	∞	∞

Table 4.4: End-to-end delay (ms) in Scenario 1 under Tout2. Reproduced from [40] with permission from De Gruyter.

	L = 0.1	L = 0.2	L = 0.4	L = 0.5	L = 0.6	L = 0.7	L = 0.8
HP(CPA1)	3.045	3.050	3.048	3.027	3.024	3.014	∞
MP(CPA1)	3.045	3.061	3.040	3.036	3.007	3.007	∞
LP(CPA1)	3.048	3.036	3.041	3.034	3.010	3.000	∞
HP(CPA2)	3.050	3.034	3.039	3.055	3.024	3.017	∞
MP(CPA2)	3.056	3.045	3.036	3.040	3.008	3.018	∞
LP(CPA2)	3.045	3.050	3.046	3.050	3.024	3.009	∞
HP(NCPA1)	3.040	3.012	∞	∞	∞	∞	∞
MP(NCPA1)	3.041	3.028	∞	∞	∞	∞	∞
LP(NCPA1)	3.022	3.004	∞	∞	∞	∞	∞
HP(NCPA2)	3.019	2.998	∞	∞	∞	∞	∞
MP(NCPA2)	3.013	3.025	∞	∞	∞	∞	∞
LP(NCPA2)	3.060	3.052	∞	∞	∞	∞	∞

Table 4.5: End-to-end delay (ms) in Scenario 1 under DA. Reproduced from [40] with permission from De Gruyter.

	L = 0.1	L = 0.2	L = 0.4	L = 0.5	L = 0.6	L = 0.7	L = 0.8
HP(DA1)	3.061	3.049	3.043	3.034	3.032	3.018	3.002
MP(DA1)	3.070	3.051	3.042	3.034	3.030	3.009	2.998
LP(DA1)	3.245	3.206	3.179	3.171	3.150	3.118	6.022
HP(DA2)	3.061	3.049	3.043	3.034	3.030	3.014	2.997
MP(DA2)	3.070	3.051	3.042	3.034	3.029	3.006	2.995
LP(DA2)	3.245	3.206	3.179	3.172	3.154	3.131	6.043

Hence, the delay of optical packets waiting in the APTB grows to infinity. Under any traffic load and Significance Set, end-to-end delay of three classes have reduced under Tout2 compared to Tout1 due to the less buffering user packets experience in ingress switches under Tout2. However, NCPA under Tout2 experiences infinite buffering delay even at small traffic loads due to the huge amount of generated optical packets that cannot be sent to the network. In any scenario, delay under small traffic loads is higher than large traffic loads. This is because at high loads, user traffic is aggregated not only when timeout happens but also when there is enough traffic in class-based buffers to create a full optical packet. DA has always a bounded delay due to $L_{OP} < 1$. The delay of LP traffic is always higher under DA compared with NCPA and CPA. However, the HP and MP delays under DA are almost close to the HP and MP delays of CPA. The actual difference between them returns to the operation of CPA and DA in creating optical packets.

- **Scenario 2**: Now, the PacNet network with diameter of five is evaluated under the CPA and DA bandwidth access schemes. Here, deflection is allowed with $D_{max} = 3$. The network parameters are: $B_c = 40$ Gbits/s, $f \times W = 24$, $\alpha_0 = 0.2$, $\alpha_1 = 0.3$, $\alpha_2 = 0.5$, N_{WC} limited-range WCs in any core switch with $d_c = 5$, and ($V_{HP} = 0.5$, $V_{MP} = 0.3$, $V_{LP} = 0.2$) for {HP, MP, LP} traffic. Timeouts of CPA are {$T_{HP} = 20\ \mu s$, $T_{MP} = 50\ \mu s$, $T_{LP} = 100\ \mu s$}, respectively, for {HP, MP, LP} traffic.

Three deflection cases are studied: (1) no deflection case (denoted by NoD in the diagrams) in which optical packets are never deflected, (2) deflection case without backward deflecting feature (denoted by DRw/oBD) in which an optical packet cannot be deflected to the link coming through it, and (3) deflection with backward deflecting feature (shown with DRwBD) in which an optical packet can be deflected to any link including the link coming through it.

Figure 4.26 illustrates network-wide user traffic loss rate (measured based on the number of user packets lost in the network; i.e., among all source–destination pairs within the network). This loss includes three loss components as shown in this figure: traffic loss due to resource lack, traffic loss due to PMD constraint, and traffic loss due to receiver blocking that happens when there are no available output ports to deliver an optical packet to a given egress switch, and there is also no way to resolve its contention either by wavelength conversion or by deflection routing. The network experiences the smallest user traffic loss rate under all traffic loads when backward deflecting feature is allowed. With backward deflection, the backward fiber link acts as an optical buffer to postpone the switching of the optical packet. In a network with either a small PMD coefficient or a small data rate, deflection scheme with backward deflecting feature could be even more effective because the PMD constraint would be less disturbing. It can be observed that CPA always outperforms DA in LP traffic loss rate, except at light traffic loads ($L \leq 0.2$). However, for HP/MP traffic loads, DA outperforms CPA at heavy traffic loads ($L \geq 0.7$). CPA provides smaller loss rate than DA at moderate traffic loads.

One can see that there is no loss due to resource lack for transit optical packets at $L \leq 0.2$ in the NoD case and at $L \leq 0.3$ in the DRw/oBD and DRwBD cases. Compared with CPA, optical packets experience lower loss under DA due to resource lack because of the smooth traffic transmission nature of DA.

There is also no loss due to the PMD constraint when deflection routing is not utilized. Under any traffic load, loss due to the PMD constraint is the highest under the DRwBD case. This is because the greater the number of optical

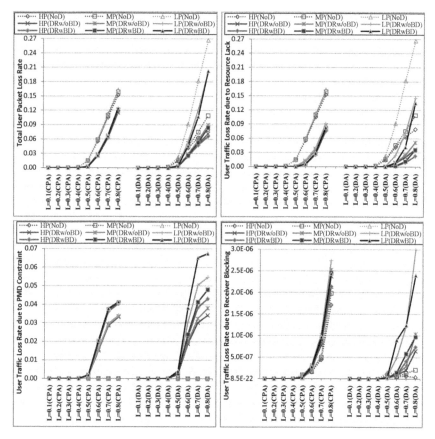

Figure 4.26: Traffic loss rate under the PA + MF + NBR + DRwBD + IAS + SHP + SPN mechanism at $f = 2$ and $N_{WC} = 10$. Reproduced from [40] with permission from De Gruyter.

packets deflected, the greater the PMD constraint disturbance. An interesting issue is that DA provides high loss because of the PMD constraint compared with CPA. This is because of the smooth traffic transmission nature of DA that allows optical packets encounter small loss due to resource lack, travel for long hops, and experience high PMD compared with CPA.

There is no loss due to receiver blocking at $L \leq 0.4$. Under deflection routing, a high chance is provided for an optical packet to pass through a core switch. Hence, the probability of collision goes up among the optical packets competing for local ports of the core switch. This results in increasing user traffic loss rate due to receiver blocking when deflection routing is used, especially at heavy traffic loads. Optical packets under DA still experience smaller loss due to receiver blocking compared with CPA because the arrival of optical packets at the local egress switch of any core switch is smooth.

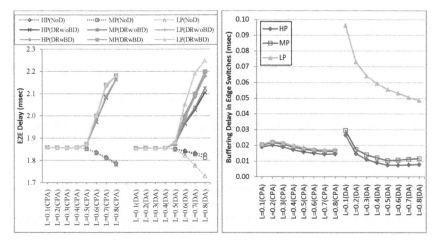

Figure 4.27: Delay under the PA + MF + NBR + DRwBD + IAS + SHP + SPN architecture at $f = 2$ and $N_{WC} = 10$. Reproduced from [40] with permission from De Gruyter.

Figure 4.27 depicts the average end-to-end delay (the duration from the time a user packet arrives at an ingress switch until it is successfully delivered to its relevant egress switch) in ms in the network, along with the average buffering delay (the duration from the time a user packet arrives at an ingress switch until it is transmitted to the OPS network inside an optical packet) in ms in ingress switches. The buffering delay goes down by increasing traffic load because of the aggregation of user traffic in optical packets, where user traffic is aggregated in optical packets with higher rate at heavy traffic loads than at light traffic loads. Under CPA, the delays among different classes are close to each other. This is due to the combined aggregation of user traffic from different classes in an optical packet. Therefore, average buffering delay goes down even for LP traffic. On the other hand, DA provides a higher buffering delay for LP traffic and lower buffering delay for HP/MP traffic compared with CPA. This returns to the nature of the scheduler used in edge switches under DA, where the scheduling of HP/MP traffic is favored over LP traffic. By this, a higher differentiation is provided for class-based traffic under DA.

By increasing traffic load under the NoD case, end-to-end delay slightly reduces due to the reduction of buffering delay at heavy traffic loads. On the other hand, end-to-end delay goes up by increasing traffic load under both DRw/oBD and DRwBD cases due to the large number of deflections that happen at high traffic loads. End-to-end delay is smaller in DRw/oBD than in DRwBD due to the smaller number of deflections under DRw/oBD.

Additional performance evaluation under DRwBD (see Table 4.6) shows that traffic loss rate reduces for all traffic classes by increasing f and N_{WC}. Note that at $L \leq 0.2$ and $N_{WC} \geq 10$, no client packet loss has been observed under both CPA and DA. Similarly under DA at $L = 0.1$ and $N_{WC} = 4$, no traffic loss has been observed. Moreover, there is no significant difference in user packet loss rate when using $N_{WC} = 30$ compared with $N_{WC} = 20$, and therefore, it is not cost-effective to use more than 20 WCs in each core switch. Furthermore, traffic loss rate of HP traffic is lower than both MP and LP traffic. Besides, loss rate is smaller for MP traffic than for LP traffic. It can be observed that LP traffic under DA experiences a higher loss rate compared with CPA. The DA mostly provides lower loss rate for both HP and MP traffic at heavy traffic loads ($L \geq 0.7$) and at light traffic loads ($L \leq 0.3$) compared with CPA. However, for moderate traffic loads, CPA mostly outperforms DA.

Table 4.6: Comparison of client traffic loss rate under DRwBD in the PA + MF + NBR + DRwBD + IAS + SHP + SPN mechanism. Reproduced from [40] with permission from De Gruyter.

Class	load	CPA						DA					
		$f=2$			$f=3$			$f=2$			$f=3$		
		$N_{WC}=4$	$N_{WC}=10$	$N_{WC}=20$	$N_{WC}=4$	$N_{WC}=10$	$N_{WC}=20$	$N_{WC}=4$	$N_{WC}=10$	$N_{WC}=20$	$N_{WC}=4$	$N_{WC}=10$	$N_{WC}=20$
HP	0.3	2.87e-4	1.38e-7	0	2.32e-5	0	0	3.10e-4	5.50e-7	0	2.19e-5	0	0
	0.4	3.29e-3	3.06e-5	1.09e-7	5.70e-4	7.84e-7	1.76e-7	3.64e-3	7.93e-5	0	5.77e-4	1.39e-6	3.24e-8
	0.5	1.65e-2	1.81e-3	1.57e-4	5.30e-7	2.88e-4	1.36e-4	1.74e-2	2.73e-3	4.42e-4	6.99e-3	5.42e-4	3.90e-4
	0.6	4.60e-2	2.42e-2	1.44e-2	6.94e-6	1.69e-2	1.35e-2	3.97e-2	2.45e-2	1.58e-2	2.72e-2	1.76e-2	1.49e-2
	0.7	9.23e-2	6.21e-2	5.05e-2	6.30e-5	5.30e-2	4.80e-2	6.54e-2	4.76e-2	4.29e-2	5.02e-2	4.27e-2	4.28e-2
	0.8	1.51e-1	1.16e-1	1.00e-1	2.75e-4	1.02e-1	9.68e-2	9.14e-2	6.59e-2	5.66e-2	6.97e-2	5.75e-2	5.52e-2
MP	0.3	3.07e-4	1.04e-7	0	2.56e-5	0	0	3.38e-4	5.13e-7	0	2.45e-5	0	0
	0.4	3.51e-3	3.28e-5	9.44e-8	6.15e-4	8.00e-7	2.02e-7	3.95e-3	8.63e-5	0	6.38e-4	1.49e-6	9.30e-9
	0.5	1.73e-2	1.92e-3	1.67e-4	1.15e-5	3.10e-4	1.48e-4	1.84e-2	2.92e-3	4.74e-4	7.50e-3	5.89e-4	4.23e-4
	0.6	4.79e-2	2.54e-2	1.54e-2	9.05e-6	1.81e-2	1.45e-2	4.19e-2	2.60e-2	1.72e-2	2.88e-2	1.91e-2	1.63e-2
	0.7	9.59e-2	6.49e-2	5.31e-2	8.59e-5	5.57e-2	5.06e-2	7.31e-2	5.27e-2	4.75e-2	5.61e-2	4.75e-2	4.75e-2
	0.8	1.57e-1	1.21e-1	1.05e-1	3.60e-4	1.08e-1	1.02e-1	1.15e-1	8.38e-2	7.36e-2	8.97e-2	7.43e-2	7.21e-2
LP	0.3	3.18e-4	1.28e-7	0	2.68e-5	0	0	5.03e-4	6.02e-7	0	4.05e-5	0	0
	0.4	3.62e-3	3.40e-5	1.15e-7	6.43e-4	8.94e-7	2.09e-7	5.29e-3	1.21e-4	0	9.10e-4	2.34e-6	6.24e-8
	0.5	1.77e-2	1.99e-3	1.71e-4	1.19e-5	3.22e-4	1.51e-4	2.30e-2	4.11e-3	6.86e-4	1.04e-2	9.22e-4	6.49e-4
	0.6	4.89e-2	2.61e-2	1.59e-2	1.12e-5	1.87e-2	1.51e-2	7.03e-2	4.64e-2	3.67e-2	5.13e-2	3.96e-2	3.65e-2
	0.7	9.77e-2	6.65e-2	5.46e-2	1.00e-4	5.72e-2	5.20e-2	1.49e-1	1.07e-1	9.96e-2	1.18e-1	1.00e-1	1.00e-1
	0.8	1.60e-1	1.24e-1	1.08e-1	4.15e-4	1.11e-1	1.05e-1	2.49e-1	2.03e-1	1.90e-1	2.18e-1	1.90e-1	1.90e-1

4.2.8 Hybrid DA + MF + NBR + SPN WC Architecture

In [30], ingress switches use the DA scheme for traffic transmission to a multi-fiber slotted OPS network. On the other hand, each core switch uses the non-blocking receiver (so that each core switch is connected to its local edge switch with six additional drop ports; i.e., eight drop ports in total) and N_{WC} Shared-Per-Node (SPN) wavelength conversion techniques for contention resolution purposes.

The performance of NCPA in ISA1 and DA in ISA2 are compared in a multi-hop all-optical network with eight core switches supporting three classes of traffic, namely EF, AF, and BE (see Section 2.1.2 for the architectures of ISA1 and ISA2). There are $f = 2$ fibers per link, $W = 4$ channels per fiber, and eight shared-per node WCs used in each core switch. To provide another scheme for contention resolution in core switches, an almost non-blocking receiver is implemented. Each core switch chooses the shortest path; i.e., the number of hops, to route the optical packets between an ingress–egress switch pair. Other network parameters are as follows: $B_C = 10\ \text{Gbits/s}$, $S_T = 9\ \mu\text{s}$, $S_O = 1\ \mu\text{s}$, frame length $F = 100$ OP sets under DA, $\sigma = 0.5$, and $\theta_{max} = F \times W = 400$. In ISA1, there are three equal-size buffers (i.e., one buffer for each class) with a total size of 216 Mbits for each egress switch. The timeout

(Tout) values for EF, AF, and BE traffic under NCPA are set to 20 μs, 40 μs, and 100 μs, respectively. Assume the optical packet transmission buffer size in ISA1 is infinite in order to avoid dropping of optical packets at ingress switches. The Pareto distribution with the probability density function $p(t) = \frac{\alpha \times b^{\alpha}}{t^{\alpha+1}}$ is used to model the inter-arrival times of client packets by using $\alpha = 1.2$. The traffic from each source to any egress switch consists of 25%, 35%, and 40% of EF, AF, and BE traffic, respectively, under any traffic load. We also have ($V_{EF} = 0.6$, $V_{AF} = 0.3$, $V_{BE} = 0.1$). Moreover, the traffic from each source and in each class is an aggregation of client packets of varying length with a distribution mentioned in [42] and with a mean of 3928 bits.

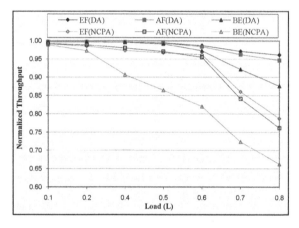

Figure 4.28: Average normalized throughput of differentiated packets under the DA + NBR + MF + SPN mechanism. Reprinted from *Computer Networks*, Vol. 53, Akbar Ghaffar Pour Rahbar and Oliver Yang, Distribution-based bandwidth access scheme in slotted all-optical packet-switched networks, 744–758, copyright 2009, with permission from Elsevier.

Figure 4.28 depicts the average normalized throughput of EF, AF, and BE client packets among all source–destination pairs within the network. It is calculated separately for each class based on the total number of client packets received in all egress switches divided by the total number of transmitted client packets from all ingress switches. Since the optical packet generation rate is higher under NCPA compared with DA (see Fig. 2.21 and Fig. 2.22) that leads to higher collision in the network, DA can provide higher throughput for all classes than NCPA. For example, consider $L = 0.7$. Under DA, the normalized throughput for EF, AF, and BE traffic are respectively (0.971, 0.963, 0.921). In the similar situation under NCPA, the normalized throughput for EF, AF, and BE traffic are respectively (0.771, 0.760, 0.688).

Figure 4.29 shows the average system delay for EF, AF, and BE client packets in ingress switches under Pareto traffic arrivals, where system delay is the duration from the time a client packet arrives at an ingress switch until it is transmitted to the OPS network within an optical packet. Since the bandwidth provisioning technique in DA may not be accurate under bursty traffic at low and moderate traffic loads due to traffic fluctuations within short intervals, NCPA achieves lower delay than DA under the Pareto traffic. Under the Poisson traffic, NCPA can provide lower delay

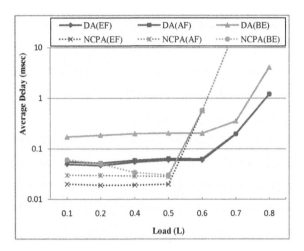

Figure 4.29: Average system delay for differentiated client packets under the DA + NBR + MF + SPN mechanism. Reprinted from *Computer Networks*, Vol. 53, Akbar Ghaffar Pour Rahbar and Oliver Yang, Distribution-based bandwidth access scheme in slotted all-optical packet-switched networks, 744–758, copyright 2009, with permission from Elsevier.

than DA at low traffic loads, whereas DA has a better (and desirable) EF and AF delay for the moderate and high traffic loads. However, the delay for all classes of client packets under NCPA goes to infinity at high traffic loads due to its instability problem. This is because of the large number of optical packets generated that results in increasing system delay. As can be observed, NCPA cannot provide much differentiation among the delay of the client packets in different classes at mid-range and high traffic loads so that the delays of EF, AF, and BE traffic become almost the same. The reason is that the volume of traffic arrival is high, and therefore the number of timeout occurrences is reduced. Hence, traffic classes with higher arrival rate can be aggregated faster than other classes with low rates. By choosing high timeout values, the delay of BE traffic will be even lower than the EF and AF packets at high traffic loads. To prevent this problem, one may decide to reduce the timeout values for EF and AF in NCPA. However, this solution increases the optical packet generation rate, thus leading to (a) high loss rate for NCPA at the core switch and (b) an unstable edge switch operation even at moderate traffic loads.

4.2.9 Smoothed Flow Decomposition (SFD)

Under SFD, proposed for slotted OPS, incoming traffic destined to a particular egress switch is smoothed at each ingress switch by a buffer scheduling algorithm and decomposed into several constant bit rate sub-flows (each with a uniform optical packet inter-arrival time), where a flow is the data stream from one ingress switch to another egress switch. On the other hand, each sub-flow emerging from a core switch also retains these traffic characteristics. Here, OPS core switches use limited FDLs and edge switches use traffic smoothing to decrease PLR. By smoothing, the effect of traffic burstiness in the OPS network reduces [43].

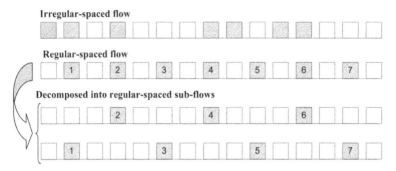

Figure 4.30: Different flows injected into a core switch: an irregular-spaced flow and a regular-spaced flow, where two decomposed sub-flows can be made from the regular-spaced flow [43]

The SFD technique proposed for single-class slotted OPS networks [43] includes two main parts:

- **Edge Traffic Smoothing:** This is the same as the smoothing mechanism detailed in Section 2.1.1.5, but without considering the network status. An ingress switch estimates traffic requirements for a time interval called renegotiation interval. It stores arriving client packets in electrical buffer, and then it injects traffic to the OPS network based on the estimated traffic for that interval. In SFD, Eq. (2.1) in Chapter 2 is only used to estimate smoothed arrival rate R_i at time unit i based on the measured arrival rate U_i during time unit i, and factor v is a constant that shows to what extent previous experience and to what extent the current measured rate must affect the new service rate. Then, the required service rate for the next negotiation interval is obtained by taking mean of R_i over all time units in the last negotiation interval. The SFD divides a smoothed flow in an ingress switch to multiple regular-spaced Constant Bit Rate (CBR) sub-flows.

Note that traffic flows could be either irregular-spaced or regular-spaced. Figure 4.30 illustrates a simple example of transmitting seven optical packets in two ways as irregular-spaced and regular-spaced flows, where the regular-spaced flow can be decomposed into two equal-rate sub-flows. In irregular-spaced flow, optical packets are transmitted within any time slot and only the total amount of traffic is controlled in each renegotiation interval. In practice, a flow may be decomposed into many sub-flows even with different rates. The egress switch finally recombines the decomposed sub-flows into the original single flow based upon the sequence information carried in each optical packet header.

The header of each generated optical packet includes a small amount of identification information to provide coordination in core switches: flow ID, sub-flow ID, and sub-flow rate. Then, each OPS core switch will check the flow-ID and sub-flow ID to determine the rate, and then it will schedule the optical packet onto the appropriate FDL on the relevant output link.

- **Flow Decomposition in Core Switches:** Assume that smoothed traffic arrives at an intermediate OPS core switch as regular-spaced flow, in which contending optical packets are resolved by simple buffering in FDLs. In this case, the buffered optical packets are transmitted by the FCFS manner, thus resulting in different amount of delays for optical packets of each flow. When the flow passes through multiple core switches, the FCFS buffering causes traffic burstiness, and therefore the flow would no longer be regular spaces. This leads to more contention and PLR. In general, optical packets in a core switch can be scheduled in three ways as (see Fig. 4.30): irregular un-smoothed optical packet arrivals; regular optical packet arrivals; or a combination of several sub-flows each with regular optical packet arrivals (i.e., decomposed sub-flows).

The SFD uses decomposed sub-flows and maintains the space of optical packets regular even after traversing through several intermediate core switches. The core switch does not use FCFS method to buffer contending optical packets. Instead, it buffers optical packets according to sub-flow rates and transmits sub-flows optical packet in a way to maintain its orders, and therefore, SFD maintains the space of the flow regular. In other words, SFD maintains the relative positions of optical packets within a sub-flow. The core switch always transmits the high-rate sub-flow optical packets before low-rate sub-flow optical packets of each flow.

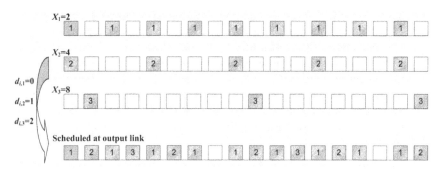

Figure 4.31: Scheduling of sub-flows on an output link [43]

Assume that after renegotiation at time T_i, a new sub-flow is introduced onto an output link of an OPS core switch. To reduce or eliminate contention, each optical packet of sub-flow j will be delayed by $d_{i,j}$ time slots, which is constant from time T_i to T_{i+1} (i.e., the next renegotiation time). Then, starting from the highest rate sub-flow and proceeding in descending order of rates, all sub-flows including the new one are scheduled on the output link. For this, sub-flow j is assigned $d_{i,j}$ in such a way that its next optical packet occupies the first time slot not occupied by already scheduled sub-flows.

Assume there are m sub-flows on an output link in the core switch and X_j is the constant inter-arrival time between optical packets in sub-flow j. Figure 4.31 displays three contending sub-flows on a wavelength destined to an output link with inter-arrival times $X_1 = 2$, $X_2 = 4$, and $X_3 = 8$. First, sub-flow 1 with the highest rate $\frac{1}{X_1} = 0.5$ is scheduled and time slots are allocated. Then, sub-flow 1 with rate $\frac{1}{X_2} = 0.25$ is scheduled in which the first empty time slot is the second time slot. Finally, sub-flow 3 with the lowest rate $\frac{1}{X_3} = 0.125$ is scheduled starting from the fourth time slot. For this purpose, SFD assigns $d_{i,1} = 0$, $d_{i,2} = 1$, and $d_{i,3} = 2$ time slot delays for scheduling the sub-flows at the output link as depicted in the lowest diagram of Fig. 4.31.

To maintain the arrival pattern of each sub-flow, two requirements must be satisfied in a core switch:

- **Requirement 1:** Optical packet inter-arrivals in each sub-flow must be constant all the time.

- **Requirement 2:** The core switch must be able to schedule each sub-flow onto a particular output link with the necessary delay in order to avoid contention.

On the other hand, the necessary conditions to keep the inter-arrival times within sub-flow X_j constant after multiplexing a number of sub-flows on an output link (even after shifting the entire sub-flow X_j in time) are:

- **Condition 1:** If $X_j < X_k$, then X_k must be an integer multiple of X_j, when $1 < j, k \leq m$. A simple case that satisfies this condition is to take $X_j = 2^i$; i.e., integral powers of 2 as in the above example. Note that no two sub-flows have the same inter-arrival time.

- **Condition 2:** The summation of sub-flow rates must not overload an output link; i.e., $\sum_{i=1}^{m} \frac{1}{X_i} \leq 1$.

The simulation results show that decomposed sub-flows can outperform regular smoothing method in all traffic conditions and SFD requires small optical buffers to have the same performance than General Traffic Smoothing (GTS) detailed in Section 2.1.1.5. Compared with GTS, the SFD does not introduce delay jitter and can reduce PLR significantly. The drawback of SFD is its requirements and conditions mentioned above.

4.3   Summary

This chapter has studied hybrid methods that can achieve an OPS network with very low optical packet loss rate. The hybrid methods could be either hybrid contention resolution schemes (such as combining different types of WCs, combining WCs and FDLs, and so on) or hybrid contention avoidance/resolution schemes that combine a number of contention avoidance schemes with a number of contention resolution schemes in the same network. Clearly, the latter schemes can provide the best performance results while reducing the network cost. Different hybrid architectures to be developed in the future could be:

- Combining the routing-based techniques with merit-based techniques, where the merits of optical packets could be a criterion for routing.

- Designing multi-fiber OPS networks that use additional wavelength channels on each fiber and use shared-per-node conversion with limited-range WCs at core switches.

- Investigating the routing-based schemes that use the "significance" parameter of an optical packet as a criterion for routing.

- Investigating other criteria for the PR technique, including (a) distance and (b) the significance of traffic carried in an optical packet.

REFERENCES

1. V. Eramo, M. Listanti, and P. Pacifici. A comparison study on the number of wavelength converters needed in synchronous and asynchronous all-optical switching architectures. *IEEE Journal of Lightwave Technology*, 21(2):340–355, Feb. 2003.

2. D. K. Hunter, M. C. Chia, and I. Andonovic. Buffering in optical packet switches. *IEEE Journal of Lightwave Technology*, 16(12):2081–2094, Dec. 1998.

3. Y. Li, G. Xiao, and H. Ghafouri-Shiraz. On the benefits of multifiber optical packet switch. *Microwave and Optical Technology Letter*, 43(5):376–378, Dec. 2004.

4. F. Xue, Z. Pan, Y. Bansal, and J. Cao. End-to-end contention resolution schemes for an optical packet switching network with enhanced edge routers. *IEEE Journal of Lightwave Technology*, 21(11):2595–2604, Nov. 2003.

5. H. L. Liu, B. Zhang, and S. L. Shi. A novel contention resolution scheme of hybrid shared wavelength conversion for optical packet switching. *IEEE Journal of Lightwave Technology*, 30(2):222–228, 2012.

6. C. Okonkwo, R. C. Almeida, R. E. Martin, and K. Guild. Performance analysis of an optical packet switch with shared parametric wavelength converters. *IEEE Communications Letters*, 12:596–598, Aug. 2008.

7. H. Liu and B. Zhang. A resolution scheme for optical packet contention based on shared limited range wavelength converters and parametric wavelength converters. *Journal of Optoelectronics Laser*, 22(3):329–332, 2011.

Quality of Service in Optical Packet Switched Networks, First Edition.
By Akbar Ghaffarpour Rahbar Copyright © 2015 IEEE. Published by John Wiley & Sons, Inc.

8. H. Liu, Y. Chen, and B. Zhang. An optical packet contention resolution supporting QoS differentiation with LRWC wavelength converters for OPS. *Photonic Network Communications*, 24:71–76, 2012.

9. C. Okonkwo, R. C. Almeida Jr., and K. M. Guild. An optical packet switch employing shared tunable parametric wavelength converters. In *International Conference on Photonics in Switching*, Sapporo, Aug. 2008.

10. Z. Zhang and Y. Yang. WDM optical interconnects with recirculating buffering and limited range wavelength conversion. *IEEE Transactions on Parallel and Distributed Systems*, 17(5):466–480, 2006.

11. V. Eramo, M. Listanti, and A. Germoni. Cost evaluation of optical packet switches equipped with limited-range and full-range converters for contention resolution. *IEEE Journal of Lightwave Technology*, 26(4):390–407, 2008.

12. H. Li and I. L. J. Thng. Cost-saving two-layer wavelength conversion in optical switching network. *IEEE Journal of Lightwave Technology*, 24(2):705–712, Feb. 2006.

13. J. J. He, D. Simeonidou, and S. Chaudhry. Contention resolution in optical packet-switching networks: under long-range dependent traffic. In *Optical Fiber Communication Conference*, pages 295–297, Baltimore, MD, 2002.

14. J. Li. Performance of an optical packet switch with limited wavelength converter. *Springer Intelligence Computation and Evolutionary Computation*, pages 223–229, 2013.

15. H. S. Hamza, T. Ismail, and K. E. Sayed. On the design of asynchronous optical packet switch architectures with shared delay lines and converters. *Photonic Network Communications*, 22:191–208, 2011.

16. S. F. Chien, C. P. Tan, A. L. Y. Low, A. H. You, and K. Takahashi. Buffering controls for IP packets in optical packet switching. In *The 9th Asia-Pacific Conference on Communications(APCC)*, 2003.

17. C. P. Tan, G. Castan, S. F. Chien, A. H. You, and A. L. Y. Low. Buffering management schemes for optical variable length packets under limited packet sorting. *Photonic Network Communications*, 12:257–268, 2006.

18. L. Xu, H. Perros, and G. Rouskas. Techniques for optical packet switching and optical burst switching. *IEEE Communications Magazine*, 39:136–142, Jan. 2001.

19. F. Callegati, W. Cerroni, and C. Raffaelli. Routing techniques in optical packet-switched networks. In *IEEE International Conference on Transparent Optical Networks (ICTON)*, Barcellona, Luglio, July 2005.

20. S. Yao, B. Mukherjee, S. J. B. Yoo, and S. Dixit. A unified study of contention-resolution schemes in optical packet-switched networks. *IEEE Journal of Lightwave Technology*, 21(3):672, 2003.

21. C. Papazoglou, G. Papadimitriou, and A. Pomportsis. Design alternatives for optical-packet-interconnection network architectures. *OSA Journal of Optical Networking*, 3(11):810–825, 2004.

22. S. Bregni and A. Pattavina. Performance evaluation of deflection routing in optical IP packet-switched networks. *Cluster Computing*, 7:239–244, 2004.

23. N. Akar, C. Raffaelli, and M. Savi. Analytical model of asynchronous shared-per-wavelength multi-fiber optical switch. In *IEEE 12th International Conference on High Performance Switching and Routing (HPSR)*, pages 264–269, Cartagena, July 2011.

24. V. Eramo, A. Germoni, C. Raffaelli, and M. Savi. Packet loss analysis of shared-per wavelength multi-fiber all-optical switch with parallel scheduling. *Computer Networks*, 53(2):202–216, Feb. 2009.

25. V. Eramo, A. Germoni, C. Raffaelli, and M. Savi. Multi-fiber shared-per-wavelength all-optical switching: architectures, control and performance. *IEEE Journal of Lightwave Technology*, 26(5):537–551, Mar. 2008.

26. V. Eramo. An analytical model for TOWC dimensioning in a multifiber optical-packet switch. *IEEE Journal of Lightwave Technology*, 24(12):4799–4810, Dec. 2006.

27. A. G. Rahbar and O. Yang. Even slot transmission in slotted optical packet-switched networks. *IEEE/OSA Journal of Optical Communications and Networking*, 1(2):A219–A235, July 2009.

28. A. G. Rahbar and O. Yang. Prioritized retransmission in slotted all-optical packet-switched networks. *Journal of Optical Networking*, 5(12):1056–1070, Dec. 2006.

29. A. G. Rahbar. Cost-effective combination of contention resolution/avoidance schemes in bufferless slotted OPS networks. *Elsevier Optics Communications*, 282(5):798–808, Mar. 2009.

30. A. G. Rahbar and O. Yang. Distribution-based bandwidth access scheme in slotted all-optical packet-switched networks. *Elsevier Computer Networks*, 53(5):744–758, Apr. 2009.

31. A. G. Rahbar. Analysis of optical packet loss rate under asymmetric traffic distribution in multi-fiber synchronous OPS switches. *OPTIK*, 125:769–772, 2013.

32. A. G. Rahbar. PTES: a new packet transmission technique in bufferless slotted OPS networks. *IEEE/OSA Journal of Optical Communications and Networking*, 4(6):490–502, June 2012.

33. Y. Li, G. Xiao, and H. Ghafouri-Shiraz. Performance evaluation of multi-fiber optical packet switches. *Computer Networks*, 51(4):995–1012, Mar. 2007.

34. A. G. Fayoumi, F. A. Al-Zahrani, A. A. Habiballa, and A. P. Jayasumana. Performance analysis of multi-fiber synchronous photonic share-per-link packet switches. In *IEEE LCN2005*, Sydney, Nov. 2005.

35. G. Muretto and C. Raffaelli. Combining contention resolution schemes in WDM optical packet switches with multifiber interfaces. *Journal of Optical Networking*, 6(1):74–89, Jan. 2007.

36. F. Callegati, W. Cerroni, C. Raffaelli, and P. Zaffoni. Wavelength and time domain exploitation for QoS management in optical packet switches. *Computer Networks*, 44:569–582, 2004.

37. A. G. Rahbar. Improving throughput of long-hop TCP connections in IP over OPS networks. *Springer Photonic Network Communications*, 17(3):226–237, June 2009.

38. T. E. Stern, G. Ellinas, and K. Bala. *Multiwavelength Optical Networks: Architectures, Design, and Control*. Second edition, Cambridge University Press, 2008.

39. G. Jeong and E. Aynoglu. Comparison of wavelength-interchanging and wavelength-selective cross-connects in multiwavelength all-optical networks. In *IEEE Infocom*, pages 156–163, 1996.

40. A. G. Rahbar. Impairment-aware merit-based scheduling in QoS-capable multi-fiber OPS networks. *Journal of Optical Communications*, 33(2):103–121, June 2012.

41. A. G. Rahbar and O. Yang. LGRR : a new packet scheduling algorithm for differentiated services packet-switched networks. *Elsevier Computer Communications*, 32(2):357–367, Feb. 2009.

42. Packet size reports, cooperative association of internet data analysis (caida), http://www.caida.org/research/traffic-analysis/aix/, 2008.

43. Z. Lu and D. K. Hunter. Optical packet contention resolution through edge smoothing into decomposed subflows. *IEEE/OSA Journal of Optical Communications and Networking*, 1(7):622–635, Dec. 2009.

CHAPTER 5

HYBRID OPS NETWORKS

Early generation optical networks such as reconfigurable optical add–drop multiplexers and all-Optical Cross Connects (OXCs) are working in circuit switching mode, and therefore they are not suitable for bursty Internet traffic since circuit switching may provide low channel utilization under bursty traffic. The major advantages of a circuit switching network are fixed end-to-end delay and loss-free traffic transmission, which are very desirable for multimedia and real-time services. On the other hand, OPS is suitable for packet-based traffic that can provide very high channel utilization by multiplexing different traffic on the same channel. The disadvantages of OPS are uneven end-to-end delay for optical packets destined to the same egress switch (due to selecting different paths with different lengths or delaying in FDLs for contention resolution) and high PLR due to the lack of enough mechanisms for contention resolution (such as FDLs and TWCs). These advantages and disadvantages have led to design hybrid optical network architectures [1].

Optical networks can be combined in different ways to make hybrid optical networks. The combination could be at the OPS network level only, where slotted OPS and asynchronous OPS networks are combined. In another way, OPS and OCS networks can be combined in three levels. At the client–server level, OPS traffic (as client traffic) is carried by an OCS network (as a server network). At the parallel

Quality of Service in Optical Packet Switched Networks, First Edition.
By Akbar Ghaffarpour Rahbar Copyright © 2015 IEEE. Published by John Wiley & Sons, Inc.

level, network resources (such as transceivers and wavelengths) are statically or dynamically allocated to each switching mechanism. For example, a number of wavelengths in a core switch are allotted to OPS traffic and the remaining wavelengths are assigned to OCS traffic. Finally, at the integrated level, network resources are completely shared among different switching mechanisms. For instance, both OPS and OCS traffic can be carried on the same wavelength channel. At the parallel and integrated levels, a core switch architecture is complex since it must be able to support the operations of different switching mechanisms. In the following, hybrid architectures proposed for OPS networks are studied in detail.

5.1 Hybrid Asynchronous and Synchronous OPS Networks

In this hybrid approach, the network operation is asynchronous while core switches operate synchronously so that the optical packets sent to the OPS network can have variable sizes. They arrive at a core switch at any time without requiring any packet realignment. However, the switching module inside an OPS core switch begins to operate only at the start of time slots, where a time slot is duration of transmitting the smallest packet size (say 40 bytes in IP networks used for TCP acknowledges). Here, large-size optical packets are transmitted in several consecutive time slots. For example, a 600-byte optical packet should be switched in 15 consecutive time slots. This approach provides the flexibility of having any packet size and makes easy contention resolution in OPS switches [2].

Contention is resolved by combining a small amount of optical buffering with wavelength conversion and, finally, with deflection routing. When contention occurs in a core switch, the switch controller first tries to switch the conflicting optical packets on different wavelengths using WCs. When all wavelengths of the target output port are busy, some optical packets are forwarded to the fiber delay lines for possible switching in a second time. Finally, when no appropriate delay line is available, a contending optical packet is deflected to a different output port.

The scheduler in a core switch decides how much delay an optical packet must tolerate before switching at its desired output port. Arriving optical packets may be switched even out of order. For example, consider S_{ts} to be the time-slot duration. Assume optical packet OP_1 with length $l_1 \times S_{ts}$ has arrived at the core switch on wavelength λ_i at time $\alpha \times S_{ts}$. Similarly, consider optical packet OP_2 with length $l_2 \times S_{ts}$ has arrived at the core switch on wavelength λ_j at time $\beta \times S_{ts}$ (where $\beta > \alpha$), from another input port, but with the same output port as OP_1. Now, assume that optical packet OP_1 must be delayed $d_1 \times S_{ts}$ in order to be switched at its desired output port; i.e., leaving the core switch at time $(\alpha + d_1) \times S_{ts}$. On the other hand, optical packet OP_2 must be delayed $d_2 \times S_{ts}$ in order to be switched at its desired output port. In this case, optical packet OP_2 can only be transmitted before optical packet OP_1, provided that $\beta + d_2 + l_2 < \alpha + d_1$, where no collision occurs between OP_1 and OP_2.

A core switch includes three units: a synchronization unit (similar to Fig. 1.11) in order to align variable length optical packets arriving at different times; a fiber delay line unit used to store optical packets and resolve contention that can provide delays from 0 to at most $B \times S_{ts}$ (similar to Fig. 3.2) for an optical packet on any wavelength channel; and a switching module that could only include WCs (as described in Section 3.1.1.2) or could combine WCs with recirculation FDLs (as stated in Section 4.1).

5.2 Hybrid OPS and OCS Networks

In this section, different hybrid architectures proposed for combination of OPS and OCS (i.e., wavelength routed optical networks) will be studied. Note that an optical network architecture is called hybrid optical network if it combines two or more basic network technologies at the same time. In other words, hybrid optical network architectures employ two or more optical networks at the same time to improve network performance by their advantages and avoiding their disadvantages. Hybrid OCS and OPS networks are very promising to fulfill future networks needs and have very good results which demands more attentions. Bringing optical packets to circuits closer together can optimize the overall network design and performance. Proposed hybrid optical networks can be classified in three categories based on the level of interaction and integration of network techniques as parallel, client–server, and integrated [1].

In QoS-capable hybrid optical networks, high-priority traffic (such as real-time streaming services) must be transmitted through wavelength routed facilities and low-priority traffic should be sent through OPS facilities. By this, real-time traffic as high-priority traffic would experience smallest delay, loss, and jitter. On the other hand, mid-priority traffic can be sent through either OPS or wavelength routed facilities.

5.2.1 Client–Server Hybrid Optical Networks

In client–server hybrid optical networks, there are two layers of optical networking in a hierarchical manner. Here, the lower layer with a physical topology operates as the server layer and provides a virtual topology for the upper layer called client layer. The virtual topology is set up by establishing a virtual direct light-path between every two optical switches according to the network's physical topology. In other words, at the absence of a direct link between every two optical switches in the physical topology, this technology establishes a virtual direct path between every two optical switches by reserving resources on the existing links between them. The layers are designed in such a way that the client layer has fine granularity (say, with OPS) and the server layer has coarse granularity (say, with wavelength routed optical network).

OPS works at the edge switches in the client layer and makes optical packets from aggregation of client packets, and then it transmits them in the server layer using already established light-paths (i.e., the virtual topology). This operation avoids intermediate switches from processing optical packets and prevents contention occurrence and optical packet loss [3, 4].

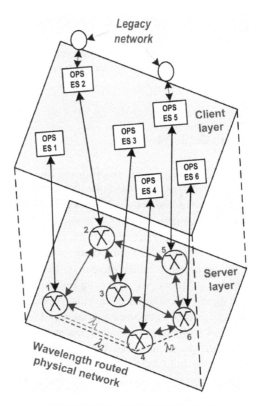

Figure 5.1: Client–server hybrid OPS network

Figure 5.1 illustrates an example for client–server hybrid OPS model with two layers. The client layer includes six OPS Edge Switches (ES) and the server layer is a wavelength routed optical network with six core switches. The server layer provides a virtual topology for the client layer in which there is a light-path setup between any pair of OPS edge switches. For instance, OP ES 1 sends its optical packets to OP ES 4 on the light-path setup on wavelength λ_1 from core switch 1 to core switch 4. For another example, OP ES 1 sends its optical packets to OP ES 6 on the light-path setup on wavelength λ_2 from core switch 1 to core switch 6, which passes through intermediate core switch 4.

Note that the server layer may also carry traffic from other networks such as SONET/SDH; i.e., coexisting OPS with SONET/SDH. In this case, some network resources are used to establish dynamic light-paths for carrying the traffic of other networks, and the remaining network resources are dedicated to the light-paths of the virtual topology for OPS traffic. In this case, an OPS edge switch is connected to one of the relevant core switch input/output ports.

The set of networks resources (such as transceivers and wavelengths) dedicated to the virtual topology is not always static, and it could be changed dynamically according to network traffic changes. For example, if traffic from other networks increases (i.e., light-path demands increase), the virtual topology provided for OPS traffic can be modified and some sets of transceivers and wavelengths are released for establishing the new light-paths.

Since there is no statistical multiplexing of different routed optical packets on the same light-path, channel utilization decreases to pure wavelength routed optical networks, as a deficiency for client–server hybrid optical networks. In other words, the client layer model needs more resources to handle all traffic available among OPS edge switches. However, it is simple in terms of core switch complexity and signaling complexity [3].

One way to increase channel utilization and reduce the required resources is to share light-paths. In this case, a full mesh virtual topology is not required in the server layer. For example, assume there is no direct light-path setup between OP ES 1 and OP ES 5 in Fig. 5.1. Instead, there is a light-path established between OP ES 1 and OP ES 2 and another light-path setup between OP ES 2 and OP ES 5. In this situation, OP ES 1 must send its optical packets destined toward both OP ES 2 and OP ES 5 on the former light-path. Then, OP ES 2 must forward the optical packets destined to OP ES 5 on the latter light-path. Clearly, in this case, OP ES 2 should convert these optical packets to the electronic domain, then save them for a while if necessary, next make new optical packets from them, and finally send the new optical packets toward OP ES 5 when appropriate. Obviously, this is not an all-optical network.

5.2.2 Parallel Hybrid Optical Networks

In parallel hybrid optical networks, two or more optical switching mechanisms operate in parallel to offer different services. There could be two major scenarios for parallel hybrid optical networks. In the first scenario, there are two separate optical networks and each network has its own dedicated resources (such as fibers, core switches). For example, in CHEETAH [5], traffic can be routed either through the Internet (as packets) or through a SONET network (as connections). In the second scenario, there is only one physical network, and resources used for transmission and switching are shared among OPS and OCS technologies. In the second scenario, the network resources (such as transceivers and wavelengths) can be statically or dynamically allocated to each switching mechanism. For instance, among W wavelengths in the network, K wavelengths are used for transmission of OPS traffic and $W - K$ wavelengths are used for transmission of OCS traffic. In both scenarios, an edge switch decides which optical networking paradigm should be used for transmission of its traffic. Note that core switches should also be able to support all switching mechanisms installed in parallel.

Figure 5.2 illustrates an ingress switch model in a parallel hybrid optical network. This is similar to the ingress switch architecture proposed for parallel hybrid OBS and OCS detailed in [6, 7]. A flow classifier in the ingress switch decides how to transmit an arriving traffic flow, and it sends client packets of flows to either the OPS section or the OCS section. It is common to send short-lived flows through OPS and long-lived flows through OCS. It is also common to send low-priority traffic through

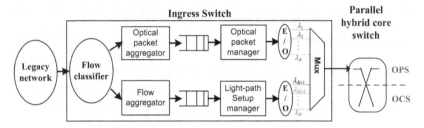

Figure 5.2: Parallel hybrid model on a single physical topology

OPS and high-priority traffic through OCS. In the OPS section, a packet aggregation manager aggregates client packets and makes aggregated units as discussed in Section 2.1.2.1. In the OCS section, light-paths are dynamically set up for new arriving traffic flows. Optical packets are sent through wavelengths λ_1 to λ_K to the core network. On the other hand, light-paths are only established on wavelengths λ_{K+1} to λ_W. All core switches manage OPS traffic only on wavelengths λ_1 to λ_K and light-paths on wavelengths λ_{K+1} to λ_W. An egress switch obtains client traffic from light-paths and from de-aggregated optical packets, and then it forwards them toward their destinations.

Contention resolution schemes can be used in a parallel hybrid optical network to improved its performance. Wavelength converters can be used for improving both OPS and OCS operations. In this case, a WC used for OPS traffic must convert a wavelength from Set 1 (i.e., $\{\lambda_1, \lambda_2, ..., \lambda_K\}$) to a wavelength in this set. Similarly, a WC used for OCS should convert a wavelength from Set 2 (i.e., $\{\lambda_{K+1}, \lambda_{K+2}, ..., \lambda_W\}$) to a wavelength in Set 2. When using the deflection routing mechanism, the same issue must be considered in which optical packets requiring deflection must be deflected only on Set 1 wavelengths and light-paths needing deflection must be deflected only on Set 2 wavelengths. In case of using FDL for resolving the contention of OPS traffic, only Set 1 wavelengths inside the FDLs must be used for delaying optical packets.

Although sharing of resources between OPS and OCS in a parallel hybrid optical network improves channel utilization, it suffers from some disadvantages. Clearly, an optical switch utilized in an OCS network does not need to be a very fast switch. However, the optical switch used in a parallel hybrid optical network must be a very fast switch for switching of OPS traffic. In addition, the complexity of control channel is higher in the parallel hybrid optical network than in the OCS network.

In the following, two cases developed for parallel hybrid optical networks are studied.

5.2.2.1 Parallel OCS/OBS/OPS
In parallel OCS/OBS/OPS, edge switches and core switches must support OCS, OBS, and OPS switching mechanisms [8]. An edge switch selects a proper mechanism for sending a traffic flow—say, based on its class of service. Then, the traffic is sent to the relevant outbound card, where there are three outbound cards for OCS traffic transmission, OPS traffic transmission, and

OBS traffic transmission. An outbound card provides bandwidth and manages QoS issues for its traffic. In each outbound card, there is an Assembly Scheduler (AS) unit that controls the traffic in the buffer of the outbound card and aggregates traffic in order to assemble a number of client packets in a big traffic unit. There is also a Resources Allocation Scheduler (RAS) unit in each outbound card that includes a Look-Up-Table (LUT) for maintaining network recourses, output buffers, and an electro-optic interface. It uses either (a) the LUTs for allocating optical recourses for transmitting assembled optical packets, (b) assembled optical bursts, or (c) long-lived wavelengths (i.e., circuits). The RAS uses the buffers for contention resolution when all wavelengths are reserved, and uses the electro-optic interface for converting traffic from electronic domain to optical domain.

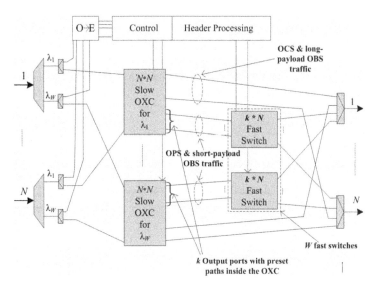

Figure 5.3: A parallel hybrid OCS/OPS/OBS core switch architecture [8]

Figure 5.3 shows a hybrid OCS/OPS/OBS core switch architecture to support applications with different bandwidth granularity requirements. Recall that OCS and long-payload OBS traffic do not need fast switching in core switches, but OPS and short-payload OBS need fast switching. To achieve this purpose, the switch architecture in Fig. 5.3 provides two different routes for switching traffic:

- Because of the high price of fast switching, W slow switches (e.g., MEMS-based switches) are used for switching of OCS and long-payload OBS traffic, where this type of traffic is routed through OXC to an appropriate output port via combiners.

- Each slow OXC is connected to a fast optical switch with k output ports. Here, there are k inputs of the OXC that are permanently connected to k output ports of the OXC as preset paths. In this case, an optical packet or a short-payload

OBS burst is sent via pre-defined paths in the OXC directly to a fast switch. Then, the fast switch switches the OPS/OBS traffic with a high speed.

The benefit of the proposed core switch architecture is to reduce the requirements on fast switches, and therefore smaller and cost efficient switches are only required.

5.2.2.2 Parallel OCS and Multi-Wavelength (MW) OPS This section details a parallel hybrid optical networking mechanism that combines OCS and MW-OPS [9, 10]. It should be noted that MW-OPS operates differently from OPS. In MW-OPS, a single optical packet is sent over v wavelengths. The header of an optical packet is sent on a specific wavelength channel, and its payload is divided into $v - 1$ segments. The segments are then encoded into $v - 1$ wavelengths and then the optical packet is sent.

In order to simultaneously provide QoS-guaranteed transport service and high bandwidth utilization, the Hybrid Optical neTwork ARchitectUre (HOTARU) is used, where traffic flows that require QoS-guaranteed transport use the OCS mechanism while the other traffic uses the MW-OPS mechanism. In HOTARU, wavelengths are shared between two switching mechanisms.

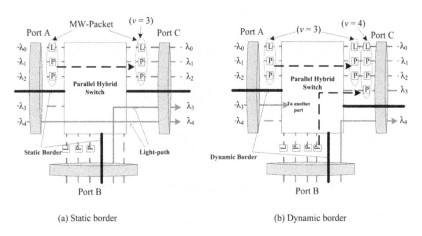

(a) Static border   (b) Dynamic border

Figure 5.4: An example for borders in hybrid OCS and MW-OPS [9]: (a) static border, (b) dynamic border

There could be two wavelength allocation models in the hybrid optical network. Figure 5.4a and Figure 5.4b display static border model and dynamic border model, respectively, where a hybrid switch has multiple fiber ports with $W = 5$ wavelengths for each fiber. One of the wavelengths is dedicated to MW-OPS headers, and the others can be used for OCS light-paths and MW-OPS payloads. In a static border model, wavelength resources in a fiber are divided into two fixed groups: one group for MW-OPS and the other group for OCS. On the other hand, in the dynamic border model, the place of the border could be changed by a request for light-path setup. In HOTARU, dynamic border model is used. Therefore, all wavelength channels can be used by MW-OPS if there is no request for a light-path. The dynamic border model could provide higher lambda utilization than the static border model. In this model,

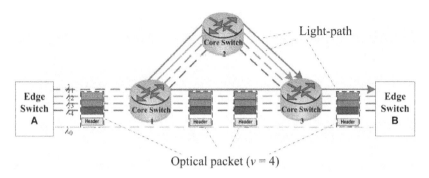

Figure 5.5: An example for sending an optical packet in hybrid OCS and MW-OPS architecture [9]

λ_0 could be dedicated to MW-OPS headers. For more explanation, consider the network depicted in Fig. 5.5. Edge switch A first determines the number of available wavelengths v to edge switch B ($v = 4$ in this example). Then, it fragments the optical packet payload into v segments, encodes the segments into different wavelengths, and sends them with a header to edge switch B. In core switches, FDLs can be used for contention resolution purposes. If FDLs are completely in use, deflection routing can be used instead.

To provide high wavelength efficiency in HOTARU, an RWA algorithm should satisfy two conditions: (1) preference for choosing low-numbered wavelengths since MW-OPS payloads can use only the high-numbered wavelengths, and (2) load balancing, which means RWA algorithms need to avoid concentrating on particular routes in order to balance wavelength utilization on each link because unbalanced wavelength utilization could obstruct a deflection route for MW-OPS packets. The first condition can be satisfied by the first-fit wavelength assignment as described in Section 1.3.1.1. The second condition can be satisfied by using three routing algorithms: Minimum Delay (MD) routing algorithm, Maximum Lambda (ML) routing algorithm, and Maximum Matched Lambda (MML) routing algorithm. The ML routing prefers a route including the links with a large number of available wavelengths regardless of their consistency along the route. The MML is similar to ML, except that it checks the consistency of wavelengths for each link on the route. The MML-FF RWA algorithm that combines the First-Fit (FF) wavelength assignment algorithm and the MML algorithm can achieve better performance in both delay ratio and blocking probability performance parameters among different RWA algorithms.

In order to evaluate wavelength availability and load balancing for MW-OPS, each ingress switch generates two types of MW optical packets to all other egress switches. One optical packet is transmitted along an MD route (called MD-Pkt route) since this route provides minimum delay. The other optical packet is transmitted along a Maximum Sequence Lambda (MSL) route (called MSL-Pkt route), which is the route with the maximum number of continuous wavelengths. Note that these routing algorithms are different from RWA routing algorithms (i.e., MD, ML, and MML). Two performance evaluation metrics are used for evaluating MW-OPS:

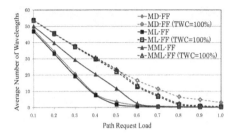

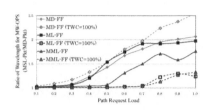

Figure 5.6: Average number of wavelengths for MW-OPS among all nodes using First-Fit [9]

Figure 5.7: Comparison of number of wavelengths for MW-OPS between MD-Pkt and MSL-Pkt, normalized by the MW-OPS in MD-Pkt [9]

- **Wavelength Utilization**: the number of available wavelengths received for MW optical packets at regular intervals. Figure 5.6 shows the average number of wavelengths that the MW-optical packets, which have successfully arrived at their destinations, have utilized with each RWA algorithm using the FF wavelength assignment algorithm. Obviously, the RWA algorithms with TWCs can provide more wavelengths for MW-OPS than for the others. The performance of MD-FF with TWCs at path request load greater than 0.6 is better than that of ML-FF with TWCs and MML-FF with TWCs since MD-FF always uses MD routes. Therefore, wavelength resources along MSL-Pkt routes can be considered to be still available. The results on ML-FF with TWCs and MML-FF with TWCs are mostly the same. On the other hand, under the three routing algorithms without TWCs, MML-FF outperforms the others for loads that range between 0.1 and 0.6. At loads higher than 0.6, it is difficult for all of the three routing algorithms to provide wavelengths for MW-OPS. At load 0.5, MML-FF has provided approximately ten wavelengths, while MD-FF and ML-FF have provided approximately less than three wavelengths. In short, if TWCs are not used, MML-FF outperforms the others in terms of the available number of wavelengths for MW-OPS, and even in transmission delay. Otherwise, ML-FF and MML-FF with TWCs both can provide the best performance.

- **Load-Balancing**: the average ratios of the number of available wavelengths for MW-payloads along MSL-Pkt routes to that along MD-Pkt routes for each RWA. If the result is close to 1.0, it shows that the difference in the number of available wavelengths between the two kinds of routes is not so large, and therefore this can be used as a criterion for load-balancing. Figure 5.7 depicts this ratio for each RWA algorithm. The ratio of MML-FF is very low at up to the load of 0.5, compared with the other two RWA algorithms in the case that TWCs are not used. In the other case, ML-FF and MML-FF balance wavelength utilization at up to the load of 0.7. Therefore, if TWCs are utilized, ML-FF and MML-FF can provide good load-balancing; otherwise MML-FF can outperform the others overall.

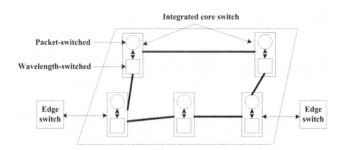

Figure 5.8: An integrated hybrid optical network model

5.2.3 Integrated Hybrid Optical Networks

Unlike the parallel hybrid optical networks, all wavelengths can be used for transmission of both OCS and OPS traffic in integrated hybrid optical networks [1]. In these networks (see Fig. 5.8), resources can be completely shared among both OPS and OCS networking techniques in the same physical network, where all core switches operate in both switching modes at the same time.

In this hybrid model, selection of the proper switching technology (either circuit switching or packet switching) is not limited to only edge switches like in parallel hybrid optical networks. Here, core switches can also make a decision about the switching mode. An arriving traffic at an ingress switch could be transmitted either in wavelength switching mode or in packet switching mode, depending on traffic service class or resource availability. Recall the traffic class of a packet could be specified either by QoS parameters available in the packet header or by explicitly requesting the ingress switch from the relevant user. A core switch or edge switch usually transmits well-behaved traffic over end-to-end light-paths in order to avoid processing in intermediate core switches, while it transmits dynamic traffic by the packet switching mode. The selection of the proper technology can also be according to QoS requirements; for example, a high-priority traffic flow is sent via light-paths and a low-priority traffic flow is sent via packet switching.

Each core switch includes both wavelength switching and packet switching devices. In case of congestion in one mode in a core switch, the core switch can change traffic transmission mode. For example, if the light-path to the desired destination in the wavelength switching mode is congested, the core switch could send the traffic by the packet switching mode. Intermediate core switches pass bit streams through light-paths in case of the wavelength switching mode and switch optical packets in the packet switching mode. Therefore, the core switch should be able to identify arriving traffic streams in order to know which mode they belong to.

Theoretically, the integrated hybrid optical networks are optimal from resource utilization point of view. However, it needs very complex switch architecture from both control and technology points of view. In the following, the techniques proposed for integrated hybrid optical networks shall be studied.

5.2.3.1 Overspill Routing in Optical Networks (ORION)   A particular example of an integrated hybrid optical network is Overspill Routing In Optical Networks (ORION) that integrates OCS and OPS [11–13]. Under ORION, the same physical resources are shared between OPS and OCS. The basic idea of ORION is to send optical packets through established light-paths. If it becomes necessary, ORION can send an optical packet on a different light-path, where the destination of the optical packet is not the same as the light-path destination. For this purpose, an intermediate core switch must access to data on a light-path under ORION. Recall in original OCS, in contrast to ORION, an intermediate core switch cannot access to the traffic on a light-path.

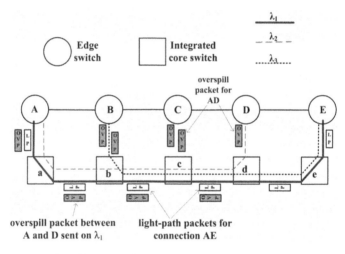

Figure 5.9: An example for integrated ORION network scenario [13]

Figure 5.9 shows a simple ORION network scenario. Suppose that we have three connections: A-a-b-c-d-e-E set up on λ_1, A-a-b-c-d-D set up on λ_2, and B-b-c-d-e-E set up on λ_3. If the capacity of light-path AD is fully utilized, this connection cannot be used for additional traffic transmission when traffic is overloaded between edge switches A and D. Now, suppose that there is a low traffic flow on light-path AE and sometimes there is no traffic carried on it. In original OCS, one cannot send the traffic to edge switch D through light-path AE. On the other hand, the ORION method can increase the bandwidth utilization by sending optical packets of connection AD through light-path AE on λ_1, where these optical packets are called overspill packets (displayed with OVP). The dominant traffic is sent in the normal mode and very few packets are sent in the overspill mode. The process of inserting optical packets belonging to a connection (say, AD) on a different light-path (say, light-path AE is set up on λ_1) is called overspilling. In overspilling, an edge switch must insert an OVP on a wavelength without disturbing the existing light-path traffic on that wavelength. Note that the optical packets sent through their right light-paths are called light-path packets (displayed with LP). One can name the out-of-order delivery of optical packets as a problem for ORION. This problem is because of high delay of overspill packets in comparison with other packets going through the normal wavelength paths.

In this scenario, each core switch that receives an OVP sends it to the relevant edge switch for further routing. In this method, it may be necessary to convert an OVP into electronic mode in the edge switch. Assume core switch b has received an OVP of connection AD. It sends the OVP to edge switch B, and the edge switch sends the OVP either through λ_1 on light-path AE or through λ_3 on light-path BE toward edge switch D again as an OVP. Note that if there was a light-path setup between edge switch B and the destination of OVP (i.e., D), the OVP would be sent as a regular traffic through the light-path BE. Core switches c and d perform the similar task, except that edge switch D stops overspilling.

In ORION, there is no way for deflection routing since an OVP should follow the same route as its light-path route, but on a different wavelength. The overspill method is a temporary solution for a transient overload traffic. If the overload traffic remains for a long period of time, new light-paths should be set up for the overload traffic.

Intermediate core switches should be able to detect the overspill packets from the light-path packets in order to handle them. For distinguishing an overspill packet from a light-path packet, a marker which could be a Frequency Shift Keying (FSK) signal is used. This marker should be recognized easily without using any O/E convertors. When a marker is detected on a wavelength, the overspill packet must be extracted, while a light-path packet must be left on the wavelength. In short, ORION needs a low packet processing capacity since it only needs to handle overspill packets, not all the packets on a wavelength.

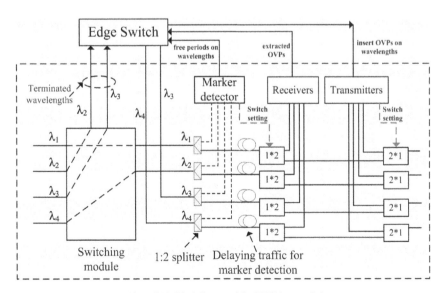

Figure 5.10: Block diagram of the ORION core switch

Figure 5.10 displays the block diagram of ORION core switch for four wavelengths. After switching the light-paths through the switching module, the packets carried on wavelengths are evaluated. To detect the marker of OVP packets, the power of each packet on a wavelength is split and processed by the marker detector unit. To provide enough time for processing of a marker, delay lines are used after splitting. If the extracted marker indicates an OVP, 1*2 fast optical switches are set to route the OVP toward the edge switch. If we assumed that the OVPs are rarely observed in a core switch, we could combine the outputs of these 1*2 switches and use only one tunable receiver. For this case, some contention resolution mechanism must be used for the case that more than one OVP enter the core switch. Otherwise, the receiver unit should use W fixed receivers. After receiving the OVP by the edge switch, the edge switch decides whether to send the OVP via the light-path originated at the edge switch toward the destination of the packet or again in the overspill mode.

The marker detector has another function as determining the start and end of occupied periods on wavelengths. Then, the information about the availability of free periods on each wavelength is sent to the edge switch so that the edge switch knows when it can send an OVP on any wavelength. For this purpose, the final stage of 2*1 switches are accordingly set.

5.2.3.2 Optical Migration Capable Network with Service Guarantees (OpMiGua)

Another integrated hybrid network has been proposed in [14, 15] that integrates OCS and OPS in which Guaranteed Service Transport (GST) packets are sent through the OCS networking on the setup light-paths without any collision with other GST packets and no header processing, and the remaining packets are sent through the OPS networking according to their attached header information. This enables high throughput for network traffic. There are network services with strict demands with no packet loss, low delay, and no jitter such as real-time streaming and remotely assisted surgery. The GST supports these types of services. The GST traffic rate is usually lower than OPS traffic, and the GST packets should receive the absolute priority over OPS packets.

In this method, resources are divided between OCS or OPS in time for achieving the increased resource utilization by using statistical multiplexing. The basic idea of this integration is similar to ORION; i.e., OPS packets are tried to be transmitted through light-paths in the light-paths' vacant spaces; in other words, the OPS packets are interleaved between the GST packets. The OPS packets in this hybrid method are divided into two classes: High Class Transport (HCT) and Normal Class Transport (NCT). HCT packets should encounter low PLR, minimum delay, and minimum jitter, while NCT packets may have moderate PLR and relaxed delay and relaxed jitter.

Figure 5.11 displays the core switch architecture in OpMiGua. At the core switch input, the class of the packets (wither GST packets or OPS packets) are specified. The packets from two classes can be divided by using FSK header labeling and fast optical switches, or by using orthogonal states of polarization labeling. At the core

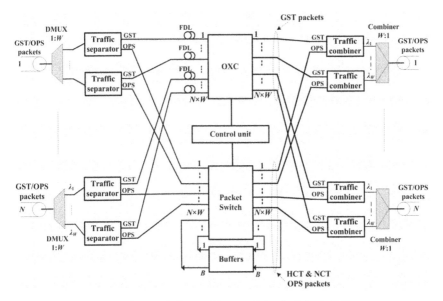

Figure 5.11: Block diagram of the core switch under OpMiGua [14]

switch output, packets from OXC and packet switch are combined, and then different combined signals are multiplexed to make a fiber output signal.

Under asynchronous OPS, an extracted OPS packet (HCT or NCT) is scheduled in the packet switching module. On the other hand, when the packet type is GST, it is delayed by an FDL for a suitable period of time and then switched in the OXC. Then, when the GST packet reaches the output port, the output will be always empty. This delaying is to ensure GST packets absolute priority. Under slotted OPS, the priority of all received packets in a time slot is checked; and if the packet is a GST packet, it will be absolutely sent. Then, the OPS packets contending with the GST packet will be possibly rejected.

Whenever a GST packet arrives at a core switch input port, the destined output port is reserved for a duration corresponding to the duration of a maximum length of HCT and NCT packets. The output may be busy scheduling a HCT/NCT packet when the GST packet arrives. Therefore, from the time the output finishes the HCT/NCT packet scheduling until the GST packet enters the output port, the core switch does not allow servicing any packets. This time gap which causes additional blocking is called Reservation Induced Blocking (RIB).

The packet switching module in Fig. 5.11 has a buffering module with B buffers for storing OPS packets, where the HCT packets have access to all the buffer inputs, and the NCT packets have access to only some inputs of the buffer. This is for service differentiation of the HCT and NCT packets implemented by a scheduling algorithm. A HCT packet should be scheduled from the buffer as soon as a wavelength becomes free to the HCT packet destination. However, this is possible for electronic buffering. There are some benefits for OpMiGua:

- GST Support: It supports GST with no loss and fixed delay while it provides high utilization by using OPS packets simultaneously.

- Reducing OPS Resource Requirements: Both OPS traffic and GST traffic can use the same wavelength channel. This bandwidth sharing reduces network resources required for OPS traffic.

- Support Migration: Old OCS networks can be supported under GST services, while new services can be supported under OPS services.

- Cost and Performance: The combination in this model can offer both lower cost and better performance.

An improved version of OpMiGua has been proposed in [16], where the performance of the HCT/NCT packets (i.e., Statistically Multiplexed (SM) class) is satisfied by applying packet-level Forward Error Correction (FEC). Two implementation strategies are proposed. The RedSM (Redundancy SM) transmits redundant optical packets as SM traffic and RedGST (Redundancy GST) transmits redundant optical packets as low-priority GST traffic. The performance evaluation results show that RedGST is the most efficient scheme if overload of GST light-paths is avoided. On the other hand, RedSM provides low sensitivity to large variations in the input traffic.

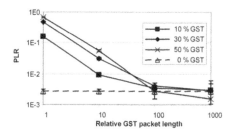 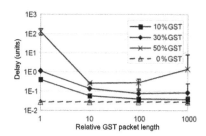

Figure 5.12: HCT PLR as a function of GST packet length for various GST traffic shares, no NCT [14]

Figure 5.13: HCT delay as a function of GST packet length for various GST traffic shares, no NCT [14]

Figure 5.12 illustrates PLR and buffering delay performances for HCT traffic as a function of GST packet length for various GST traffic shares (0% to 50%) and no NCT traffic. In the diagrams, the GST packet length is fixed and its value is set to 1 to 1000 times of the average HCT/NCT packet length in the horizontal axis. The GST traffic share is varied, while total traffic load is kept constant. For short GST packets, the impact of RIB is high and the PLR of HCT packets is high under all GST traffic shares, but the PLR increases with GST traffic share, as expected. For long GST packets, the RIB is reduced, and the GST traffic share has less impact on the resulting PLR. For very long GST packets, increased GST traffic improves the PLR. Here, RIB is negligible, and the high GST traffic shares cause a relative increase in available buffering resources, causing the HCT PLR reduction. The RIB also influences the buffering delay performance in core switches (see Fig. 5.13). For short GST packet lengths, RIB is high, and therefore the probability for contention at the outputs will be high, and buffered HCT packets stay in the buffer for a long time. On the other hand, when the GST packet length increases, the time each GST packet occupies the switch outputs will increase, and therefore the time that buffered packets stay in the buffer before an output becomes idle will increase.

5.2.3.3 3-Level Integrated Hybrid Optical Network (3LIHON)   The 3LIHON [17–19] is an extension of OpMiGua with the following advantages compared with OpMiGua : (1) better QoS support for short packets with high real-time demands, (2) better utilization of network resources, and (3) cheaper core switch architecture. There are three classes in 3LIHON:

- **Guaranteed Service Type (GST)**: GST resembles an optical circuit switched service with no traffic loss inside the network. It is suitable for video/music services and applications generating large information units for transport.

- **Statistically Multiplexed Real Time (SM/RT)**: This class is used for packet switched services with contention for bandwidth and possible packet loss inside core switches. This class must ensure no delay or very limited delay for optical packets inside core switches. It is suitable for voice conversational, high real-time interactive messaging services, and low bandwidth services (say music streaming). Note that GST traffic travels through the network without any loss, while SM/RT traffic is sent in gaps in-between GST packets.

- **Statistically Multiplexed Best Effort (SM/BE)**: This class is used for packet switched services with very small overall packet loss (that can be retransmitted later on), but with no guaranteed delay inside core switches. It is suitable for (a) all applications/services that generate non-large information units for transport, and (b) low real-time interactive messaging services.

Figure 5.14 displays the core switch architecture under 3LIHON, where there are three switching modules inside the core switch: one all-optical OXC with optical buffers at its input side for switching of GST traffic (where a GST packet is forwarded to its pre-established wavelength on the output fiber), one OPS switch with no (or limited) FDL buffering and wavelength conversion for switching of SM/RT optical packets, and the other one for switching of SM/BE optical packets. The last switching module could be either an Electronic Packet Switch (EPS), i.e. a store-and-forward electronic router, or a hybrid OPS with electronic buffering. Both of the second and third modules are responsible for resolving the contention of optical packets. The detect packet type module used for each wavelength channel separates the type of arriving traffic and forwards it to the appropriate switching module.

Since three different traffic types may compete to use the same wavelength channel on a core switch output link, collision avoidance must be managed. Maximum priority is assigned to GST packets, which can always be forwarded without loss through pre-established virtual circuits. Consider a SM/RT packet is being sent out on wavelength channel j at output port k. Similar to OpMiGua, GST packets should

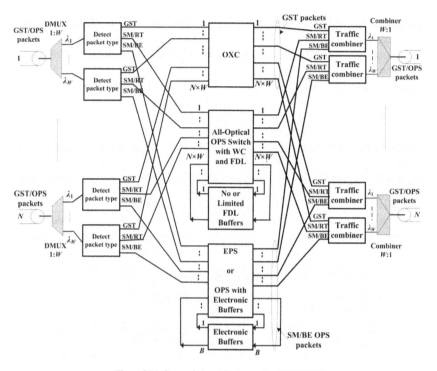

Figure 5.14: Core switch architecture under 3LIHON [17]

enter a fixed-length FDL (with length equal to the maximum SM/RT packet length) in order to allow the current SM/RT packet to finish its switching. Therefore, there is no collision among GST transport class and SM/RT transport class as GST has a non-preemptive priority over the SM/RT. However, GST packets can preempt SM/BE packets. Therefore, when a GST packet arrives, the EPS/OPS switching module must immediately stop sending out an SM/BE packet under transmission. Note that SM/RT traffic can be switched when it finds a free wavelength. It can also preempt SM/BE packets under transmission.

Define b as the ratio between the GST burst length and the SM/RT average packet length. Choosing high (low) values of b results in a small (large) number of GST bursts, but each with large size (small size). Consider traffic mix 60% GST, 10% SM/RT, and 30% SM/BE. Figure 5.15 displays packet loss probability of SM/RT traffic as a function of parameter b, varying the total load on output wavelength channels. Due to the higher priority of GST traffic, SM/RT packets experience loss, which decreases by b. This is because, for a given GST load, the higher the GST burst length (compared with the average SM/RT), the lower the Reservation Induced Blocking (RIB) (see Section 5.2.3.2). In addition, this means that the probability for an arriving SM/RT packet to find an available wavelength increases. For different

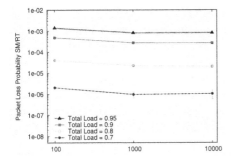

Figure 5.15: Packet loss probability of SM/RT traffic as a function of length ratio b [18]

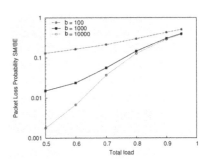

Figure 5.16: Packet loss probability of SM/BE traffic as a function of total load on output wavelengths [18]

tested values of b, the difference is not high, and thereforethe values higher than $b = 100$ ensures system efficiency.

Figure 5.16 illustrates packet loss probability of SM/BE traffic, where traffic loss in this case is due to preemption caused by higher-priority traffic (both GST and SM/RT). When traffic load is not too high, loss is lower at higher values of b since there are less GST bursts (but large size) in the network. In this case, it is less likely that an arriving GST burst will interrupt SM/BE packets.

5.2.3.4 Integrated Hybrid Optical Switching (HOS)
The integrated Hybrid Optical Switching (HOS) integrates OCS, OBS, and OPS [20–24]. This method is based on a unified control packet for all incoming data types. There is a field in the header by which a core switch realizes whether the control packet is a light-path setup request, a burst control packet, or an optical packet header. Then, the core switch uses a suitable scheduling algorithm according to the type of the data. In this integration, OCS traffic has the highest priority and OPS traffic has the lowest priority. Here, some wavelength channels may operate in time-slotted mode and some other in asynchronous mode.

Client packets arriving at the edge switches of HOS are mapped to four traffic types in optical domain as circuit, packet, short-burst, and long-burst defined in the following:

- **Circuits as OCS Traffic**: This traffic type is used for transporting the services that have the most stringent requirements in terms of QoS characterized by large and stable flows. Data are queued at an ingress switch until a circuit has been set up through the network using a two-way reservation mechanism. After establishing the circuit, data are carried without loss, jitter, and delay. A circuit reserves one wavelength on each link along the path that connects the source and destination nodes. Note that in a TDM circuit, time is divided into frames with a fixed number of time slots. The nodes along the circuit path can fill unused TDM slots of a circuit with their optical packets with suitable length that have the same destination as the circuit. Telephony and multimedia conferencing traffic are two types of client packets that can be mapped to circuits.

- **Long Bursts as OBS Traffic**: This traffic type is appropriate for applications that generate long flows with not stringent requirements in terms of delay (such as broadcast video traffic, low latency data, and high throughput data). Because of long aggregation time, long bursts experience a relatively high delay. On the other hand, offset times for long bursts are relatively high so that long bursts experience low loss rates. This is because long offset times provide some sort of prioritized handling for long bursts compared with short bursts and packets [25]. Long bursts are sent on asynchronous wavelength channels.

- **Short Bursts as OBS Traffic**: This traffic type is suitable for applications that require a minimum guaranteed level of QoS in terms of both loss and delay when traffic changes rapidly (such as real-time interactive and multimedia streaming traffic). The First Fit Unscheduled Channel with Void Filling (FFUC-VF) and the Best Fit with Void Filling (BF-VF) algorithms can be used for scheduling of both long-burst and short-burst traffic flows on asynchronous wavelength channels.

- **Packets as OPS Traffic**: This traffic type is used to transport services that have high tolerances to loss, delay, and jitter (such as low-priority data). When a client packet reaches an edge switch, the switch tries to send it via a light-path within a free time slot. When there is no free time slot in the light-path, the packet will be sent on the first idle asynchronous wavelength channel. A core switch can also fill the unused time slots in a frame with OPS packets of suitable length, but with the same destination as the light-path destination.

On a time slotted wavelength channel, in order to obtain a time slot in competition between a circuit, a burst, and a packet, the winner will be the circuit since the circuit has the highest priority and it can be blocked only by other circuits. Consequently, in case of high circuit request rate on the same output buffer, the bursts and the packets on that fiber would be blocked. Therefore, there must be a limit on the number of circuit requests in each fiber.

Notice that HOS is different from both ORION and OpMiGua in which an optical packet could be sent via a light-path even though its destination is not the same as the light-path end. Under HOS, an optical packet scheduled in a time slot of a light-path does not consume new network resources, thus leaving more space available for other traffic flows.

Figure 5.17 illustrates a HOS core switch architecture. The control information associated with each wavelength channel is extracted by the Control Information Extractor (CIE) unit and then sent to the controller, while the corresponding data are routed toward the switching module. The control unit converts the control signal into the electronic domain and processes it in order to choose the proper output port for the data. The switching module includes (a) a fast optical switch fabric based on the fast SOA technology and (b) a slow optical switch fabric based on the slow MEMS technology. The fast optical switch is used for switching of the short-burst and packet traffic, and the slow optical switch is used for switching of the long-burst and circuit traffic. Wavelength converters and FDLs can be used for contention resolution at

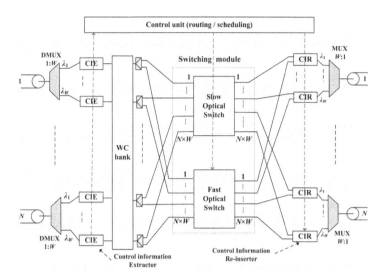

Figure 5.17: Core switch architecture in HOS [20, 21]

output ports. After switching of network traffic, the header information is inserted in optical data in CIR units.

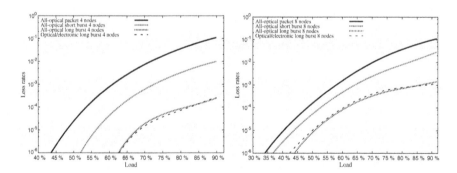

Figure 5.18: Data loss rate as a function of the input load in each link under HOS [21]

Figure 5.18 illustrates average data loss rates of all-optical packets, short bursts, and long bursts as a function of the input traffic load in two network topologies with four and eight cascaded core switches. In the all-optical HOS network with four cascaded core switches, optical packets have the highest loss probability. Due to using shorter offset times, short bursts experience higher loss rates than long bursts, but always lower than the optical packet loss rates, thus providing a minimum level of QoS. Long bursts have the lowest loss rates, always lower than 10^{-3} even at very high input traffic loads. For eight cascaded core switches, the same trend can be observed, but with higher loss probabilities compared with four cascaded core switches. It should be noted that the short and long bursts suffer too much from the increased number of cascaded core switches. On the other hand, optical packets do not suffer too much from the increased number of cascaded core switches since they can be inserted into the unused TDM slots of circuits.

The diagrams in Fig. 5.18 also show data loss rate for optical/electronic long bursts. Note that in the optical/ electronic HOS network, the packets and short bursts are transmitted using the fast electronic switch and electronic buffers are used to solve contentions in core switches. Therefore, it can be assumed that their loss rates are negligible. Long bursts that are forwarded using slow optical switching elements show almost the same loss rates as in the all-optical HOS network. Note that the circuit setup blocking probability is always zero in both network topologies and under both the all-optical and optical/electronic mechanisms.

The proposed hybrid model is able to provide an efficient handling of data packets, bursts, and circuits directly in the optical domain. The performance analysis shows that the PLR remains lower than 10^{-8} up to almost 60% of offered load, while the burst loss rate is lower than 10^{-8} for almost 75% of traffic load. Furthermore, the packet loss decreases when increasing the percentage of traffic reserved for circuits. This is because of the possibility for scheduling of optical packets on suitable unused TDM slots in established circuits. In contrast, the burst loss increases by increasing the circuit traffic. It is worthy to mention that using fiber delay lines for packet contention resolution implies a significant increase in the node performance, and using a more efficient burst scheduling algorithm can reduce the burst loss probability.

5.3 Comparison of Hybrid OPS Schemes

Hybrid optical networks can improve the overall network design by bringing heterogeneous networks together. One can compare the hybrid networks from three aspects as resource requirement, technology complexity, and control complexity. In case of resource requirement, integrated networks could be better than both parallel and client–server networks as integrated networks can interleave OPS traffic within the already established light-paths.

If we compare the hybrid networks according to their technology and control complexity, client–server networks perform better than both integrated and parallel networks since in client–server networks edge switches and core switches do not take part in the selection of the proper network technology (OCS/OBS/OPS) for sending a packet. Comparing different types of hybrid networks, it is obvious that client–server networks are preferred to integrated and parallel networks in both technology and control complexity; but in case of resource requirements, integrated networks can provide the best performance. While the performance of integrated hybrid optical networks is superior to other hybrid optical networks, physical implications of integrated hybrid optical networks are very challenging.

The summary of these comparisons are presented as follows [1]:

- **Resource Requirement**: integrated < parallel < client–server

- **Technology Complexity**: client–server < parallel < integrated
- **Control Complexity**: client–server < parallel < integrated

5.4 Summary

Hybrid OCS/OPS networks are very promising for fulfilling future networks demands with very good performance results. Hybrid optical networks are classified into three classes based on their degrees of interaction and integration: client–server, parallel, and integrated. In the client–server network class, the server layer offers virtual topology for the OPS client layer. In the parallel network class, only edge switches select a proper switching mode for sending received traffic flows. In the integrated network class, both edge nodes and core nodes could select a proper switching mode for sending received traffic flows. In this chapter, each class has been reviewed in detail and some examples have been provided for each of them. Comparing different hybrid networks, it is obvious that client–server networks are preferable to the integrated and parallel networks in both technology and control complexities; but in case of resource requirements, integrated networks can provide the best performance.

REFERENCES

1. C. M. Gauger, P. J. Kuhn, E. V. Breusegem, M. Pickavet, and P. Demeester. Hybrid optical network architectures: bringing packets and circuits together. *IEEE Communications Magazine*, 44(8):36–42, 2006.

2. A. Pattavina. Architectures and performance of optical packet switching nodes for IP networks. *IEEE/OSA Journal of Lightwave Technology*, 23(3):1023–1032, Mar. 2005.

3. M. OMahony, D. Simeonidou, D. Hunter, and A. Tzanakaki. The application of optical packet switching in future communications networks. *IEEE Communications Magazine*, 39(3):128–135, Mar. 2001.

4. I. Miguel, J. C. Gonzlez, T. Koonen, R. Durn, P. Fernndez, and I. T. Monroy. Polymorphic architectures for optical networks and their seamless evolution towards next generation networks. *Photonic Network Communications*, 8(2):177–189, Sept. 2004.

5. X. Zheng, M. Veeraraghavan, N. S. V. Rao, Q. Wu, and M. Zhu. CHEETAH: circuit-switched high-speed end-to-end transport architecture testbed. *IEEE Communications Magazine*, 43(8):11–17, 2005.

6. C. Xin, C. Qiao, Y. Ye, and S. Dixit. A hybrid optical switching approach. In *IEEE GLOBECOM*, San Francisco, CA, Dec. 2003.

7. G. M. Lee, B. Wydrowski, M. Zukerman, J. K. Choi, and C. H. Foh. Performance evaluation of an optical hybrid switching system. In *IEEE Golbecom*, San Francisco, CA, Dec. 2003.

8. D. Simeonidou, G. Zervas, and R. Nejabati. Design considerations for photonic routers supporting application-driven bandwidth reservations at sub-wavelength granularity. In

Quality of Service in Optical Packet Switched Networks, First Edition.
By Akbar Ghaffarpour Rahbar Copyright © 2015 IEEE. Published by John Wiley & Sons, Inc.

The Third International Conference on IEEE in Broadband Communications, Networks and Systems, San Jose, CA, Oct. 2006.

9. K. Machida, H. Imaizumi, H. Morikawa, and J. Murai. A RWA performance comparison for hybrid optical networks combining circuit and multi-wavelength packet switching. In *International Conference on Optical Network Design and Modeling (ONDM)*, Braunschweig, Germany, Feb. 2009.

10. K. Watabe, T. Saito, N. Matsumoto, T. Tanemura, H. Imaizumi, A. A. Amin, M. Takenaka, Y. Nakano, and H. Morikawa. 80Gb/s multi-wavelength optical packet switching using PLZT switch. In *Optical Network Design and Modeling*, pages 11–20, Athens, Greece, May 2007.

11. E. V. Breusegem, J. Cheyns, D. D. Winter, D. Colle, M. Pickavet, P. Demeester, and J. Moreau. A broad view on overspill routing in optical networks: a real synthesis of packet and circuit switching? *Optical Switching and Networking*, 1(1):51–64, Jan. 2005.

12. J. Cheyns, E. V. Breusegem, D. C. UGent, M. P. UGent, and P. Demeester. ORION: a novel hybrid network concept: overspill routing in optical networks. In *IEEE International Conference on Transparent Optical Networks*, pages 144–147, Warsaw, Poland, June 2003.

13. K. Christodoulopoulos, E. Van Breusegem, K. Vlachos, M. Pickavet, E. Varvarigos, and D. Colle. Performance evaluation of overspill routing in optical networks. In *IEEE International Conference on Communications*, pages 2538–2543, Istanbul, Turkey, June 2006.

14. S. Bjornstad, D. R. Hjelme, and N. Stol. A packet-switched hybrid optical network with service guarantees. *IEEE Journal on Selected Areas in Communications*, 24(8):97–107, Aug. 2006.

15. M. Nord, S. Bjornstad, O. Austad, V. L. Tuft, D. R. Hjelme, A. S. Sudb, and L. E. Eriksen. OpMiGua hybrid circuit- and packet-switched test-bed demonstration and performance. *IEEE Photonics Technology Letters*, 18(24):2692–2694, Dec. 2006.

16. A. Kimsas, S. Bjornstad, H. Øverby, and N. Stol. Improving performance in the OpMiGua hybrid network employing the network layer packet redundancy scheme. *IET Communications*, 4(3):334–342, 2010.

17. N. Stol, M. Savi, and C. Raffaelli. 3-level integrated hybrid optical network (3LIHON) to meet future QoS requirements. In *IEEEE Globecom*, Houston, TX, 2011.

18. G. Leli, C. Raffaelli, M. Savi, and N. Stol. Performance assessment of congestion resolution scheduling in asynchronous 3-level integrated hybrid optical network (A-3LIHON). In *International Telecommunications Network Strategy and Planning Symposium (NETWORKS)*, Rome, Italy, Oct. 2012.

19. S. Yang and N. Stol. Architecture and performance evaluation of the edge router for an integrated hybrid optical network. In *IEEE International Conference on Communications in China (ICCC)*, 2012.

20. M. Fiorani, M. Casoni, and S. Aleksic. Performance and power consumption analysis of a hybrid optical core switch. *IEEE Journal of Optical Communications and Networking*, 3(6):502–513, June 2011.

21. M. Fiorani, M. Casoni, and S. Aleksic. Hybrid optical switching for energy-efficiency and QoS differentiation in core networks. *Journal of Optical Communications and Networking*, 5(5):484–497, May 2013.

22. M. Fiorani, M. Casoni, and S. Aleksic. Large data center interconnects employing hybrid optical switching. In *The 8-th Conference on Network and Optical Communications (NOC), 2013 18th European Conference on and Optical Cabling and Infrastructure (OC&i)*, Graz, 2013.

23. M. Fiorani, S. Aleksic, and M. Casoni. Hybrid optical switching for data center networks. *Journal of Electrical and Computer Engineering*, 2014.

24. M. Fiorani, M. Casoni, and S. Aleksic. Hybrid optical switching for an energy-efficient Internet core. *IEEE Internet Computing*, 17(1):14–22, Jan.-Feb. 2013.

25. M. Yoo, C. Qiao, and S. Dixit. QoS performance of optical burst switching in IP-Over-WDM networks. *IEEE Journal on Selected Areas in Communications*, 18:2062–2071, 2000.

CHAPTER 6

METRO OPS NETWORKS

Future metro networks should provide more capacity in order to cope not only with the current traffic loads but also with the unexpected future demand growth. The next generation all-optical metro networks should be agile, where network bandwidth should be dynamically allocated to traffic demands. Due to the high traffic dynamics, OPS is necessary for metro networks in order to use network resources efficiently. Therefore, all-optical OPS appears to be the sole approach to provide such capacity [1–3].

In this chapter, recent architectures of OPS suitable for metro networks (with ring and star topologies) are studied. Unlike mesh networks, configuration and management of the ring and star architectures is simple. Note that OPS switches used in this chapter are active optical switches, not passive switches.

6.1 OPS in Star Topology

One common network topology for a metro network is the single-hop star topology with active optical switches, where the operation of each core switch in switching traffic is independent of other core switches. This means that the overlaid topology

Quality of Service in Optical Packet Switched Networks, First Edition.
By Akbar Ghaffarpour Rahbar Copyright © 2015 IEEE. Published by John Wiley & Sons, Inc.

can be divided into independent single-star networks. One advantage of the overlaid star topology is to support load balancing by equally dividing its network traffic among the several parallel (overlaid) single-star networks. Overlaid OPS networks will be studied in this section.

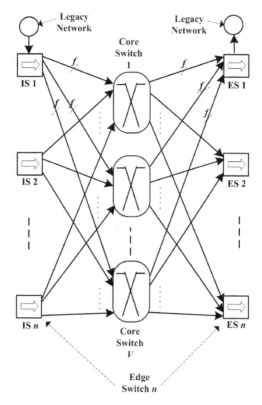

Figure 6.1: Basic single-hop overlaid OPS network model

6.1.1 Overlaid OPS Architectures

Figures 6.1 and 6.2 illustrate the overlaid single-hop OPS network model. This network overlays V OPS core switches, where n edge switches are connected to each core switch. The core switches may be either co-located or located at different places. Each core switch is an all-optical wavelength selective and buffer-less optical switch. In overlaid OPS, the core switch has $N = n$ input ports/links and $N = n$ output ports/links; i.e., an $n * n$ switch. Note that term input/output port is used in single-fiber networks, and term input/output link is used in multi-fiber networks, where each link has f ports.

In the basic architecture depicted in Fig. 6.1, each edge switch is connected to the core switch with f fibers and each fiber includes W wavelengths and one control

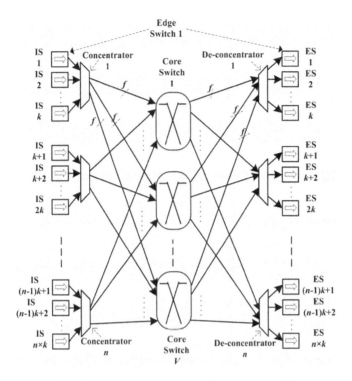

Figure 6.2: Alternative single-hop overlaid OPS network model

channel. Each core switch has $n \times f$ input ports and $n \times f$ output ports. Transmission of control information between edge switches and a core switch is carried out on the control channel. Since ingress switches can divide their traffic uniformly among the core switches, the overlaid topology can be divided into V independent single-hop networks. As can be observed, edge switch i has two parts illustrated separately: (a) ingress switch i for transmitting the traffic received from legacy networks to the OPS network on f fibers and (b) egress switch i for receiving the traffic from the OPS network on f fibers and delivering it to the legacy networks. Note that there is no FDL in any core switch, and edge switches use electronic buffers to save client traffic and send the traffic to the network at appropriate times.

An alternative overlaid OPS architecture could be based on using concentrators and de-concentrators to increase the number of edge switches [4, 5]. Figure 6.2 displays the multi-fiber architecture for this alternative overlaid OPS, where each core switch has $n \times f$ input ports and $n \times f$ output ports, but the number of edge switches is increased to $n \times k$; i.e., k edge switches are managed to be connected to each port of the core switch. In this architecture, the number of input/output links in core switches reduces significantly. A concentrator used between k ingress switches and V core switches selects the traffic from the desired ingress switch and sends it to the core switches. At the other side, a de-concentrator used between V core switches and k egress switches selects the traffic received from the core switches and sends it to the relevant egress switch. In this architecture, two mechanisms can be used for sharing bandwidth between k ingress switches. One choice is to assign some fibers/wavelengths to each ingress switch. In this case, a concentrator could be f passive couplers and a de-concentrator could be f passive splitters. In another choice when the network is slotted OPS and fibers/wavelengths can be used by all k ingress switches, a concentrator should assign time slots on wavelength channels to k ingress switches by a scheduling mechanism such as round robin.

6.1.2 Advantages of Overlaid OPS Architectures

The overlaid-star topology has the following merits in terms of topological design in OPS networks [4]:

- Since the transmission path between two edge switches passes through only one core switch, one can simply use pure optical switching in the core switch and use electronic buffers at the edge switches.

- Overlaid OPS is simple in routing, configuration, and control plane.

- Apart from the design simplicity, the synchronization required in a slotted OPS network can be easily achieved by the overlaid OPS network since time-slotted multi-hop OPS networks need precise time coordination among core switches [2].

- Overlaid OPS encounters a reduced switch crosstalk because of having only one core switch along a transmission path compared to ring and mesh topologies [2, 4].

- Since a star network suffers from the central node failure problem, the overlaid star topology can provide robustness and reliability. This is because of using overlaid core switches so that when a core switch fails, edge switches use another core switch for traffic transmission in the network [4, 6]. This is why we should have at least $V = 2$. Path reliability has been studied in [7], and the traffic protection problem has been studied in [8]. By monitoring optical signals at the receiver side, a core switch or link failure can be detected. Some spare capacity should be provided on each link in order to reroute the traffic of a failed link. Compared with the ordinary protection mechanisms in mesh networks, the protection and restoration in overlaid OPS is simple, flexible, and efficient when dealing with failures at different levels (such as wavelength level, link level, or core switch level).

- Overlaid OPS can provide a good scalability as the network size grows physically because it is easy to add an edge switch to the network, especially in the alternative overlaid OPS architecture (see Fig. 6.2).

- Overlaid OPS can reduce the big concern for telecommunication carriers; i.e., operating expenses including network planning, network management (such as installation, path provisioning, and alarm collection), network expansion (such as system capacity and number of switches), and system maintenance (such as hardware and software systems, and different platforms).

6.1.3 Reservation-Based Traffic Transmission in Overlaid OPS

A loss-free OPS is much desired in overlaid-star OPS networks, where no traffic is lost in the network. In reservation-based traffic transmission mechanisms, pro-posed mainly for time slotted networks, core switches are responsible for scheduling of edge switches traffic in a way that no optical packet contention occurs in core switches. Under reservation-based traffic transmission, the overlaid OPS architec-ture does not need any contention resolution mechanism such as wavelength con-version, optical buffering, or deflection routing in core switches. Reservation-based techniques are contention avoidance schemes that provide a loss-free OPS network. They perform well in terms of packet loss and bandwidth utilization to meet real–time QoS requirements.

Since the distance between edge switches to a core switch is usually short in a metro area, the reservation-based scheduling schemes can be applied to star topology because the operation of reservation scheduling will not lead to a long delay. Under reservation-based traffic transmission, the scheduling is executed periodically at a core switch in response to traffic demands from edge switches. These techniques are complex and impose a large processing burden on each core switch to schedule traffic demands of all edge switches, especially when the number of edge switches is high.

The idea behind the scheduling in a core switch can be taken from the scheduling in input-queued switches that map the scheduling problem to the bipartite match-ing problem. The scheduling in input-queued switches can be either slot-by-slot or frame-based, where a frame includes F time slots. In slot-by-slot reservation, the scheduling is carried out at time slot level. For example, the slot-by-slot schedul-ing [6, 9, 10] is used to perform the scheduling in overlaid star networks. Here, for each traffic demand requested by ingress switches, only the $n \times W$ time slots within a time-slot period are reserved by the core switch. The slot-by-slot design could not be suitable for a metro topology because of the performance degradation when edge switches are located at long distances from the core switch. On the other hand, in frame-based scheduling such as [11–14], the scheduling algorithm is run once within a frame time to obtain the switch configuration in each time slot inside the frame. A frame-based scheduling performs the scheduling for a number of time slots at the same time and can decrease the frequency of matching computation [12] compared with the slot-by-slot algorithms. In addition, a frame-based scheduling can reduce the communication overhead during scheduling [15] and can accommodate more scheduled slots within a frame with width F time slots than F times slot-by-slot schedulings [16], where there is only one communication channel in the network.

Due to its suitability, the frame-based scheduling is studied in this section. The Birkhoff von Neumann (BvN) scheduling [17, 18] could be used at the OPS core switches of an overlaid network to schedule lossless traffic transmission among edge switches. Since ingress switches can divide their traffic uniformly among the core switches, only one star network is studied in the following discussions. Using the BvN can guarantee the best scheduling for the required traffic between any source–destination pair, but at the expense of high complexity to perform perfect matching. Using heuristic matching algorithms such as [19], the complexity of the scheduling can be reduced, but at the expense of bandwidth wastage. Let us consider an example for frame-based scheduling based on BvN. Each edge switch sends its traffic demand to other edge switches in the network toward the core switch. Based on the traffic demands received from all edge switches, the core switch creates a traffic demand matrix. This traffic demand matrix must first be converted to a doubly stochastic matrix at the core switch. Then, the demanded traffic can be scheduled within a frame by using the BvN decomposition algorithm that can use either the bipartite matching algorithm or the maximum flow problem as its core function. Then, the edge switches are informed on which time slots they can send their traffic.

It should be mentioned that the scheduling algorithms for input-queued switches (including the BvN scheduling) are mainly defined for a network with single channel only. The BvN has also been used in a symmetric single-channel optical star network, where symmetric means that all edge switches are located at the same distance from the core switch and propagation delay on each link is an integer number of time-slot periods [13]. However, in the following, BvN is detailed for an asymmetric multi-fiber–multi-wavelength overlaid-star OPS network.

To use BvN in multi-fiber–multi-wavelength networks, four approaches could be utilized:

- One way to schedule the network traffic using the BvN scheduling algorithm is to divide each request equally among $f \times W$ wavelength channels. Then, the core switch uses $f \times W$ separate schedulers to carry out BvN scheduling on $f \times W$ wavelength channels. However, this needs $f \times W$ processors in the core switch to perform the $f \times W$ schedulings, thus resulting in an expensive network. This approach is named Multi-Processor BvN (MPBvN) from now on [11].

- Another BvN method to schedule traffic separately for each wavelength channel is called Separated BvN (SBvN) scheduling. In SBvN, after dividing each request equally among $f \times W$ wavelength channels, the core switch uses only one processor and then performs $f \times W$ consecutive BvN schedulings. However, this approach may be time-consuming [20].

- The third approach, Efficient BvN (EBvN), uses one processor to schedule traffic in multi-fiber–multi-wavelength networks more efficient than SBvN [20].

- Finally, EBvN with Filing Empty Cells (EBvN_FEC) is an improved version of EBvN that allocates empty cells in a scheduled frame so that more traffic can be scheduled than EBvN [20].

6.1.3.1 Some Definitions for Frame-Based Scheduling

Consider a time-slotted overlaid-star OPS network, where each star network includes n edge switches connected to a core switch, where each switch is connected to the core switch with f fibers and each fiber includes W data wavelengths (i.e., multi-fiber–multi-wavelength architecture). A number of definitions are necessary to be stated first before detailing the reservations-based scheduling schemes:

Table 6.1: An example for a scheduled frame at $F = 8$, $f = 2$, and $W = 3$ in ingress switch 3. Reprinted from Akbar Ghaffar Pour Rahbar, EBvN: efficient BvN in multi-fiber/multi-wavelength overlaid-star optical networks, *Journal of Annals of Telecommunications* 2012, 67:575–588, Table 1, with kind permission from Springer Science and Business Media.

Wavelength	Fiber	Slot Set1	Slot Set2	Slot Set3	Slot Set4	Slot Set5	Slot Set6	Slot Set7	Slot Set8
λ_1	Fiber$_0$	6	2	1	2	2	5	2	-
	Fiber$_1$	4	2	[5]	6	6	5	2	2
λ_2	Fiber$_0$	-	6	1	5	4	5	5	2
	Fiber$_1$	6	1	6	-	2	-	2	4
λ_3	Fiber$_0$	6	2	4	2	2	5	-	-
	Fiber$_1$	5	2	1	2	2	1	2	-

- **Frame:** A collection of $F \times f \times W$ time slots from all wavelengths on all fibers over a fixed duration of F time slots of an edge switch is called frame. A frame can be visualized as a table with F columns and $f \times W$ rows mapped to $f \times W$ wavelengths, where a number in a frame cell denotes an egress switch address. For example, Table 6.1 shows a frame with $F = 8$ columns and $f \times W = 6$ rows, where the first to the sixth rows denote the schedulings, respectively, on wavelength λ_1 on Fiber$_0$, wavelength λ_1 on Fiber$_1$, wavelength λ_2 on Fiber$_0$, wavelength λ_2 on Fiber$_1$, wavelength λ_3 on Fiber$_0$, and wavelength λ_3 on Fiber$_1$. In this table, the cell with bold number inside brackets shows that ingress switch 3 should send its traffic to egress switch 5 on wavelength λ_1 of Fiber$_1$ at the third column of the frame. As can be observed, there is no traffic destined to egress switch 3 in the table because an edge switch never sends traffic to itself.

- **Slot set:** Set of $f \times W$ rows in a column of a frame is called a slot set (see Table 6.1).

- **OBM (Optical Bandwidth Manager):** There is an OBM unit in each ingress switch that allocates the bandwidth of wavelengths/fibers to different torrents (where the traffic going to the same egress switch is called a torrents) in an integer number of time slots. The OBM should know in advance which torrent should send its traffic in which particular time slot of which slot set.

- **Traffic Demand Array in Ingress Switch i:** The number of time slots required to carry traffic from ingress switch i to any egress switch j (where $1 \leq j \leq n$ and $i \neq j$) is called traffic demand between them. For $i = j$, the number of required slots is set to 0. These n time slot requests build a traffic demand array in ingress switch i that sent to the core switch at an appropriate time so that the core switch will receive all traffic demand arrays from all ingress switches at the same time.

- **Traffic Demand Matrix in the Core Switch:** Upon receiving traffic demand arrays from ingress switches at a frame boundary, the core switch makes a matrix that includes traffic demands between any ingress–egress pair. This matrix is denoted by $A_{n,n}$. An entry a_{ij} of traffic demand matrix $A_{n,n}$ shows that ingress switch i needs a_{ij} time slots to send its traffic to egress switch j. Therefore, a_{ij} is an integer number.

- **Permutation Matrix:** A matrix with entries of only 0 and 1 in which the summation of each row and summation of each column is always equal to 1.

- **Doubly Sub-stochastic Matrix:** A matrix in which the summation of each row and summation of each column of a traffic matrix is less than or equal to 1.0. This matrix is shown with $R'_{n,n}$ in the following.

- **Doubly Stochastic Matrix:** A matrix in which the summation of each row and summation of each column exactly equals 1.0. This matrix is denoted by $R_{n,n}$. To provide permutation matrices from a traffic demand matrix, the BvN scheduling algorithm (see Algorithm 2 in [17, 18]) needs a doubly stochastic matrix as its input. Hence, before running the BvN scheduling, a doubly sub-stochastic traffic demand matrix must first be converted to a doubly stochastic traffic demand matrix.

- **Scheduled Matrix:** The matrix resulted from the BvN scheduling is denoted by $S_{n,n}$. An entry s_{ij} of scheduling matrix $S_{n,n}$ shows that only s_{ij} time slots have been assigned to ingress switch i in order to carry its traffic toward egress switch j.

- **Traffic Load:** Load L shows the number of requested time slots by an edge switch normalized with respect to total number of available time slots for an edge switch within a frame (i.e., $F \times f \times W$).

- **Scheduling Time:** The time from receiving a traffic demand matrix by a core switch until making ready the scheduling for all edge switches is called scheduling time. The worst-case scheduling time for a traffic demand matrix is named maximum scheduling time.

- **Residual Traffic Percentage** (R_p): Percentage of traffic that cannot be scheduled (i.e., blocked). This parameter is obtained based on original traffic demand matrix A and scheduled matrix S entries by

$$R_p\% = 100 \times \frac{\sum_{i=1}^{n}\sum_{j=1}^{n}(a_{ij} - s_{ij})}{\sum_{i=1}^{n}\sum_{j=1}^{n}a_{ij}}. \tag{6.1}$$

6.1.3.2 Common Steps in Frame-Based Scheduling The operations of MP-BvN, SBvN, EBvN, and BvN_FEC only differ in the operations of the core switch. Other steps stated in the following are the same for all of them. Before detailing the steps, the general operations of the BvN frame-based scheduling are illustrated with an example.

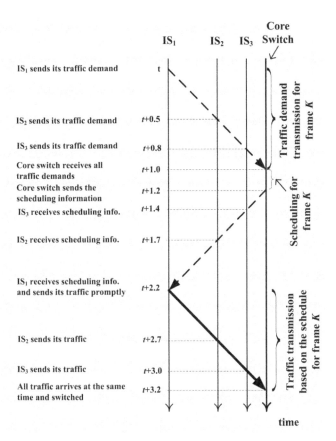

Figure 6.3: The framework of reservation-based scheduling in overlaid-star OPS. Reprinted from *Computer Networks*, Vol. 54, Akbar Ghaffarpour Rahbar and Oliver Yang, Agile bandwidth management techniques in slotted all-optical packet interconnection networks, 387–403, copyright 2010, with permission from Elsevier.

Example: Figure 6.3 shows the time evolution of the BvN operation in a core switch for frame K. The propagation delays from IS_1 (Ingress Switch 1), IS_2, and IS_3 to the core switch are 1.0 ms, 0.5 ms, and 0.2 ms, respectively. Assume the scheduling time for a frame to be 0.2 ms. The transmission time for either the traffic demand information or scheduling information between an ingress switch and the core switch is negligible (and therefore not shown) due to the high data rate of wavelengths. Here, IS_1, IS_2, and IS_3 send their traffic demands at times $t, t + 0.5, t + 0.8$, respectively, so that all traffic demands arrive at the core switch at time $t + 1.0$. Then, the core switch schedules traffic demands and determines a scheduling pattern to be sent back to the ingress switches. As illustrated, IS_3, IS_2, and IS_1 will receive this pattern at $t + 1.4$, $t + 1.7$, and $t + 2.2$, respectively. By this way, they will know when, on which wavelength, and to which egress switch they can send their traffic. As displayed, IS_1 can send its traffic as soon as the scheduling pattern arrives. However, IS_2 and IS_3 must wait and send their traffic at times $t + 2.7$ and $t + 3.0$, respectively. Traffic from all ingress switches will arrive at the core switch starting from $t + 3.2$. Note that this figure displays only the BvN time evolution for one frame. Similar time evolution is applied for other frames. Let a frame duration be T_f. Then, IS_1, IS_2, and IS_3 will send their traffic demands in future frames at times $(t + T_f, t + 2 \times T_f, t + 3 \times T_f, \ldots)$, $(t + 0.5 + T_f, t + 0.5 + 2 \times T_f, t + 0.5 + 3 \times T_f, \ldots)$, and $(t + 0.8 + T_f, t + 0.8 + 2 \times T_f, t + 0.8 + 3 \times T_f, \ldots)$, respectively. Effectively, the BvN framework functions like a pipeline processing in which each ingress switch provides its traffic demand information to the core switch at fixed frame intervals and the core switch prepares a timetable to schedule the traffic of all ingress switches. The steps of scheduling in any BvN are detailed in the following:

- **Synchronization at Frame Level:** Since the arrival of time slots at the core switch from different edges must be synchronized at the frame level in order to respect the collision-free schedule, an edge switch with the longest propagation delay has the most impact on transmission time of other edge switches. Define $T_{s,i}$ to be the time offset that edge switch i must start transmitting its traffic to the core switch after initialization, $D_{P,i}$ to be the propagation delay from edge switch i to the core switch, and $D_{P,max}$ to be the propagation delay from the furthest edge switch. Let z be the furthest edge switch, then we must have $T_{s,i} + D_{P,i} = T_{s,z} + D_{P,max}$ (where $i = 1, 2, \ldots, n$) to provide the synchronized arrival. If we use edge switch z as a reference point and set $T_{s,z} = 0$, we obtain $T_{s,i} = D_{P,max} - D_{P,i}$ for any other edge switches.

- **Constructing Traffic Demands in Edge Switches:** The OBM in an ingress switch calculates the required number of time slots to carry traffic to each egress switch within a frame period at each frame boundary. Define $N_{RS,ij}$ to be the number of required time slots to carry the traffic from ingress switch i to egress switch j (where $1 \leq i, j \leq n$ and $i \neq j$) that has arrived during the previous frame, d'_{ij} to be the demand of required time slots (an integer number) between ingress switch i and egress switch j, and d_{ij} to be the number of required time slots between ingress switch i and egress switch j that OBM reports to the core switch. Now, we say the demand of required time slots between ingress switch i and egress switch j is "granted" if $d_{ij} = d'_{ij}$. Otherwise, $N_{NG,ij} = d'_{ij} - d_{ij}$ is the number of required time slots that are not granted between ingress switch i and egress switch j during the previous frame.

Whenever the total number of time slots demanded to all egress switches exceeds the total number of available time slots within a frame period (with $F \times f \times W$ time slots), we need to reduce the required number of time slots allocated to each egress switch proportionally by a scaling factor γ_i as defined in Fig. 6.4. If $\gamma_i < 1$, the number of non-granted required time slots is updated

Algorithm: Constructing Traffic Demand in Ingress Switch i

1: $d'_{total} = 0$ {initialize total number of demands to be requested in ingress switch i}
2: **for** $j = 1$ to n , where $j \neq i$ {for all egress switches} **do**
3:    $d'_{ij} = N_{NG,ij} + N_{RS,ij}$
4:    $d'_{total} = d'_{total} + d'_{ij}$
5: **end for**

6: $\gamma_i = \min \left(1, \frac{F \times f \times W}{d'_{total}}\right)$ {compute scaling factor in ingress switch i}
7: **for** $j = 1$ to n , where $j \neq i$ {for all egress switches} **do**
8:    $d_{ij} = \lfloor \gamma_i \times d'_{ij} \rfloor$ {scale the volume of demand}
9:    $N_{NG,ij} = d'_{ij} - d_{ij}$ {update non-granted traffic demand}
10: **end for**
11: Send traffic demand information (d_{ij}) to the core switch at the first frame boundary
12: END

Figure 6.4: Construction of traffic demand information in an ingress switch. Reprinted from *Computer Networks*, Vol. 54, Akbar Ghaffarpour Rahbar and Oliver Yang, Agile bandwidth management techniques in slotted all-optical packet interconnection networks, 387–403, copyright 2010, with permission from Elsevier.

for each egress switch. If $\gamma_i = 1$, then the requested time slots to each egress switch can be completely granted, where we have $N_{NG,ij} = 0$. The parameter γ_i is updated in each frame. For example, consider $f = 1$, $n = 4$, $W = 8$, and $F = 40$. In ingress switch $i = 1$, set $N_{RS,12} = 175$, $N_{RS,13} = 140$, $N_{RS,14} = 135$, $N_{NG,12} = 15$, $N_{NG,13} = 10$, and $N_{NG,14} = 20$. We obtain $d'_{12} = 190$, $d'_{13} = 150$, and $d'_{14} = 155$. Based on these demands, we have $\gamma_1 = 0.6465$ from Fig. 6.4. We obtain traffic demand entries by $d_{12} = 122$, $d_{13} = 96$, and $d_{14} = 100$. Then, the non-granted required time slots for the next frame are $N_{NG,12} = 68$, $N_{NG,13} = 54$, and $N_{NG,14} = 55$.

- **Scheduling Traffic Demand Matrix in the Core Switch:** As stated, the operations of MPBvN, SBvN, EBvN, and BvN_FEC differ only in this step that will be discussed in the following subsections.

- **Transmitting Scheduled Traffic by Edge Switches:** Ingress switch i must always wait for synchronization period of $2(D_{P,max} - D_{P,i})$ before transmitting its traffic based on the new scheduling (see Fig. 6.3). As displayed, IS_1 has the maximum propagation delay of $D_{P,max} = D_{P,1} = 1.0$ ms. We have $D_{P,2} = 0.5$ ms and $D_{P,3} = 0.2$ ms for IS_2 and IS_3, respectively. Based on Fig. 6.3, IS_1, IS_2, and IS_3 must wait 0.0 ms, 1.0 ms, and 1.6 ms, respectively, for the synchronization period. For instance, IS_3 receives scheduling information at $t + 1.4$, but it sends its traffic at time $t + 3.0$ (i.e., a waiting time of 1.6 ms). However, there is nothing to stop an ingress switch from sending its traffic from the previous scheduling while waiting for the synchronization of the new scheduling. This overlapping of operation allows us to increase channel utilization while reducing the waiting time.

6.1.3.3 Multi-Processor BvN (MPBvN)   The MPBvN [11, 21] is based on reservation, where the reservation is executed periodically at a core switch in response

to traffic demands of edge switches. The MPBvN uses the idea of scheduling in input-queued switches. As aforementioned, existing scheduling algorithms for input-queued electronic switches (such as BvN) are designed for single-channel scheduling in electronic switches, where both electronic input buffers and the scheduler are located inside the input-queued electronic switches. In OPS model, however, input buffers are located in edge switches to benefit from their electronic buffers, and the core switch is responsible for scheduling. Hence, the scheduler should schedule a delayed version of traffic from edge switches.

The MPBvN modifies BvN in electronic switches in such a way that it can be used to schedule traffic in asymmetric multi-fiber–multi-wavelength star OPS. Upon receiving traffic requests from ingress switches at a frame boundary, the core switch divides each request equally among $f \times W$ wavelengths. Then, the core switch uses $f \times W$ schedulers to carry out BvN scheduling on $f \times W$ wavelength channels, where each scheduling is run on a separate processor.

Let the divided demand between ingress switch i and egress switch j in the core switch for a given wavelength be a_{ij}. Define $N_{RG,ij}$ to be the difference between the number of requested and granted time slots for the traffic demand between ingress switch i and egress switch j in the previous frame.

```
Algorithm:  The MPBvN Scheduling in Core Switch

1:  for i = 1 to n {scan all ingress switches} do
2:      for j = 1 to n {scan all egress switches} do
3:          a_ij = a_ij + max(0,N_RG,ij)
4:      end for
5:  end for

6:  while TRUE do
7:      Create traffic demand matrix A_n,n = [a_ij]_n,n
8:      r_max=maximum of all row summations in A_n,n
9:      c_max=maximum of all column summations in A_n,n
10:     u = max(r_max,c_max)
11:     if u > F then
12:         Adjust each entry in A_n,n by a_ij = F/u × a_ij
13:     else
14:         Convert doubly sub-stochastic matrix R' = [a_ij/u] to doubly stochastic matrix R
15:         Decompose R into R = Σ_{i=1}^k v_i × P_i
16:         Obtain the schedule of each edge switch from S = u × Σ_{i=1}^k v_i × P_i
17:         Exit loop
18:     end if
19: end while
20: END
```

Figure 6.5: The MPBvN scheduling in the core switch. Reprinted from *Computer Networks*, Vol. 54, Akbar Ghaffarpour Rahbar and Oliver Yang, Agile bandwidth management techniques in slotted all-optical packet interconnection networks, 387–403, copyright 2010, with permission from Elsevier.

Figure 6.5 details the scheduling process for a given wavelength. Here, traffic demand matrix A is constructed based on the updated values of a_{ij}. Let u be the largest value between the maximum of all row summations in A and the maximum of all column summations in A. Parameter u is used to limit the summation in each row and in each column to F. Two cases may arise for $N_{RG,ij}$:

- $N_{RG,ij} > 0$: This case arises when we have $u > F$. Then, the core switch needs to scale down traffic demand entries proportionally by granting less time-slot scheduling to any ingress switch, thus leading to a positive $N_{RG,ij}$.

- $N_{RG,ij} \leq 0$: This arises when the procedure of converting the demand matrix to a doubly stochastic matrix grants more time slots than requested for the traffic between ingress switch i and egress switch j. Having the additional granted time slots, ingress switch i finds more opportunity to send its traffic to egress switch j.

Based on u, traffic demand matrix A is converted to doubly sub-stochastic matrix $R' = \left[\frac{a_{ij}}{u}\right]$ in which the summation of each row and each column is less than or equal to 1.0. The algorithm in [18] is used for converting R' to doubly stochastic matrix R in which the summation of each row and summation of each column equals 1.0; otherwise the traffic matrix will not be decomposable [22]. The BvN applies a decomposition algorithm on R from which the permutation matrices for each wavelength are obtained. In an iteration, a bipartite maximum matching algorithm is performed using the maximum flow algorithm [23]. Then, each permutation P_i receives a weight v_i where the summation of these weights equals 1.0. Hence, we obtain $R = \sum_{i=1}^{k} v_i \times P_i$ with k permutation matrices. Finally, scheduling matrix S is computed by multiplying u by each term $v_i \times P_i$. The pseudo code in Fig. 6.6 shows the BvN decomposition for a given wavelength:

```
Algorithm: BvN Decomposition on Doubly Stochastic Matrix R
```

1: $i = 1$
2: **while** TRUE **do**
3:     Find matching P_i corresponding to R
4:     Set v_i to the minimum value of R corresponding to P_i
5:     $B = R - v_i \times P_i$
6:     **if** $B = [0]$ {when all items in matrix B is zero} **then**
7:         Exit loop
8:     **end if**
9:     $R = \frac{B}{(1-v_i)}$
10:     $i = i + 1$
11: **end while**
12: $k = i$
13: $S = u \times v_1 \times P_1 + u \times v_2 \times P_2 + \cdots + u \times v_k \times P_k$
14: END

Figure 6.6: BvN decomposition in the core switch. Reprinted from *Computer Networks*, Vol. 54, Akbar Ghaffarpour Rahbar and Oliver Yang, Agile bandwidth management techniques in slotted all-optical packet interconnection networks, 387–403, copyright 2010, with permission from Elsevier.

For the same traffic demand matrix among $f \times W$ wavelengths in the best case, the above steps will be executed once for all $f \times W$ demand matrices. However, in the worst-case, these steps may have to be executed for each one of the $f \times W$ demand matrices. After specifying the scheduling of ingress switch i, the core switch will announce the ingress switch its scheduling on each wavelength channel and fiber.

Example: Figure 6.7 shows a 4×4 traffic demand matrix on a given wavelength in a network at $n = 4$. As shown, IS_2 needs 4, 3, and 2 time slots (see the second row of matrix A) to send its traffic to egress switches 1, 3, and 4, respectively, within a

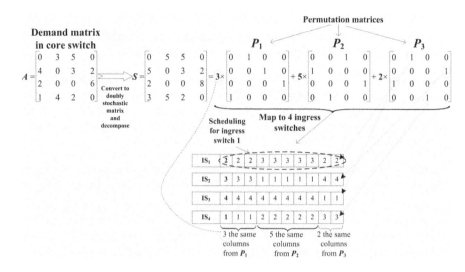

Figure 6.7: A 4×4 traffic demand matrix, its decomposition, and the scheduling map to ingress switches at $F = 10$. Reprinted from *Computer Networks*, Vol. 54, Akbar Ghaffarpour Rahbar and Oliver Yang, Agile bandwidth management techniques in slotted all-optical packet interconnection networks, 387–403, copyright 2010, with permission from Elsevier.

frame. Here, we have $u = 10$. The decomposed matrix is $S = 3 \times P_1 + 5 \times P_2 + 2 \times P_3$, where $u \times v_1 = 3$, $u \times v_2 = 5$, and $u \times v_3 = 2$. Figure 6.7 also displays the frame of traffic transmission for each ingress switch, where a number shown in a frame cell is an egress switch address. The first three columns have the same scheduling because 3 is the coefficient of P_1. Likewise, since the coefficients for P_2 and P_3 are 5 and 2, respectively, we have the same scheduling for the next five columns and the next two columns. Given a permutation matrix $P = p_{ij}$, an ingress switch i has the scheduling to egress switch j if $p_{ij} = 1$. For example, consider the first column extracted from P_1 in the frame. In matrix P_1, we have $p_{12} = p_{23} = p_{34} = p_{41} = 1$, and therefore IS_1, IS_2, IS_3, and IS_4 can send one optical packet to egress switches 2, 3, 4, and 1, respectively.

6.1.3.4 The Efficient BvN (EBvN)

The EBvN modifies the BvN to efficiently schedule traffic in overlaid star OPS networks with multi-fiber–multi-wavelength architecture. Instead of using a dedicated processor to schedule traffic on each wavelength channel in an OPS core switch, EBvN uses only one processor to schedule all traffic demands on all fibers/wavelength channels at the same time [20]. The following steps show the procedure of EBvN in a core switch:

- **Step 1:** The core switch makes original traffic demand matrix $A = [a_{ij}]_{n \times n}$ from traffic demand arrays received from all ingress switches at its input ports at the same time. Note that all entries of A are integer numbers. The following steps work on temporary traffic demand matrix D initially set to $D = A = [a_{ij}]_{n \times n}$.

- **Step 2:** The summation of each row (r_i, where $1 \leq i \leq n$) and each column (c_j, where $1 \leq j \leq n$) of traffic demand matrix D are computed. Let r_m denote maximum of row summations and c_m be the maximum of column summations in D. This procedure is called max-rows-columns. Then, the core switch computes matrix scale down factor $v = \max(r_m, c_m)$.

- **Step 3:** Summation of entries in each row and each column of D are adjusted in a way to be less than $F \times f \times W$. For this, the scale-down factor v must be smaller than the total number of available time slots in one frame period; i.e., $F \times f \times W$. If this condition is not satisfied and $v > F \times f \times W$, then each entry of matrix D is multiplied by $\frac{F \times f \times W}{v}$; i.e., $a_{ij} = \lfloor a_{ij} \times \frac{F \times f \times W}{v} \rfloor$, so that all entries of D are still integer. Then, the max-rows-columns procedure in Step 2 is repeated again and a new value is obtained for v so that we have $\forall i \sum_j a_{ij} \leq v$ and $\forall j \sum_i a_{ij} \leq v$.

- **Step 4:** A doubly sub-stochastic matrix $R' = [r'_{i,j}]_{n \times n}$ is obtained by dividing all entries of D over v, where the summation of each row and summation of each column in matrix R' is less than or equal to 1.0. On the other hand, every entry of R' is a multiple of $\frac{1}{v}$.

- **Step 5:** Matrix $R' = [r'_{i,j}]_{n \times n}$ is scaled to doubly stochastic matrix $R = [r_{i,j}]_{n \times n}$ using the Algorithm 1 stated in [17, 18] that boosts some entries in such a way that $r'_{i,j} \leq r_{i,j}$, where this scaling can be performed in $O(n^3)$. Based on this algorithm, entry (i,j) is found in a way that $g_1 = \sum_m r'_{m,j} < 1$ and $g_2 = \sum_n r'_{i,n} < 1$. Then, boost parameter $\varepsilon_{i,j}$ for entry (i,j) is computed by $\varepsilon_{i,j} = 1 - \max(g_1, g_2)$, and then $r'_{i,j} = r'_{i,j} + \varepsilon_{i,j}$. This process is continued until the required constraint for having a doubly stochastic matrix is met.

  Since both g_1 and g_2 are multiples of $\frac{1}{v}$, parameter $\varepsilon_{i,j}$ is a multiple of $\frac{1}{v}$ as well. Therefore, a selected entry is boosted by a multiple of $\frac{1}{v}$. Thus, after conversion, every entry of R is a multiple of $\frac{1}{v}$ as well.

- **Step 6:** After obtaining doubly stochastic matrix R, the BvN scheduling is applied on this matrix. The BvN decomposes matrix R to k permutation matrices $P_1, P_2, \ldots, P_k$ so that we have $R = \sum_{i=1}^{k} w_i \times P_i$, where summation of weights w_i is 1.0 (i.e., $\sum_{i=1}^{k} w_i = 1$). During decomposition, weights are extracted in a way to be a multiple of $\frac{1}{v}$ (see Algorithm 2 in [17, 18]).

- **Step 7:** In order to obtain integer weights, the weights of k permutation matrices are multiplied by v. By this, the scheduled matrix is given by $S = (v \times w_1) \times P_1 + (v \times w_2) \times P_2 + (v \times w_3) \times P_3 + \cdots + (v \times w_k) \times P_k$.

- **Step 8:** Using the permutation matrices with integer weights obtained in Step 7, the scheduling is assigned to the time slots within the frame and among multiple channels/multiple fibers of different edge switches. The pseudo code for the time-slot assignment process is displayed in Fig. 6.8, where matrix sch has four

Algorithm: Time-Slot Assignment in EBvN

1: $n_W = n_f = slotPtr = w_s = x = 0$ {initializations}
2: $sch = EMPTY$ {set all entries in sch to EMPTY}
3: **for** $j = 1$ to k {compute total number of permutation matrices} **do**
4:    $w_s = w_s + v \times w_j$ {summation of integer weights of k permutation matrices}
5: **end for**
6: $\Delta = \frac{F \times f \times W}{w_s}$ {compute average steps between slot sets within the frame}

7: **for** $j = 1$ to k {for each permutation matrix P_j in S} **do**
8:    **for** $i = 1$ to $v \times w_j$ {scan all weights of the permutation matrix P_j} **do**
9:       **for** $IS = 1$ to n {for each ingress switch IS} **do**
10:          $sch[IS][slotPtr][n_W][n_f] = GetCol(P_j, IS)$ {get egress switch address}
11:       **end for**
12:       $n_f = (n_f + 1) mod\ f$ {update fiber number related to each ingress switch}
13:       **if** $n_f = 0$ **then**
14:          {if we have obtained f numbers from the permutation matrices}
15:          $n_W = 1 + (n_W + 1) mod\ W$ {update wavelength number $(\lambda_1 - \lambda_W)$, where λ_0 is control channel}
16:          **if** $n_W = 1$ **then**
17:             {update slot number after getting $f \times W$ numbers from permutation matrices}
18:             $x = x + \Delta$
19:             $step = \lfloor x \rfloor$
20:             $x = x - step$ {decease the given step}
21:             $slotPtr = slotPtr + step$ {point to the next slot set in the frame}
22:          **end if**
23:       **end if**
24:    **end for**
25: **end for**
26: END

Figure 6.8: Pseudo code for time-slot assignment from scheduling matrix S with k permutation matrices in EBvN. Reprinted from Akbar Ghaffarpour Rahbar, EBvN: efficient BvN in multi-fiber/multi-wavelength overlaid-star optical networks, *Journal of Annals of Telecommunications* 2012, 67:575–588, Figure 2, with kind permission from Springer Science and Business Media.

dimensions: fiber number (n_f), wavelength number (n_W), slot-set pointer within the frame ($slotPtr$), and ingress switch number (IS). It is noted that the function $GetCol(P_j, IS)$ returns the column in which there is 1 at row IS of permutation matrix P_j, where the returned value shows an egress switch address. Notice that there is only one individual frame for each ingress switch obtained from the pseudo code.

The EBvN breaks the weight of each permutation matrix down into at most F slot set partitions. Each slot set partition includes at most W wavelength partitions, and each wavelength partition consists of at most f fiber partitions. Each slot set partition represents one slot set within the frame, and each wavelength partition indicates the configuration on each wavelength. Note that in a multi-fiber network, f optical packets can be switched at each output port at the same time. Therefore, up to f optical packets with the same destination never collide on the same wavelength from different fibers. Recall that each time slot carries only one optical packet. This is why we have f fiber partitions in each wavelength partition.

For illustration purposes, consider a scenario in which we have $F = 4, f = 2, W = 3$ and a decomposed matrix of $S = 9P_1 + P_2 + 4P_3 + 5P_4 + 4P_5$. This decomposition (see Fig. 6.9) is broken down into $F = 4$ slot set partitions (each slot set partition is depicted within a square bracket pair), each slot set parti-

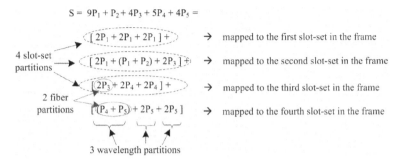

Figure 6.9: An example for mapping of permutation matrices to ingress switches. Reprinted from Akbar Ghaffar-pour Rahbar, EBvN: efficient BvN in multi-fiber/multi-wavelength overlaid-star optical networks, *Journal of Annals of Telecommunications* 2012, 67:575–588, with kind permission from Springer Science and Business Media.

tion with $W = 3$ wavelength partitions (i.e., three terms inside a square bracket) and each wavelength partition with at most $f = 2$ fiber partitions (i.e., coefficient of each term inside a square bracket). In this mapping, the first slot set partition denotes the configuration for the first slot set within the frame. The first wavelength partition of the first slot set partition (i.e., $2P_1$) shows that two egress switch addresses must be obtained from permutation matrix P_1 (for each ingress switch) for λ_1 on Fiber$_0$ and λ_1 on Fiber$_1$, respectively. As another example, the second wavelength partition of the second slot set partition for the second slot set (i.e., $(P_1 + P_2)$) shows that one egress switch address must be obtained from permutation matrix P_1 (for each ingress switch) for λ_2 on Fiber$_0$, and the other egress switch address should be obtained from permutation matrix P_2 (for each ingress switch) for λ_2 on Fiber$_1$. Table 6.2 shows the mapping of the equation in Fig. 6.9 to the frame relevant to ingress switch m, where P_i^m denotes egress switch address obtained from permutation matrix P_i for ingress switch m using the *GetCol* function. Note that applying *GetCol* on row m of permutation matrix P_i returns the column number in which there is a 1, where the returned column number shows egress switch address.

Table 6.2: Mapping permutation matrices to a frame for ingress switch m. Reprinted from Akbar Ghaffarpour Rahbar, EBvN: efficient BvN in multi-fiber/multi-wavelength overlaid-star optical networks, *Journal of Annals of Telecommunications* 2012, 67:575–588, Table 2, with kind permission from Springer Science and Business Media.

Slot Set	Wavelength λ_1		Wavelength λ_2		Wavelength λ_3	
	Fiber$_0$	Fiber$_1$	Fiber$_0$	Fiber$_1$	Fiber$_0$	Fiber$_1$
slot set1	P_1^m	P_1^m	P_1^m	P_1^m	P_1^m	P_1^m
slot set2	P_1^m	P_1^m	P_1^m	P_2^m	P_3^m	P_3^m
slot set3	P_3^m	P_3^m	P_4^m	P_4^m	P_4^m	P_4^m
slot set4	P_4^m	P_5^m	P_5^m	P_5^m	P_5^m	-

The proposed EBvN tries to evenly assign time slots within the frame to ingress switches. This is done by computing total weight of permutation matrices $w_s = \sum_{i=1}^{k} v \times w_i$ from k permutation matrices. Then, total number of scheduled time slots within a frame for all n edge switches is at most $n \times w_s$ since each permutation matrix includes the scheduling for at most n egress switches. For even traffic transmission, optimum distance between two consecutive slot sets within a frame should be set to $\Delta = \frac{n \times F \times f \times W}{n \times w_s} = \frac{F \times f \times W}{w_s}$ slot sets, where the numerator shows total number of available time slots within a frame for all n edge switches. However, Δ may not be an integer value in practice. Hence, parameter x (where $0 \le x < 1$) is introduced to provide an optimum distance between slot sets. Considering the current slot-set pointer of $slotPtr$, the next slot set for occupation of scheduled time slots should be $slotPtr + step$, where $step = \lfloor x + \Delta \rfloor$. Parameter x is then updated by $x = x + \Delta - step$.

- **Step 9:** After computing scheduling frame for ingress switch m, the core switch sends this scheduling back to ingress switch m. Then, ingress switch m uses its scheduling and sends traffic toward specified egress switches determined by the scheduling in reserved time slots/wavelengths/fibers.

For example, ingress switch 3 follows the scheduling in Fig. 6.1 and transmits traffic at the first time slot to egress switches 6, 4, 6, 6, and 5, respectively on λ_1 of Fiber$_0$, λ_1 of Fiber$_1$, λ_2 of Fiber$_1$, λ_3 of Fiber$_0$, and λ_3 of Fiber$_1$. In the first time slot, no traffic should be sent to any egress switch on λ_2 of Fiber$_0$.

Respecting the provided schedulings, there will be no collision on the traffic sent by different ingress switches within the same time slot. The reason for this issue returns back to the operation of Line 10 in Fig. 6.8, which extracts different egress switch addresses for the same slot set, fiber, and wavelength (to determine a specific cell in the frame) from permutation matrix P_j by calling the *GetCol* function and then assigns egress switch addresses to ingress switches.

6.1.3.5 The Separated BvN (SBvN)

6.1.3.5 The Separated BvN (SBvN) Under SBvN, traffic demand matrix entries should be equally divided among $f \times W$ wavelength channels. However, entry a_{ij} may not be divisible by $f \times W$. Therefore, the algorithm displayed in Fig. 6.10 is used to divide entry a_{ij} among $f \times W$ wavelengths and assign divided parts to traffic demand matrix entries on wavelength λ_l, which is denoted by b_{lij}. After this, each traffic demand matrix is scheduled similar to the mechanism detailed in Section 6.1.3.4, but with considering $f \times W = 1$ in all steps provided in Section 6.1.3.4.

Algorithm: Division of Traffic Demand among Wavelength Channels in SBvN

1: $\Delta = \frac{a_{ij}}{f \times W}$ {compute average steps between slot sets within the frame}
2: **for** $l = 1$ to $f \times W$ {for all wavelengths} **do**
3:   $x = x + \Delta$
4:   $step = \lfloor x \rfloor$
5:   $x = x - step$ {decease the given step}
6:   $b_{lij} = step$ {assign to traffic demand matrix entry on wavelength λ_l}
7: **end for**
8: END

Figure 6.10: Pseudo code for dividing an entry of traffic demand matrix among wavelength channels under SBvN. Reprinted from Akbar Ghaffarpour Rahbar, EBvN: efficient BvN in multi-fiber/multi-wavelength overlaid-star optical networks, *Journal of Annals of Telecommunications* 2012, 67:575–588, Figure 3, with kind permission from Springer Science and Business Media.

6.1.3.6 The EBvN with Filling Empty Cells (EBvN _FEC) The EBvN can successfully schedule all requested traffic under uniform traffic even at load $L = 1.0$. However, some traffic requests cannot be scheduled under the severe non-uniform traffic distribution. Some of this unscheduled traffic is due to the cells that are left empty in frames after scheduling. In addition, during conversion from a sub-stochastic matrix to a doubly stochastic matrix, some entries in the sub-stochastic matrix may be increased, even at entry (i,i), where $i \in [1,n]$. Hence, after scheduling, some allocated bandwidth may be useless; whereas some traffic requests may not be granted.

As a tradeoff between scheduling time and residual traffic, a new scheduling algorithm based on EBvN, called EBvN with Filling Empty Cells (EBvN_FEC), has been proposed to reduce residual traffic, but at the expense of slightly increasing scheduling time. The EBvN_FEC is more effective than EBvN under non-uniform traffic distribution [20]. The EBvN_FEC differs from EBvN only in extracting scheduling information from permutation matrices and building the scheduling frames. Figure 6.11 illustrates the initialization phase of EBvN_FEC and Figure 6.12 displays the algorithm for building the scheduling frames for ingress switches in EBvN_FEC.

```
Algorithm:  Initialization Phase of EBvN_FEC

1:  n_W = n_f = slotPtr = w_s = x = 0  {initializations}

2:  for i = 1 to n {for each edge switch} do
3:      for s = 1 to F {for each slot set in frame} do
4:          for j = 1 to W {for each wavelength} do
5:              C[s][j][i] = f {set the capacity of each item in matrix C to f}
6:              for l = 1 to f {for each fiber} do
7:                  sch[i][s][j][l] = EMPTY {initialize scheduling matrix}
8:              end for
9:          end for
10:     end for
11: end for

12: for i = 1 to n {for each ingress switch} do
13:     for dst = 1 to n {for each egress switch} do
14:         E_{i,dst} = ∅ {initialize the set of empty cells }
15:     end for
16: end for

17: for j = 1 to k {compute total number of permutation matrices} do
18:     w_s = w_s + v × w_j {summation of integer weights of k permutation matrices}
19: end for
20: Δ = (F × f × W)/w_s  {compute average steps between slot sets within the frame}
21: END
```

Figure 6.11: Pseudo code for initialization phase of EBvN_FEC. Reprinted from Akbar Ghaffarpour Rahbar, EBvN: efficient BvN in multi-fiber/multi-wavelength overlaid-star optical networks, *Journal of Annals of Telecommunications* 2012, 67:575–588, Figure 9, with kind permission from Springer Science and Business Media.

In Fig. 6.11, Lines 1–11 perform initialization of scheduling matrix *sch* and capacity matrix C. The capacity matrix C shows available capacity on each wavelength of every slot set to any egress switch. Every entry of matrix C is set to f because there could be at most f optical packets that can be switched on any wavelength at a given slot set destined to a given egress switch in the core switch. Lines 12–16 initialize sets of available empty time slots dedicated to each ingress/egress switch pair within a frame. For even traffic transmission, the optimum distance between two consecutive slot sets within a frame is set to $\Delta = \frac{F \times f \times W}{w_\xi}$ slot sets, where the numerator shows the total number of available time slots within a frame for all n edge switches (see Lines 17–20).

In Fig. 6.12, Lines 1–25 extract scheduling information from permutation matrices and make scheduling frames. In addition, capacity matrix C and original demand matrix A are updated accordingly. Note that the condition in Line 25 does not allow additional bandwidth to be allocated to the demand between ingress switch IS and egress switch dst. Lines 26–42 build sets of available empty cells for each ingress/egress switch pair. The location of empty cells (determined with slot set number, wavelength number, and fiber number) that can be used between ingress switch i and egress switch dst are added to set $E_{i,dst}$. Lines 43–55 allocate empty cells obtained from sets $E_{i,dst}$ to unscheduled demands.

6.1.3.7 Comparison of Reservation-Based Traffic Transmission in Overlaid OPS

The computational complexities of EBvN, SBvN, and EBvN_FEC are [20]:

$$C_{EBvN} = O(n^{4.5}) + O(n \times F^2 \times f^2 \times W^2 \times L^2).$$
$$C_{SBvN} = O(f \times W \times n^{4.5}) + O(n \times F^2 \times f \times W \times L^2).$$
$$C_{EBvN\_FEC} = O(n^{4.5}) + O(n \times F^2 \times f^2 \times W2 \times L^2) + O(n^2 \times F \times f \times W).$$

The complexity of EBvN is higher than SBvN only when we have $f = 1$, $F = 1$, and $n = 1$. In this case, the complexity of EBvN is $O(W^2 \times L^2)$ and the complexity of SBvN is $O(W \times L^2)$. However, in practice, we always have $F > 1$ in frame-based scheduling algorithms and $n > 1$ in a star network. On the other hand, the complexity of both EBvN and SBvN will be equal to $O(n^{4.5}) + O(n \times F^2 \times L^2)$ when we have $f = 1$ and $W = 1$. Nevertheless, the proposed algorithms are designed for multi-fiber–multi-wavelength overlaid OPS networks; i.e., $f > 1$ and $W > 1$, where the complexity of EBvN becomes smaller than SBvN. Finally, the computational complexity of EBvN_FEC is higher than the complexity of EBvN. Nevertheless, the EBvN_FEC is more efficient in the scheduling of demanded traffic compared with EBvN.

6.1.3.7.1 Performance Comparison of SBvN and EBvN

Assume an overlay star-based OPS network with $f = 2$ fibers per link, $n = 16$ edge switches, and frame size $F = 16$. In each simulation replication, 10,000 traffic demand matrices are randomly generated based on normalized traffic load L, where L is the number of requested time slots by an edge switch normalized with respect to total number of available time slots for an edge switch within a frame (i.e., $F \times f \times W$). Two traffic scenarios are evaluated as follows:

- **Uniform Traffic Distribution:** Here, traffic sent by every ingress switch to each one of $n - 1$ egress switches is uniform. The entry a_{ij} of each generated traffic demand matrix $n \times n$ is set in a way that:

Algorithm: Time-Slot Assignment in EBvN_FEC

1: **for** $j = 1$ to k {compute total number of permutation matrices} **do**
2:     **for** $i = 1$ to $v \times w_j$ {scan all weights of the permutation matrix P_j} **do**
3:         **for** $IS = 1$ to n {for each ingress switch} **do**
4:             $dst = GetCol(P_j, IS)$ {get scheduled egress switch address}
5:             **if** $A[IS][dst] \neq 0$ **then**
6:                 {if still granted time slots required in original traffic demand matrix A}
7:                 $sch[IS][slotPtr][n_W][n_f] = dst$ {set scheduling item}
8:                 dec $A[ingressNode][dst]$ {update number of granted time slots in original demand matrix A}
9:                 dec $C[slotPtr][n_W][dst]$ {update capacity of each destination in capacity matrix C}
10:             **end if**
11:         **end for**
12:         $n_f = (n_f + 1) mod\ f$ {update fiber number related to each ingress switch}
13:         **if** $n_f = 0$ **then**
14:             {if we have obtained f numbers from the permutation matrices}
15:             $n_W = 1 + (n_W + 1) mod\ W$ {update wavelength number $(\lambda_1 to \lambda_W)$}
16:             **if** $n_W = 1$ **then**
17:                 {update slot number after getting $f \times W$ numbers from permutation matrices}
18:                 $x = x + \Delta$
19:                 $step = \lfloor x \rfloor$
20:                 $x = x - step$ {decease the given step}
21:                 $slotPtr = slotPtr + step$ {point to the next slot set in the frame}
22:             **end if**
23:         **end if**
24:     **end for**
25: **end for**
26: **for** $i = 1$ to n {for each edge switch} **do**
27:     **for** $s = 1$ to F {for each slot set in frame} **do**
28:         **for** $j = 1$ to W {for each wavelength} **do**
29:             **for** $l = 1$ to f {for each fiber} **do**
30:                 **if** $sch[i][s][j][l] = EMPTY$ **then**
31:                     **for** $dst = 1$ to n {for each egress switch} **do**
32:                         **if** $C[s][j][dst] > 0$ and $A[i][dst] > 0$ **then**
33:                             {if bandwidth is required between ingress switch i and egress switch dst}
34:                             append (s, j, l) to $E_{i,dst}$ {append the location of empty time slot to set $E_{i,dst}$}
35:                             dec $C[s][j][dst]$ {update capacity}
36:                         **end if**
37:                     **end for**
38:                 **end if**
39:             **end for**
40:         **end for**
41:     **end for**
42: **end for**
43: **for** $i = 1$ to n {for each ingress switch} **do**
44:     **for** $dst = 1$ to n {for each egress switch} **do**
45:         **while** $A[i][dst] > 0$ **do**
46:             {evaluate residual traffic in original matrix A}
47:             **if** $E_{i,dst} = \varnothing$ **then**
48:                 Exit loop {exit while loop if no free time slot}
49:             **end if**
50:             $(s, j, l) = extract\ and\ remove\ head\ of\ E_{i,dst}$ {extract components of the first item of $E_{i,dst}$, then remove}
51:             $sch[i][s][j][l] = dst$ {schedule dst in the determined position}
52:             dec $A[i][dst]$ {update demand matrix}
53:         **end while**
54:     **end for**
55: **end for**
56: END

Figure 6.12: Pseudo code for time-slot assignment from the scheduling matrix S with k permutation matrices with filling empty slots. Reprinted from Akbar Ghaffarpour Rahbar, EBvN: efficient BvN in multi-fiber/multi-wavelength overlaid-star optical networks, *Journal of Annals of Telecommunications* 2012, 67:575–588, Figure 9, with kind permission from Springer Science and Business Media.

- The value of a_{ij} is a random integer number since it shows number of required time slots.

- When $i = j$, we have $a_{ij} = 0$ since a given edge switch never requests to send traffic to itself.

- At traffic load L, ingress switch i sets $a_{ij} = \lfloor \frac{F \times f \times W \times L}{n-1} \rfloor$ for any $j \neq i$. This leads to uniform traffic distribution from ingress switch i to $n-1$ egress switches. Note that $F \times f \times W \times L$ shows the maximum number of time slots that can be requested by ingress switch i for sending its traffic to $n-1$ egress switches.

- **Non-Uniform Traffic Distribution:** Here, traffic transmitted from every ingress switch to $n-1$ egress switches is non-uniform. The entry a_{ij} of each generated traffic demand matrix $n \times n$ has the following characteristics:

  - The number of required time slots a_{ij} is a random integer number.

  - When $i = j$, we have $a_{ij} = 0$.

  - Ingress switch i chooses a_{ij} (where $j \neq i$) as a random number uniformly distributed between 0 and $\frac{2 \times F \times f \times W \times L}{n-1}$. This causes a severe non-uniform traffic distribution from ingress switch i to $n-1$ egress switches.

  - Summation of all time slots requested by ingress switch i to all $n-1$ egress switches (i.e., the summation in each row of the traffic demand matrix) is $\sum_{i=1}^{n} a_{ij} = F \times f \times W \times L$. However, note that the summation of entries in any column j of the traffic demand matrix, (i.e., $\sum_{j=1}^{n} a_{ij}$) varies between 0 and $2 \times F \times f \times W \times L$ since a column may have at most $n-1$ entries each with $\frac{2 \times F \times f \times W \times L}{n-1}$ time slot requests in the worst-case, thus leading to a severe non-uniformity.

All simulations in this section are run in a DELL Inspiron 6,400 laptop with Intel core 2 Duo CPU at 2.0 GHz, 2 MB cache memory, and 1 GB memory under 32-bit Windows 7 Ultimate operating system. Three different performance parameters are evaluated in the following:

- **Residual Traffic Performance:** This performance is evaluated under two traffic distributions:

  - **Uniform traffic distribution:** There is no residual traffic observed under EBvN for any W and L. Under SBvN, there is no residual traffic in the core switch at any W and $L \in [0.4, 0.6, 0.8]$. However, 5.8% of requested traffic cannot be scheduled in the core switch at $L = 1.0$ under SBvN. Hence, EBvN is more efficient in scheduling demanded traffic.

  - **Non-uniform traffic distribution:** Figure 6.13 depicts average percentage of residual traffic under non-uniform traffic, where EBvN is more efficient in scheduling of demanded traffic compared with SBvN. This is because EBvN schedules all traffic at the same time on all wavelength channels, but

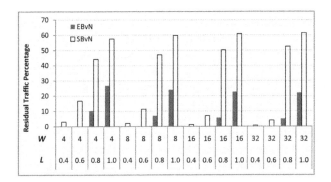

Figure 6.13: Average residual traffic percentage under non-uniform traffic distribution. Reprinted from Akbar Ghaffar-pour Rahbar, EBvN: efficient BvN in multi-fiber/multi-wavelength overlaid-star optical networks, *Journal of Annals of Telecommunications* 2012, 67:575–588, Figure 6, with kind permission from Springer Science and Business Media.

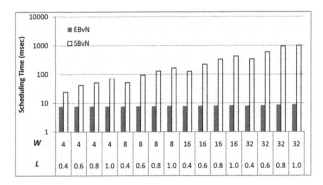

Figure 6.14: Average scheduling time under uniform traffic distribution. Reprinted from Akbar Ghaffarpour Rahbar, EBvN: efficient BvN in multi-fiber/multi-wavelength overlaid-star optical networks, *Journal of Annals of Telecommunications* 2012, 67:575–588, Figure 4, with kind permission from Springer Science and Business Media.

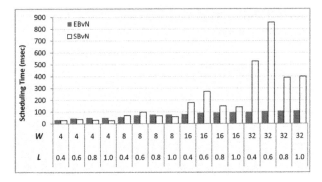

Figure 6.15: Average scheduling time under non-uniform traffic distribution. Reprinted from Akbar Ghaffarpour Rahbar, EBvN: efficient BvN in multi-fiber/multi-wavelength overlaid-star optical networks, *Journal of Annals of Telecommunications* 2012, 67:575–588, Figure 7, with kind permission from Springer Science and Business Media.

SBvN separately schedules traffic of each wavelength channel. Note that most of the traffic cannot be scheduled at high traffic loads $L = 0.8$ and $L = 1.0$ under SBvN because of the initial process required for converting a traffic demand matrix to a doubly stochastic matrix in which entries of the traffic demand matrix may be reduced during this step because of severe non-uniform traffic distribution, where some requested time slots between any ingress–egress switch pair may not be granted. At the same traffic load, under SBvN, residual traffic goes up by increasing W. On the other hand, residual traffic goes down by increasing W under EBvN. This shows that EBvN is more efficient than SBvN, especially when there are many wavelength channels in the network.

- **Average Scheduling Time Performance:** The performance of scheduling time is evaluated under two traffic distributions:

  - **Uniform traffic distribution:** Figure 6.14 displays average time (in ms) necessary for performing scheduling (in logarithmic scale) under uniform traffic distribution, where EBvN significantly outperforms SBvN. Average scheduling time varies between 7.55 ms to 8.91 ms under EBvN; whereas this range is between 24.3 ms and 1002 ms under SBvN.

  - **Non-uniform traffic distribution:** Figure 6.15 illustrates the average time required for performing scheduling (in ms) under both EBvN and SBvN. As can be observed, using a small number of wavelengths, say $W = 4$, SBvN can always be performed faster than EBvN. However, at $W > 8$, EBvN always outperforms SBvN. At $W = 8$ and average traffic loads, EBvN is executed faster than SBvN. At load $L > 0.6$, execution time under SBvN has been reduced for any W. However, this reduction is at the expense of increasing residual traffic (see Fig. 6.13).

- **Maximum Scheduling Time Performance:** This performance is evaluated under two traffic distributions:

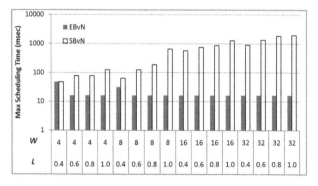

Figure 6.16: Maximum scheduling time under uniform traffic distribution. Reprinted from Akbar Ghaffarpour Rahbar, EBvN: efficient BvN in multi-fiber/multi-wavelength overlaid-star optical networks, *Journal of Annals of Telecommunications* 2012, 67:575–588, Figure 5, with kind permission from Springer Science and Business Media.

- **Uniform traffic distribution:** Figure 6.16 depicts the worst-case scheduling time (in ms) in logarithmic scale, obtained among all simulations for different traffic demand matrices, traffic loads, and wavelengths. According to this diagram, EBvN and SBvN have encountered the maximum worst-case scheduling time of 47 ms and 1966 ms, respectively.

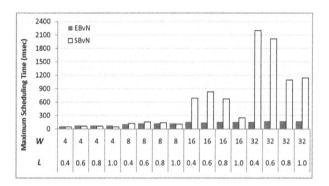

Figure 6.17: Maximum scheduling time under non-uniform traffic distribution. Reprinted from Akbar Ghaffarpour Rahbar, EBvN: efficient BvN in multi-fiber/multi-wavelength overlaid-star optical networks, *Journal of Annals of Telecommunications* 2012, 67:575–588, Figure 8, with kind permission from Springer Science and Business Media.

- **Non-uniform traffic distribution:** Figure 6.17 illustrates the worst-case scheduling time (in ms), obtained among all simulations at different traffic demand matrices, traffic loads, and wavelengths. According to this diagram, applying SBvN to a traffic demand matrix may need much more running time, as high as 2200 ms at $L = 0.4$ and $W = 32$, especially at $W > 8$. On the other hand, the worst-case scheduling time under EBvN is about 156 ms, which occurs at $L = 1.0$ and $W = 32$.

Performance evaluation results under both uniform and non-uniform traffic distributions show that more traffic demands can be scheduled under EBvN compared with SBvN. In addition, scheduling speed of EBvN is mostly faster than SBvN. Finally, EBvN can provide a bound on the maximum scheduling time. The EBvN can successfully schedule all requested bandwidth under uniform traffic load even at load 1.0. However, the problem of residual traffic observed even for EBvN at high loads is due to the severe non-uniform traffic distribution. The EBvN can be easily used in a network with many edge switches and many wavelengths per fiber. In such a network, the performance of EBvN becomes comparable with SBvN.

6.1.3.7.2 Performance Comparison of EBvN and EBvN_FEC Here, the performances of EBvN and EBvN_FEC are compared under the non-uniform traffic distribution detailed in Section 6.1.3.7.1. Figure 6.18 illustrates average percentage of residual traffic that cannot be scheduled in the core switch, where EBvN_FEC is more efficient in scheduling of demanded traffic compared with EBvN. This is because of filling empty cells (time slots) after performing the scheduling. The reduction in residual traffic varies between 20% and 32%. For example, residual traffic percentage has been reduced from 22% to 17.2% at $L = 1.0$ and $W = 32$.

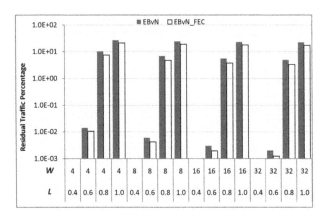

Figure 6.18: Average residual traffic under non-uniform traffic in EBvN and EBvN_FEC. Reprinted from Akbar Ghaffar-pour Rahbar, EBvN: efficient BvN in multi-fiber/multi-wavelength overlaid-star optical networks, *Journal of Annals of Telecommunications* 2012, 67:575–588, Figure 10, with kind permission from Springer Science and Business Media.

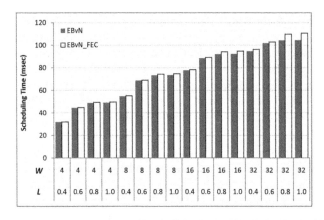

Figure 6.19: Average scheduling time under non-uniform traffic in EBvN and EBvN_FEC. Reprinted from Akbar Ghaf-farpour Rahbar, EBvN: efficient BvN in multi-fiber/multi-wavelength overlaid-star optical networks, *Journal of Annals of Telecommunications* 2012, 67:575–588, Figure 11, with kind permission from Springer Science and Business Media.

Figure 6.19 displays average time required for performing scheduling (in ms), where EBvN_FEC experiences a little higher scheduling time than EBvN, as expected. However, this increase varies from 0.7% up to 6% that could be ignored. Figure 6.20 shows the worst-case scheduling time (in ms), where EBvN_FEC experiences higher worst-case delay than EBvN. The worst-case scheduling time is at most 187ms under EBvN_FEC.

Clearly, the complexity of EBvN_FEC is higher than the complexity of EBvN. However, compared with EBvN, EBvN_FEC can reduce residual traffic by at most

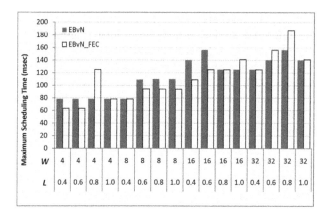

Figure 6.20: Maximum scheduling time under non-uniform traffic in EBvN and EBvN_FEC. Reprinted from Akbar Ghaffarpour Rahbar, EBvN: efficient BvN in multi-fiber/multi-wavelength overlaid-star optical networks, *Journal of Annals of Telecommunications* 2012, 67:575–588, Figure 12, with kind permission from Springer Science and Business Media.

32%, but at the expense of increasing scheduling time by at most 6%. EBvN_FEC is more effective than EBvN under non-uniform traffic distribution.

6.1.4 Contention-Based Traffic Transmission in Overlaid OPS

A number of contention-based techniques proposed for traffic transmission in an overlaid-star OPS network are studied in this section. In these techniques, no reservation is made for transmission of optical packets. Recall that there are n edge switches in this network and the core switch has $N = n$ input ports/links, and $N = n$ output ports/links; i.e., an $n * n$ core switch. Recall that the term input/output port is used in single-fiber networks, and the term input/output link is used in multi-fiber networks, where each link has f ports.

6.1.4.1 Hybrid DA + MF + PR + SPN WC + COPS Type 3 Architecture   In an overlaid-star OPS network, the DA contention avoidance scheme can be combined with prioritized retransmission, multi-fiber architecture, SPN wavelength conversion, and composite optical packet scheduling [11]. Here, there are three classes for client packets as HP, MP, and LP. Slot-level synchronization is adequate for DA in an overlaid star OPS network so that the start time at ingress switch i is given by $T_{s,i} = D_{P,i} - S_{ts} \times \left\lceil \frac{D_{P,i}}{S_{ts}} \right\rceil$, where $D_{P,i}$ is the propagation delay from edge switch i to the core switch in the network.

The composite optical packet scheduling, COPS Type 3 (see Section 3.2.3.3), is used in the core switch. Define "significance" parameter V_c for class c to show how valuable the traffic of class c is when carried in an optical packet within a time slot. Recall that DA aggregates client packets from different classes as an optical packet and sends the optical packet in a time slot. The network manager sets V_c, where $0 < V_c < 1$ and $V_{HP} + V_{MP} + V_{LP} = 1.0$. These values are used for contention resolution to privilege some optical packets with high-value traffic over other optical packets with low-value traffic. Recall that by assigning a high "significance" parameter to a class, the traffic of that class is given a high chance to pass through any core switch in the network, and therefore one can expect a high throughput for that class.

Let the set $\Theta_{l,i} = \{OP_0, OP_1, \ldots, OP_{N_c-1} | 0 < N_c \leq n \times f\}$ denote the subset of N_c contending optical packets on wavelength λ_i ($i = 0, \ldots, W-1$) at output link l. Since each link has f fibers, a total of f optical packets can be switched on wavelength λ_i of the output link. Let vector $(A_j, ID_j, \delta_j, n_{HP,j}, n_{MP,j}, n_{LP,j})$ denote traffic parameters for optical packet OP_j recorded in the header coming from ingress switch A_j, where ID_j is the identification code of optical packet j; δ_j denotes the prioritization factor of optical packet j; and $n_{HP,j}$, $n_{MP,j}$, and $n_{LP,j}$ denote respectively the number of client packets in class HP, class MP, and class LP carried in optical packet j. Recall that DA collects client packets from different classes as an optical packet in a time slot, and therefore we have $n_{HP,j} \geq 0$, $n_{MP,j} \geq 0$, and $n_{LP,j} \geq 0$. The core switch resolves the contention by the following procedure (see Fig. 6.21) on wavelength λ_i at output link l. After determining the unresolved collisions, the remaining $N_c - \min(f, N_c)$ optical packets in $\Theta_{l,i}$ should be resolved using SPN wavelength converters. However, the contention resolution mechanism should consider all output links because WCs are shared-per-node.

Algorithm: Contention Resolution in Core Switch under DA+MF+PR+SPN+COPS Type 3

1: Compute traffic value parameter V_j for optical packet OP_j by $V_j = 1000 \times \delta_j + V_{HP} \times n_{HP,j} + V_{MP} \times n_{MP,j} + V_{LP} \times n_{LP,j}$.
2: Sort the index of optical packets in set $\Theta_{l,i}$ in a descending order based on their V values.
3: Select the top $\min(f, N_c)$ optical packets from the sorted list for transmission on wavelength λ_i of f fibers (where output fibers are randomly chosen).
4: Collect unused wavelengths in set U_l
5: Combine all sets U_l into set U
6: Randomize set U
7: **for** each item in set U **do**
8:     Obtain link m' and unused wavelength λ'
9:     Find list $\Theta_{m',a'}$ with a number of unresolved optical packets
10:     Schedule an optical packet from the top of list $\Theta_{m',a'}$ for transmission on wavelength λ' after conversion from wavelength a'
11:     **if** No WC available **then**
12:         Exit loop
13:     **end if**
14: **end for**
15: Drop the optical packets that cannot be resolved
16: Send NACK commands encapsulated with IDs of dropped optical packets back to their source ingress switches
17: END

Figure 6.21: Pseudo code of contention resolution in DA + MF + PR + SPN + COPS Type 3. Reprinted from *Computer Networks*, Vol. 54, Akbar Ghaffar Pour Rahbar and Oliver Yang, Agile bandwidth management techniques in slotted all-optical packet interconnection networks, 387–403, copyright 2010, with permission from Elsevier.

6.1.4.2 Packet Transmission Based on Scheduling of Empty Time Slots (PTES)

The Packet Transmission based on the scheduling of Empty time Slots (PTES) as a software-based contention reduction scheme is proposed, which is suitable for overlaid star topology. Under PTES, edge switches are not completely coordinated with each other in order to have a loss-free traffic transmission as in centralized scheduling. Instead, all edge switches first send their traffic loads to the core switch at some regular intervals, and the core switch computes average traffic load Λ among all edge switches and sends Λ back to all edge switches. Then, ingress switch i executes PTES (in a distributed manner) within a frame interval to provide a scheduling of empty time slots. After scheduling, some time slots are scheduled as empty time slots and the remaining time slots are scheduled as non-empty time slots. This scheduling is used for optical packet transmission. Then, ingress switch i must avoid sending its traffic in empty time slots; i.e., an empty time slot has no traffic to any egress switch. Instead, a non-empty time slot can be used for an optical packet transmission. With respect to this scheduling, the variance of the number of non-empty time slots that carry traffic from all edge switches is minimized in the core switch at each time slot, thus reducing traffic loss. It is analyzed that this issue can lead to a lower traffic loss than Immediate Packet Transmission (IPT), and therefore it is counted as a software-based contention reduction scheme. The PTES shapes traffic transmission to the OPS network, and therefore leads to low traffic loss, especially at light and moderate traffic loads compared with IPT. The PTES can be combined with any contention reduction scheme and any hardware-based contention reduction scheme [24–26].

There are some definitions and assumptions that must be addressed before detailing the operation of PTES:

- **Optical Packet (OP):** An optical packet may include a number of client packets.

- **Time-Slot Status:** The status of a time slot is either "empty" or "non-empty". An ingress switch must not send any optical packet in an empty time slot. On the other hand, if the ingress switch has traffic, it can send only one optical packet in a non-empty time slot.

- **Core Switch:** Each core switch has $n \times f$ input ports and $n \times f$ output ports. There are n input links from n edge switches to the core switch, as well as n output links from the core switch to n edge switches.

- **Control Channel:** There are $W + 1$ wavelength channels used on every fiber in the network. One wavelength channel is used for controlling purposes (called control channel), and W data channels are used for traffic transmission.

- **Distance:** The propagating delay from edge switch i to any core switch is D_i time slots, where D_i is an integer value. Note all core switches are co-located. The largest propagation delay is denoted by D_{max}.

- **Traffic Transmission:** When sending traffic, an edge switch uses the round-robin technique to serve its electronic buffers. When serving buffer j, a user packet is removed from head of buffer j and then sent toward egress switch j within a time slot.

- **Slot Set:** A set of $f \times W$ optical packets (OPs) simultaneously sent by an edge switch on $f \times W$ wavelengths. Before sending a slot set, a header is made based on the optical packets in a slot set and sent on the control channel. The information carried in the header of a slot set includes egress/ingress addresses of its OPs.

- **nf-slot-set**: The set of $n \times f$ optical packets sent from n ingress switches on the same wavelength channel and on f fibers within a given time slot. Within each nf-slot-set, at most $n \times f$ non-empty time slots including $n \times f$ optical packets may arrive at the core switch on the same wavelength channel from n edge switches on f fiber links. For instance, the set of optical packets received on λ_2 at the input ports of the core switch from $n \times f$ links is called nf-slot-set on wavelength λ_2. The number of non-empty time slots in an nf-slot-set is denoted by N_{OP}.

- **Frame:** Since PTES is frame-based, a frame is defined on each wavelength λ_i with $n \times n$ time slots, where its width has the size of n time slots and its height is relevant to n edge switches so that traffic on row r is sent by edge switch r on λ_i. As an attribute, "number of empty time slots" E is assigned to each frame so that E empty time slots should always appear in each column of the frame. This frame is shown with $F(n,E,w)$. Totally, there are W frames used for packet transmission on W wavelength channels.

- **Multi-frame:** A set of K consecutive frames is called a multi-frame. There are W multi-frames for W wavelength channels.

- **Scheduling Interval:** The interval S_I within which the PTES scheduling is performed once. Considering K frames within a multi-frame and each frame with n time slots, we could have $S_I = K \times n$ time slots in a scheduling interval.

Table 6.3: Distribution of empty time slots on wavelength λ_w at $n = 8$, $E = 3$, and $b_w = 4$, where the left column shows ingress switch numbers (from 0 to 7) [26]

		A Frame (8 × 8)								
		7	6	5	4	3	2	1	0	
0	...		e		e		e			...
1	...	e		e			e			...
2	...	e			e	e				...
3	...	e		e		e				...
⇒4	...			e		e		e		...
5	...			e	e			e		...
6	...		e		e			e		...
7	...		e		e		e			...

Due to independent stars in an overlaid topology, PTES is stated only for one star network. The functions of PTES protocol are distributed over the core switch and edge switches. Before the beginning of a multi-frame, each ingress switch runs PTES scheduling once for providing scheduling of empty time slots on W frames since there are W wavelengths in each edge switch. Schedulings for the first W frames in W multi-frames are performed by edge switches at suitable times so that they can be

Table 6.4: Distribution of empty time slots on wavelength λ_w at $n = 14$, $E = 4$, and $b_w = 0$, where the left column shows ingress switch numbers (from 0 to 13) [26]

		A Frame(14 × 14)													
		13	12	11	10	9	8	7	6	5	4	3	2	1	0	
⇒0	...			e				e				e			e	...
1	...			e				e				e			e	...
2	...		e					e		e				e		...
3	...		e					e		e				e		...
4	...		e			e				e			e			...
5	...		e			e				e			e			...
6	...	e				e			e				e			...
7	...	e				e			e				e			...
8	...	e			e				e			e				...
9	...	e			e				e			e				...
10	...	e			e			e				e				...
11	...	e			e			e				e				...
12	...	e		e				e			e					...
13	...	e		e				e			e					...

usable at the beginning of all W multi-frames. There may be at most two different "number of empty time slots", say E_1 and E_2, to be assigned to the frames within a multi-frame (see Section 6.1.4.2.4). Hence, an edge switch may need to run the PTES scheduling at most twice for all the frames within a multi-frame, one based on E_1 for some frames and the other based on E_2 for the remaining frames within the multi-frame. In parallel to traffic transmission from the scheduling done based on E_1, an edge switch may run PTES for computing the second scheduling based on E_2 within a multi-frame. For example, Table 6.3 and Table 6.4 show distribution of empty time slots (shown with 'e') for a frame on wavelength channel λ_w, respectively, in a network with eight edge switches and a network with 14 edge switches. Scheduling at row 4 in Table 6.3 is provided by edge switch 4.

6.1.4.2.1 Edge Switch Functions under PTES

- Edge switch r sends its up-to-date traffic load to the core switch at time $T_{s,r}$ (see Section 6.1.4.2.6 for computing $T_{s,r}$).

- Edge switch r receives average load Λ from the core switch at time $t_a + 1 + D_r$. Based on Λ, PTES in edge switch r assigns a specific "number of empty time slots" to each frame within a multi-frame (see Section 6.1.4.2.4 for computing the number of empty time slots).

- For each frame $F(n,E,w)$, where $0 \leq w \leq W - 1$, edge switch r runs PTES and schedules time slots at row r of the frame as either empty or non-empty (see Section 6.1.4.2.5 for scheduling procedure). Edge switch r makes this scheduling independent from other edge switches.

- To efficiently utilize WCs in the core switch, distribution of empty time slots within a frame must begin from different edge switches. Let b_w be the edge switch number from which the scheduling starts on wavelength channel λ_w. For example, we have $b_w = 4$ and $b_w = 0$ in Table 6.3 and Table 6.4, respectively. Since the scheduling of empty time slots is performed in a distributed manner, each edge switch should compute b_w for frame $F(n, E, w)$ using

$$b_w = (w \times E) \bmod n . \tag{6.2}$$

For instance, assume $n = 5$, $W = 3$, and $E = 2$. For distribution of empty time slots on wavelengths λ_0, λ_1, λ_2, and λ_3, we should respectively have $b_0 = 0$, $b_1 = 2$, $b_2 = 4$, and $b_3 = 1$.

- When sending traffic, edge switch r transmits its optical packets to the core switch based on its provided scheduling, but only in non-empty time slots. For example, from right to left in Table 6.3, edge switch 4 must first avoid sending any optical packet. Then, it can send one optical packet in the next time slot. Then, it must again avoid sending traffic. Next, it can send two optical packets in two consecutive non-empty time slots.

6.1.4.2.2 Core Switch Functions under PTES
The functions of the core switch are as following:

- The core switch receives all up-to-date traffic loads from all edge switches at time t_a.

- Based on these traffic loads, it computes average traffic load Λ among all edge switches and sends Λ back to each edge switch at time $t_a + 1$.

- The core switch should receive traffic from all edge switches from its input ports, starting from the head of frames, at the same time. After arriving the relevant header for a slot set and during the time gap between two time slots, the core switch evaluates potential contention, then resolves the contention (say using r_{wc} shared-per-link WCs at any output link), and finally makes its switching structure ready to switch incoming optical packets toward their desired egress switches. Note that the arrival of optical packets in time slots from different edge switches on the same wavelength channel at the core switch must be in a way that the core switch must exactly see $E \times f$ empty time slots at each nf-slot-set.

6.1.4.2.3 Computing Normalized Traffic Load in the Core Switch
Edge switches and the core switch cooperate with each other to obtain average normalized traffic load Λ (where $0 \leq \Lambda \leq 1$) in network, which is the same on each channel for all edge switches. At time $T_{s,i}$, edge switch i sends its current traffic load Λ_i to the core switch within the header sent for the last slot set. Define H_i to be the size of traffic (in bits) arrived at edge switch i during the last scheduling interval. Let $\overline{H_i}$ be average size

of the traffic arrived at edge switch i until $T_{s,i}$, given by $\overline{H_i} = \delta \times \overline{H_i} + (1 - \delta) \times H_i$, where $\delta \in [0,1]$ is a coefficient. Define Q_i to be the size of residual un-transmitted traffic (in bits) in edge switch i. Edge switch i computes $\Lambda_i = \frac{\overline{H_i} + Q_i}{B_a \times S_I \times S_{ts}}$, where S_{ts} is time-slot length in seconds and B_a is wavelength channel bandwidth. The denominator shows total bandwidth available in bits for edge switch i within scheduling interval S_I. Here, $\overline{H_i}$ estimates traffic arrival at the next scheduling interval; therefore, numerator is the size of total traffic that will be at edge switch i within the next scheduling interval.

Assume that at time slot t_a the core switch receives all up-to-date traffic loads from all edge switches. It then computes average traffic load Λ among all edge switches according to $\Lambda = \frac{1}{n} \sum_{i=1}^{n} \Lambda_i$ and finally sends Λ back to all edge switches.

```
Algorithm: Uniform Distribution of E Values

1:  s=0.5 {initialization}
2:  E_s = ∅ {empty set}
3:  q = E_a/f − ⌊E_a/f⌋
4:  ζ = ⌊q×K⌋/K
5:  for i = 1 to K do
6:      s = s + ζ {update s}
7:      if s < 1 then
8:          append ⌊E_a/f⌋ to E_s
9:      else
10:         append ⌈E_a/f⌉ to E_s
11:         s = s − 1 {update s}
12:     end if
13: end for
14: END
```

Figure 6.22: Pseudo code for uniform distribution of E values inside set E_s [26]

6.1.4.2.4 Computing the Number of Empty Time Slots in an Edge Switch

After receiving Λ, an edge switch should compute "number of empty time slots" required for each frame within a given multi-frame. First, define E_a to be average "number of empty time slots" out of $n \times f$ time slots in an nf-slot-set; i.e., on $n \times f$ the same wavelength channels, where $0 \le E_a \le n \times f$. Parameter E_a is given by $E_a = n \times f \times (1 - \Lambda)$. The $n \times f$ time slots in an nf-slot-set may all be full (i.e., $E_a = 0$), or there may be some empty time slots in an nf-slot-set (i.e., $E_a > 0$). Since E_a may not be an integer number, let us define another parameter E to be an integer number of empty time slots within a frame for a given wavelength channel on the same fibers of all edge switches (say on wavelength channel λ_2 of fiber 3 of all n edge switches).

Let E_s denote the set of K "number of empty time slots" assigned to K consecutive frames within a multi-frame, where these numbers are integer values. It is better to assign these values in a uniform manner to set E_s in order to have a uniform traffic transmission within a multi-frame period. The pseudo code of this uniform distribution is shown in Fig. 6.22. After initialization, the fractional part of $\frac{E_a}{f}$ is rounded off in such a way that it becomes dividable by $\frac{1}{K}$, and the result is called ζ. For example, when $\frac{E_a}{f} = 6.509$ and $K = 5$, the fractional part $\frac{E_a}{f}$ is rounded down to

$\zeta = 0.4$, where 0.4 is dividable by $\frac{1}{K} = 0.2$. Then, the loop at Line 5 is repeated for K times.

Each time, according to the value of s, either $\left\lfloor \frac{E_a}{f} \right\rfloor$ or $\left\lceil \frac{E_a}{f} \right\rceil$ is appended to set E_s (see Lines 8 and 10). When we have $s \geq 1.0$, the value of s is reduced by 1.0 (see Line 11). The pseudo code computes $\phi = K \times \zeta$. In set E_s, to have E_a empty time slots in an interval of K frames, $K - \phi$ frames must have $E_1 = \left\lfloor \frac{E_a}{f} \right\rfloor$ empty time slots, and the remaining ϕ frames must have $E_2 = \left\lceil \frac{E_a}{f} \right\rceil$ empty time slots on the same wavelength channel of the same fibers. For example, at $f = 2$, $\Lambda = 0.47$, $K = 10$, and $n = 10$, we obtain $E_a = 10.6$. These parameters lead to $\zeta = 0.3$ and $\phi = 3$. Then, we have $E_1 = 5$ for $K - \phi = 7$ frames, and $E_2 = 6$ for $\phi = 3$ frames. Therefore, we have $E_a = \frac{2 \times (7 \times 5 + 3 \times 6)}{K} = 10.6$ empty time slots in an S_I interval on the same wavelength channels of two fibers. Using the distribution algorithm, we find $E_s = [5, 6, 5, 5, 6, 5, 5, 5, 6, 5]$ for $K = 10$ consecutive frames.

6.1.4.2.5 Scheduling of Empty Time Slots in an Edge Switch After obtaining the "number of empty time slots" per frame, an edge switch can start the scheduling process. Here, the scheduling is stated for frame $F(n, E, w)$, where $0 \leq w \leq W - 1$. First, the following issues should be considered:

- PTES is presented for wavelength λ_w on a given fiber. The same algorithm is applied for other wavelengths, but with different starting points b_w given in Eq. (6.2).

- Since each edge switch is assigned a unique index r ranging from 0 to $n - 1$, edge switch r runs the PTES scheduling only for row r of the frame.

- In edge switch r, the scheduling of empty time slots on the same wavelength channel of different fibers is the same, e.g., scheduling on λ_1 is the same for all f channels λ_1 on all f fibers.

- To keep the scheduling information of empty time slots in edge switch r for a period of one frame on channel λ_w of fiber f_v (where $0 \leq v \leq f - 1$), an Empty time Slot Scheduling (ESS) array $S_{v,w,r}$ with size n is used. The location *col* of this array (where $0 \leq col \leq n - 1$) is denoted by $S_{v,w,r,col}$ that includes either "e" to show an empty time slot or an egress switch number.

- Consider a frame $F(n, E, w)$ for wavelength λ_w on fiber f_v. This frame is mapped to matrix $S_{v,w}$ of size $n \times n$ and the scheduling is performed in matrix $S_{v,w}$, where each row of $S_{v,w}$ is an ESS array defined above. For example, Fig. 6.23 illustrates the scheduling of empty time slots within frame $F(17, 7, w)$, where $S_{v,w}$ is considered as a 17×17 matrix. The elements of this matrix are assigned in a distributed manner so that edge switch r computes its own scheduling at row r of $S_{v,w}$ (i.e., in $S_{v,w,r}$).

- Scheduling matrix $S_{v,w}$ is divided into E regions (see Fig. 6.23 with regions R_0 to R_6 separated with bold lines). This is because the number of columns needed

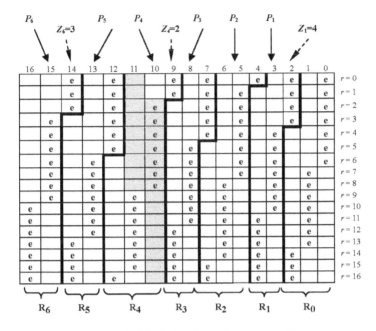

Figure 6.23: An example of distribution of empty time slots at $n = 17$ and $E = 7$ [26]

to distribute n of "e" empty time slots is $a = \frac{n}{E}$. Hence, the required number of regions equals frame length n divided by a; i.e., $\frac{n}{a} = E$. The column located at intersection of regions R_i and R_{i+1} is denoted as shared column i, e.g., column 4 in Fig. 6.23 is shared between regions R_1 and R_2. If distribution of empty time slots can be completed in an integer number of columns before region R_{i+1}, we have no shared column between regions R_i and R_{i+1}.

- Scheduling of empty time slots in matrix $S_{v,w}$ is mapped to the scheduling in one-dimensional array $A_{v,w}$ with size $n \times (n + E - 1)$, used for wavelength λ_w. In this array, n sets each with E empty time slots are distributed in sequence starting from positions $j \times (n + E)$, where $j = 0, 1, ..., n - 1$. Here, the difference between the start of two consecutive sets of distributed empty time slots is always $n + E$. By this, the status of position m in array $A_{v,w}$ is set to empty if g defined by Eq. (6.3) belongs to $[0, 1, ..., E - 1]$.

$$g = m \bmod (n + E) . \tag{6.3}$$

- Since the same algorithm is executed in all edge switches (independent from each other), the scheduling algorithm is presented for edge switch r in Fig. 6.24. Lines 1 to 12 perform initializations. At Line 2, all elements in row r of matrix $S_{v,w}$ are initialized to *NON_EMPTY*. At Line 4, the scheduling algorithm is over when $E = 0$.

Algorithm: The PTES Scheduling

```
1: for col = 0 to n − 1 do
2:     S_{v,w,r,col} = NON_EMPTY
3: end for
4: if E = 0 then
5:     return {no scheduling required}
6: end if

7: k = E − 1 {region number}
8: for i = 1 to k do
9:     P_i = ⌈i×n/E⌉
10:    y = i × n − E × ⌊i×n/E⌋
11:    Z_i = (E − y) mod E
12: end for

13: for col = n − 1 downto 0 do
14:    if r ≥ b_w then
15:        m = n × col + r − b_w
16:    else
17:        m = n × col + r − b_w + n
18:    end if
19:    if k > 0 and (m ≥ n × P_k or n × (P_k − 1) ≤ m < n × (P_k − 1) + Z_k) then
20:        g = (m + k × n) mod (n + E)
21:    else
22:        if k > 0 then
23:            k = k − 1
24:        end if
25:        g = (m + k × n) mod (n + E)
26:    end if
27:    if 0 ≤ g < E then
28:        S_{v,w,r,col} = EMPTY
29:    end if
30: end for
31: END
```

Figure 6.24: Pseudo code of scheduling in edge switch r for frame $F(n,E,w)$ [26]

Since there are E regions, region number k is set to $E - 1$ at Line 7 to point to the last region. Line 9 provides pointer P_i (where $i > 0$) that points to the start of the first column in region R_i, where there is no distributed empty time slot in this column remaining from region R_{i-1}. Figure 6.23 shows pointers P_1 to P_6 for regions R_1 to R_6, respectively. For example, we have $P_3 = 8$ that points to the first column in R_3, where this column is separated from the previous column in region R_2. To compute P_i, one should first note that there are exactly n empty time slots distributed in each region, and E empty time slots distributed in each column. Therefore, to find the maximum number of columns required to distribute E empty time slots in each column of regions R_0 to R_{i-1}, we can use $P_i = \lceil \frac{i \times n}{E} \rceil$.

At Line 11, the number of empty time slots remaining from one region to be distributed at the beginning of the immediate next region is computed, where Z_i is the number of empty time slots remaining from region R_{i-1} that must be distributed at the beginning of region R_i. To compute Z_i (where $i > 0$), note that there are $i \times n$ empty time slots distributed in regions R_0 to R_{i-1}. Of these empty time slots, $E \times \lfloor \frac{i \times n}{E} \rfloor$ empty time slots are distributed before the shared column $i - 1$. Then, in this column, the remaining $y = i \times n - E \times \lfloor \frac{i \times n}{E} \rfloor$ empty time slots are distributed in region R_{i-1}, and finally $Z_i = (E - y) \bmod E$ empty time slots need to be distributed in region R_i. Note that when $i \times n$ is divisible by E, there is no shared column, and therefore we have $Z_i = 0$. This is why the *mod* function must be used in calculation of Z_i. For example, $Z_4 = 2$ in Fig. 6.23 shows that there are two empty time slots remaining from region R_3 which need to be distributed at the beginning of region R_4 (see column 9 in Fig. 6.23). Similarly, we have $Z_1 = 4$ and $Z_6 = 3$ that show the number of empty time slots distributed at the start of columns 2 and 14 (located in R_1 and R_6), respectively.

The loop starting from Line 13 until Line 30 schedules empty time slots within row r of matrix $S_{v,w}$ (i.e., for edge switch r), where *col* is the position of a time slot at row r. At Lines 14 to 18, the two-dimensional position of time slot (r, col) in matrix $S_{v,w}$ is mapped to one-dimensional position m in array $A_{v,w}$. The offset b_w is also involved in the mapping process.

Lines 19 to 26 specify to which region point m belongs, and then g is appropriately set. Here, the scheduling of empty time slots in matrix $S_{v,w}$ is mapped to the scheduling of empty time slots in array $A_{v,w}$. Note that the parameter *col* starts from the last of array $S_{v,w,r}$ in order to enable us to easily find the region of point m in matrix $S_{v,w}$. Point m belongs to region R_k (when $k > 0$) if either point m is not located in shared column $k - 1$ (i.e., $m \geq n \times P_k$), or this point is located in shared column $k - 1$ (i.e., $n \times (P_k - 1) \leq m < n \times (P_k - 1) + Z_k$). Otherwise, parameter k is updated in order to point to the next region. Note that at $k = 0$, point m only belongs to region R_0. One can see from Fig. 6.23 that the distance between the beginning of two sets of empty time slots distributed in two consecutive columns within a region is $n + E$. For example, in region R_4 the distance between two sets of distributed empty time slots starting at position $(r = 2, col = 10)$ and $(r = 9, col = 11)$ is 17+7 = 24 (shown with gray time slots in columns 10 and 11 of Fig. 6.23).

However, distance $n + E$ reduces to E from one region to its neighbor region (see the starting points at columns 7 and 8). For mapping the status of a time slot from matrix $S_{v,w}$ to array $A_{v,w}$, value of $k \times n$ is added to m at region R_k to compensate k of n time slots differences created between two neighbor regions. Then, g is computed similar to Eq. (6.3), but considering offset $k \times n$. At Line 28, if $g \in [0, E - 1]$, then the status of the time slot at position (r, col) in matrix $S_{v,w}$ is set to "e"; otherwise, it is left as a non-empty time slot.

6.1.4.2.6 Timing Issues in PTES

Since the network is heterogeneous due to different distances between edge switches and the core switch, transmission times of either traffic load values or optical packets (by different edge switches) must be in a way that the core switch receives all of them at the same time. This section details when edge switches should start sending their information to the core switch in order to compensate for different links delays. Assume the start of each scheduling interval should be at time slot $T_l = l \times S_l$ (where $l = 1, 2, 3, \ldots$), at which traffic arrives at the core switch based on the new schedules from all edge switches.

Recall that edge switch i should send its current traffic load Λ_i to the core switch at time slot $T_{s,i}$. Then, at time slot $t_a = T_{s,i} + D_i$, the core switch must receive all traffic loads from all edge switches. Assuming that one time slot is required for computing average traffic load Λ, the core switch will send Λ back to all edge switches at time slot $t_a + 1$.

Then, at time slot $t_a + 1 + D_i$, parameter Λ arrives at edge switch i, and it starts the assignment of "number of empty time slots" to the frames in a multi-frame and scheduling of empty time slots based on Λ. Let schedulings for W frames (to be sent on W wavelength channels) take at most τ_s time slots. Since the head of frames have offset with respect to each other, the transmission of traffic based on the new scheduling must be in such a way that the head of frames from different edge switches arrive at the core switch at the same time. To do this, edge switch i must use the new scheduling of empty time slots after $t_{w,i}$ time slots. In other words, edge switch i should start sending its traffic based on the new scheduling at time slot $t_a + 1 + D_i + t_{w,i}$, and this traffic will arrive at the core switch at time slot $t_a + 1 + 2 \times D_i + t_{w,i}$. However, the furthest edge switch can send its traffic based on the new scheduling right after finishing the scheduling at time slot $t_a + 1 + D_{max} + \tau_s$. Then, traffic from the furthest edge switch will start arriving at the core switch at time slot $t_a + 1 + 2 \times D_{max} + \tau_s$. Since the arrival of traffic at the core switch from all edge switches must be at the same time, we have $t_a + 1 + 2 \times D_i + t_{w,i} = t_a + 1 + 2 \times D_{max} + \tau_s$. Hence, we have $t_{w,i} = 2 \times (D_{max} - D_i) + \tau_s$. As assumed, traffic sent based on the new scheduling must arrive at the core switch at time slots T_l. Hence, we have $t_a + 1 + 2 \times D_i + t_{w,i} = T_l$, or $T_{s,i} = T_l - (D_i + 2 \times D_{max} + \tau_s + 1)$.

6.1.4.2.7 Discussion on the Features of PTES   Important features of PTES are:

- PTES can reduce the variance of optical packets sent in non-empty time slots by all edge switches at each time slot, thus reducing traffic loss. What matters in PTES is to have exactly $E \times f$ empty time slots among the time slots in each nf-slot-set. This could be realized as a kind of traffic shaping. Under IPT, for example, at $n = 10$, $f = 1$, and $\Lambda = 0.8$, there may be different number of non-empty time slots between 0 and 10 at each nf-slot-set. However, long-term average number of non-empty time slots at $nf - slot - sets$ is 8. Under PTES, there are exactly eight non-empty time slots (or two empty time slots) at each nf-slot-set. Note that the number of non-empty time slots and empty time slots are respectively computed by $n \times f \times \Lambda$ and $n \times f \times (1 - \Lambda)$. As another example, at $n = 10$, $f = 2$ and $\Lambda = 0.37$, edge switches under PTES must send their optical packets in time slots in a way that there are either 8 or 7 non-empty time slots (equivalent to 2 or 3 empty time slots) at each nf-slot-set.

- PTES scheduling is simple. Complexity of the first to the third loops in Fig. 6.24 are $O(n)$, $O(E)$, and $O(n)$, respectively. Since $E \leq n$, PTES complexity in each edge switch is $O(n)$ for a frame with size $n \times n$.

- PTES is fair in distribution of empty time slots. Since n empty time slots are distributed in region R_i, there is only one empty time slot distributed within a part of a row of matrix $S_{v,w}$ that is located in region R_i. On the other hand, there are E regions in matrix $S_{v,w}$. Hence, there are exactly E empty time slots (and $n - E$ non-empty time slots) in each row of matrix $S_{v,w}$. In short, PTES is fair in distributing both empty and non-empty time slots among edge switches because an equal number of empty time slots are assigned to each edge switch within any frame period.

- The distance between any two consecutive empty time slots distributed in any row of matrix $S_{v,w}$ is either $\lfloor \frac{n}{E} \rfloor$ or $\lceil \frac{n}{E} \rceil$ (see Fig. 6.23, where the distance is either 2 or 3 for any edge switch). This is because between any two consecutive empty time slots at row r of matrix $S_{v,w}$, there are exactly n empty time slots to be distributed. In the best case, these n empty time slots may require $\lfloor \frac{n}{E} \rfloor$ columns, whereas in the worst-case they may need $\lceil \frac{n}{E} \rceil$ columns to be distributed. Hence, PTES can provide a uniform distribution of empty time slots within any row of matrix $S_{v,w}$.

- By distributing empty time slots on different wavelength channels starting from different edge switches, WCs inside the core switch can be efficiently utilized because wavelength channels can be used in a uniform manner and overwhelming the same wavelength channel can be avoided. Note that at any time slot, the core switch can use all wavelengths at its output ports to send optical packets toward their egress switches.

6.1.4.2.8 Packet Loss Rate Analysis in PTES In the following computations, PLR is obtained for a given output link (referred to as the tagged output link) within a time slot in steady state, where there are r_{wc} shared-per-link WCs located at each output link. Uniform traffic transmission to all output links is assumed in the following computations.

Define the c-optical packet collision as an event where there is a collision at the core switch from any c optical packets (out of the N_{OP} optical packets in non-empty time slots) going to the tagged output link on the same wavelength λ_w in such a way that only f optical packets are allowed to be switched on wavelength λ_w, and thus, $\max(c - f, 0)$ optical packets cannot be switched on this wavelength channel. Clearly, if no contention resolution scheme is used at the core switch, these $\max(c - f, 0)$ optical packets will be dropped. For example, Fig. 6.25 shows $c = 3$ optical packets destined to tagged output link 2.

Consider an nf-*slot-set* with $n \times f$ optical packets on wavelength channel λ_w from all input ports of the core switch. These $n \times f$ optical packets can be grouped into three types of optical packets as depicted with an example in Fig. 6.25 on tagged output link 2 (where a number inside a time slot carrying an optical packet indicates an output link number for the optical packet):

- There are exactly $E \times f$ empty time slots among them ($E \times f = 2$ empty time slots in the example). Therefore, there are $N_{OP} = n \times f - E \times f$ optical packets in N_{OP} non-empty time slots, each destined to one of n output links of the core switch.

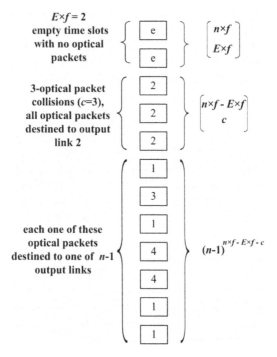

Figure 6.25: An *nf-slot-set* with $n = 6$, $f = 2$, and $E = 1$ on wavelength λ_w [26]

- There are exactly c optical packets among the N_{OP} optical packets that are destined to the tagged output link (three optical packets destined to output link 2 in the example).

- There are exactly $N_{OP} - c$ optical packets that are destined to other output links of the core switch (seven optical packets destined to other output links (1, 3, or 4) in the example).

Total possible number of combinations for an event of c-optical packets collision, n_c, is given by (see Fig. 6.25 for illustration)

$$
n_c = \begin{cases} \left(\begin{array}{c} n \times f \\ E \times f \end{array} \right) \left(\begin{array}{c} N_{OP} \\ c \end{array} \right) (n-1)^{N_{OP}-c} & \text{for} \quad N_{OP} \geq c, \\ 0 & \text{for} \quad N_{OP} < c, \end{cases} \tag{6.4}
$$

where the first term is the combination of empty time slots in an *nf-slot-set*; the second term is the combination of c-optical packets collisions among N_{OP} optical packets destined to the tagged output link; and the third term is the combinations for the remaining $N_{OP} - c$ optical packets, where each optical packet can be destined to one of $n - 1$ output links (notice the tagged output link is excluded). On the other hand, total number of output link combinations in the *nf-slot-set*, n_t, is given by

$$n_t = \left(\begin{array}{c} n \times f \\ E \times f \end{array} \right) n^{N_{OP}}, \tag{6.5}$$

where the first term is the combination of the empty time slots in the nf-$slot$-set; and the second term denotes the combination of N_{OP} optical packets in the nf-$slot$-set, where each optical packet can be destined to one of the n output links of the core switch.

Define random variable p_x to be the probability of having a x-optical packets collision at the tagged output link, where $0 \le x \le N_{OP}$. Considering uniform optical packet arrivals at the core, p_x is given by

$$p_x = \frac{n_x}{n_t} \qquad \text{for } 0 \le x \le N_{OP} . \tag{6.6}$$

A set of f the same wavelength channels (say, f of wavelength channels λ_1) at the tagged output link is referred to as the same-wavelength group from now on. Therefore, there are W the same-wavelength groups at the tagged output link. In the following analysis, W the same-wavelength groups can be divided into two categories:

- j the same-wavelength groups on which all arriving time slots at any of the wavelengths of each group are empty; and

- $u = W - j$ the same-wavelength groups on which there is at least one non-empty time slot on one wavelength channel of each group. For simplicity, the wavelength channels in the u groups are assigned alias names from λ'_1 to λ'_u.

Define random variable X to be the number of the same-wavelength groups on which all time slots at any of the wavelength channels of each group are empty. Thus, probability of having $X = j$ is distributed based on a binomial distribution

$$\text{Prob.}\{X = j\} = \left(\begin{array}{c} W \\ j \end{array} \right) p_0^j (1 - p_0)^{W-j} , \tag{6.7}$$

where $j \in [0, 1, 2, .., W]$, and p_0 computed using Eq. (6.6) is the probability of having no non-empty time slots (i.e., $x = 0$) at the channels of a given same-wavelength group.

Define random variable Y_i to be the number of optical packets at the channels of the same-wavelength group i (i.e., the group of f wavelength channels λ'_i). Computing p_0 and p_y from Eq. (6.6), the probability of having $Y_i = y$ is given by

$$\text{Prob.}\{Y_i = y\} = \frac{p_y}{1 - p_0} \qquad \text{for } y > 0 . \tag{6.8}$$

The probability of having $y_1 > 0$ optical packets arriving on f channels of the same-wavelength group 1, along with $y_2 > 0$ optical packets on f channels of the same-wavelength group 2, till $y_u > 0$ optical packets on f channels of the same-wavelength group u can be computed by

$$\prod_{k=1}^{u} \text{Prob.}\{Y_k = y_k\}. \tag{6.9}$$

Among y_k optical packets arriving on f wavelengths of the same-wavelength group k, $\min(f, y_k)$ of them are easily switched on f channels λ'_k. In other words, of $\sum_{k=1}^{u} y_k$ optical packets arriving at the tagged output link, $\sum_{k=1}^{u} \min(f, y_k)$ optical packets can be switched. In addition, some of the remaining optical packets can then be switched after using r_{wc} WCs located at the tagged output link. First, recall that there are $f \times j$ wavelength channels on which all time slots at the tagged output link are empty. In addition, among the u same-wavelength groups, there are $\eta = \sum_{k=1}^{u} \max(0, f - y_k)$ wavelengths on which we have empty time slots. In total, $\beta = \min(r_{wc}, f \times j + \eta)$ optical packets can be switched using r_{wc} WCs. Thus, for a given state $(j, y_1, y_2, ..., y_u)$, the number of optical packets that cannot be switched is given by

$$\alpha_{(j,y_1,y_2,...,y_u)} = \max\left(\left(\sum_{k=1}^{u} (y_k - \min(f, y_k)) \right) - \beta, 0 \right). \tag{6.10}$$

Considering the above discussions, the average number of lost optical packets at the tagged output link is computed by

$$n_{loss} = \sum_{j=0}^{W} \left(\text{Prob.}\{X = j\} \times \sum_{\substack{1 \leq y_1 \leq N_{OP} \\ 1 \leq y_u \leq N_{OP}}} \left(\prod_{k=1}^{u} \text{Prob.}\{Y_k = y_k\} \times \alpha_{(j,y_1,y_2,...,y_u)} \right) \right), \tag{6.11}$$

or

$$n_{loss} = \sum_{j=0}^{W} \left(\text{Prob.}\{X = j\} \times \sum_{\substack{1 \leq y_1 \leq N_{OP} \\ 1 \leq y_u \leq N_{OP}}} \frac{p_{y_1} \times p_{y_2} \cdots \times p_{y_u}}{(1 - p_0)^u} \times \alpha_{(j,y_1,y_2,...,y_u)} \right), \tag{6.12}$$

where the upper limit N_{OP} for y_k shows total number of optical packets in $nf\text{-}slot\text{-}set$ that can be destined to the tagged output link on f wavelengths λ'_k in the worst-case.

Of N_{OP} optical packets arriving at the same wavelength at n input links of the core switch, $n_{dlv} = \frac{N_{OP} \times W}{n}$ optical packets on average are delivered to $f \times W$ channels of the tagged output link. Hence, average PLR at the tagged output link among the frames each with $E \times f$ empty time slots (where E is an integer number) is given by

$$Loss(n, f, W, E) = \frac{n_{loss}}{n_{dlv}} . \tag{6.13}$$

To compute PLR for the case that we have E_a empty time slots on average in the network, where E_a is a real number, we consider the K values of E from set E_s as described in Section 6.1.4.2.4. Since there are $(K - \phi)$ frames with $E = E_1 = \left\lfloor \frac{E_a}{f} \right\rfloor$ empty time slots, and ϕ frames with $E = E_2 = \left\lceil \frac{E_a}{f} \right\rceil$ empty time slots in a multi-frame (among K frames), the average PLR with E_a empty time slots can be calculated by

$$PLR(n, f, W, E_a) = \frac{(K - \phi) \times Loss(n, f, W, E_1) + \phi \times Loss(n, f, W, E_2)}{K} . \tag{6.14}$$

Due to uniform traffic transmission to all output links, the same PLR is obtained for the core switch.

6.1.4.2.9 *Packet Loss Rate Performance Evaluation in PTES*   Effectiveness of PTES over IPT is evaluated by analysis and simulation under overlaid network model with $n = 8$, f fibers per link, r_{wc} converters per output link, and $W = \frac{8}{f}$ wavelength channels per fiber. Each optical packet carries only one client packet. Propagation delays between edge switches and core switches are uniformly set from 10 to 100 time slots (2 to 20 km). Since 10% of bandwidth is used for time gap, load $L = 0.9$ means full channel utilization under the failure case. Hence, results are depicted until $L = 0.8$.

Client packet inter-arrival times from legacy networks at edge switches are modeled as Poisson process. Figure 6.26 depicts analytical comparison of PLR under both IPT and PTES and simulation results under PTES at $f = 2$. The PTES always leads to a lower PLR than IPT. At light loads, PTES can result in a PLR significantly lower than IPT, e.g., PTES leads to no traffic loss at $L = 0.1$. However, PLR performances of IPT and PTES become closer to each other by increasing load so that an almost equal PLR can be observed for both IPT and PTES at high traffic loads. According to the diagram, at a given r_{wc}, IPT analysis results are always higher than PTES analysis results, and PTES analysis results are always higher than PTES simulation results. Simulation results of PTES match with analytical results, but with some difference due to the fact that traffic loss analysis always assumes that there are exactly $E \times f$ empty time slots in an $nf\text{-}slot\text{-}set$. However, the number of empty time slots may be more than $E \times f$ within simulation. This is because there may be no optical packet to be sent in some non-empty time slots, and therefore PLR

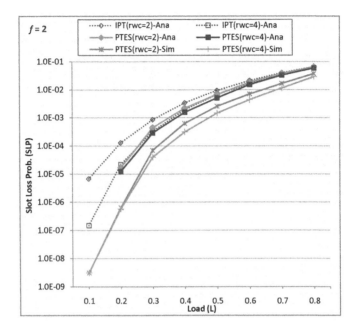

Figure 6.26: PLR in IPT and PTES under Poisson traffic at $f = 2$ [26]

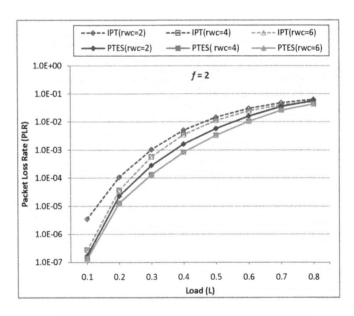

Figure 6.27: PLR in IPT and PTES under Pareto traffic at $f = 2$ [26]

is computed smaller in simulation. In other words, the PTES analysis provides an upper-bound on PLR. Additional performance analysis and simulation results show that under any f and L, PLR decreases by increasing the number of WCs. On the other hand, PLR goes down by increasing the number of fibers under any r_{wc} and L. However, there is a noticeable difference between PLR from $f = 1$ to $f = 2$ and PLR from $f = 2$ to $f = 4$.

Additional performance evaluation under Pareto traffic arrivals also shows that PTES always provides a lower PLR compared with IPT at $f = 2$ (see Fig. 6.27). By increasing f or r_{wc}, PLR decreases under both schemes. At $f = 4$, there is almost no difference between PLR performances under $r_{wc} = 2$ and $r_{wc} = 4$ because there are only two wavelength channels on each fiber at $f = 4$; therefore, employing additional WCs is useless.

6.1.4.3 Hybrid Additional Wavelengths + Prioritized Retransmission Architecture

The combination of using additional wavelength channels as a contention avoidance scheme and prioritized retransmission as a contention resolution scheme has been evaluated in overlaid slotted single-hop OPS networks [27].

Assume that each connection link needs W wavelengths to carry all of its traffic. However, additional W_a wavelengths are used on each fiber so that an ingress switch can have up to $W + W_a + 1$ wavelengths on a fiber, where one wavelength is used for transmitting the header of optical packets (see Aggregated Header in Section 1.4.4). Each ingress switch can transmit up to $W + W_a$ optical packets (called an OP set) at the same time to the core switch. Each core switch is an $n * n$ optical switch. Define L to be the normalized traffic load, normalized with respect to the channel bandwidth, on each wavelength channel when no additional wavelength channel is used on a fiber. Define L_e to be the effective normalized traffic load, normalized with respect to the channel bandwidth, on each wavelength channel when there are W_a additional wavelength channels on a fiber. Let n-OP-set be the set of n optical packets from n ingress switches on the same wavelength in a given time slot. At each n-OP-set, at most n optical packets may arrive at the core switch on the same wavelength channel. Therefore, there are $Z = n \times L_e$ optical packets in an n-OP-set. Let H be the maximum number of retransmissions required to transmit a tagged optical packet successfully so that an optical packet with priority H is dropped with a probability of less than rare event probability X, where $X \ll 1$.

The analysis of the drop probability starts from level H. At level j, we refer to the level-j retransmitted optical packets as *tagged-priority optical packets*, and all the other optical packets in the n-OP-set are referred to as *non-tagged-priority optical packets*. Define n_r to be the average number of non-empty tagged-priority optical packets at any level. Clearly, the remaining $Z - n_r$ optical packets are the average number of non-tagged-priority optical packets. Define PLR $R_L\{n_r, k\}$ to be the drop probability among n_r tagged-priority optical packets destined to the tagged output link given that only k optical packets capacity ($k = 0$ or 1) are available at the tagged output link defined as follows [28]:

$$R_L\{n_r,k\} = \frac{\sum_{c=k+1}^{n_r} (c-k) \times \binom{n_r}{c} \times (n-1)^{n_r-c}}{n^{n_r-1} \times n_r} \qquad \text{if } n_r \geq 1, \qquad (6.15)$$

$$R_L\{n_r,k\} = 0 \quad \text{otherwise.}$$

Since parameter n_r in Eq. (6.15) is not necessarily an integer value (due to multiplying Z by a probability), the Gamma function should be used to compute $x!$; i.e., $x! = \Gamma(x+1)$ should be used to compute $\binom{n_r}{c}$.

Using W_a additional wavelengths, there are $W + W_a$ time slots for the transmission of W optical packets in each OP set in an ingress switch. Let optical packets be uniformly distributed over $W + W_a$ wavelengths. Therefore, the probability of having one optical packet on a wavelength channel is $\frac{W}{W+W_a}$. Considering the normalized traffic load L in an ingress switch, the effective normalized traffic load (L_e) on each wavelength channel is given by $L_e = \frac{W}{W+W_a} \times L$ when using W_a additional wavelengths.

Let $\Pi_H = \{\pi_0 \ \pi_1 \ \pi_2 \ ... \ \pi_H \ \pi_{H+1}\}$ denote the probability vector that at most H retransmissions are required in an n-OP-set to send an optical packet to the tagged output link in steady state, where π_0 is the transmission probability of new optical packets; and π_i is the probability that an optical packet is retransmitted for the ith time. Let $P_{i,drop}$ denote the probability of dropping an optical packet at the ith retransmission level. Similar to [28], one can derive the following equations:

$$\pi_i = P_{i-1,drop}, \quad \text{where} \quad 1 \leq i \leq H \,,$$

$$\pi_{H+1} = P_{H,drop}, \quad \text{where} \quad P_{H,drop} < X \,, \qquad (6.16)$$

$$\pi_0 = 1 - \sum_{j=1}^{H} \pi_j \,.$$

To solve Eq. (6.16), $P_{i,drop}$ must be found which considers the probability that the retransmitted optical packets at levels of greater than or equal to $i+1$ have already occupied an output port in the tagged output link. Let N_k denote the number of optical packet arrivals at the tagged output link at a retransmission level greater than or equal to k, and define $P_a\{N_k = \alpha\}$ to be the corresponding probability of having α optical packet arrival ($\alpha = 0$ or 1) such that α ports of the tagged output link are occupied. Using the assumptions of uniform optical packet arrivals at each output link and equal normalized traffic load L_e on all wavelength channels, one can obtain

$$P_a\{N_k = \alpha\} = \begin{cases} 1 - L_e \times \pi_k & \text{if} \quad \alpha = 0 \,, \\ L_e \times \pi_k & \text{if} \quad \alpha = 1 \,, \end{cases} \qquad (6.17)$$

where $L_e \times \pi_k$ is the probability of one optical packet arrival (and occupation) to the tagged output link at retransmission level of greater than or equal to k. Let $P_{i,\alpha,drop}$ be the probability that a tagged-priority optical packet is dropped at retransmission level i, provided that α optical packets have already arrived and occupied the tagged-output link ports at retransmission level $i + 1$. Then,

$$P_{i,\alpha,drop} = \frac{n_{r,i} \times P_a\{N_{i+1} = \alpha\} \times R_L\{n_{r,i}, 1 - \alpha\}}{Z} , \qquad (6.18)$$

where the numerator represents the average number of dropped optical packets among $n_{r,i} = Z \times \pi_i$ tagged-priority optical packets with priority i in the n-OP-set. The second factor in the numerator indicates the probability of α optical packets arrivals at levels of greater than or equal to $i + 1$. The third term represents the drop probability among $n_{r,i}$ tagged-priority optical packets given that a capacity of only $1 - \alpha$ optical packets is available at the tagged output link. The denominator is the average number of Z optical packets in the n-OP-set. The marginal probability $P_{i,drop}$ is obtained as

$$\pi_{i+1} = P_{i,drop} = P_{i,0,drop} + P_{i,1,drop} =$$
$$\pi_i \times \left((1 - L_e \times \pi_{i+1}) \times R_L\{n \times L_e \times \pi_i, 1\} + L_e \times \pi_{i+1} \times R_L\{n \times L_e \times \pi_i, 0\}\right) . \qquad (6.19)$$

For comparison with the Random Retransmission (RR) detailed in 3.2.1.3.1, one needs to obtain π_i (where $i = 0, 1, 2, \ldots$). Since there is no difference between a newly transmitted optical packet and a previously retransmitted optical packet in RR, the probability of k transmissions until a success is computed by

$$P\{k \ transmissions \ until \ success\} = \pi_{k-1} = \pi_0 \times (1 - \pi_0)^{k-1} , \qquad (6.20)$$

where π_0, the probability of a successful transmission, is obtained from $\pi_0 = 1 - R_L\{Z, 1\} = 1 - R_L\{n \times L_e, 1\}$.

Solving the $H + 1$ equations in Eq. (6.16) in the worst-case (at $L = 1$, $W_a = 0$, $X = 10^{-7}$, and $n = 1000$ as the largest possible core switch dimension), one can find that $\pi_5 \ll X$, and therefore we have $H = 4$. Thus, for any other combination, one need only to solve at most five equations in Eq. (6.16).

Table 6.5 compares the performances of PR and RR at $L = 0.7$, $W = 4$, and $n = 32$. This table verifies the correctness of the analysis. Both the analysis and simulation results indicate that most parts of the dropped optical packets can pass through the core switch within two retransmissions under PR, and only less than 0.2% would require the third retransmission. The number of retransmissions is not limited in RR; as observed, an optical packet may have been retransmitted for the 18-th time. The volume of traffic retransmission at each retransmission level and under the same retransmission scheme is lower at $W_a = 2$ and $W_a = 4$ than at $W_a = 0$. For example, using additional wavelength channels, more traffic can pass through the core switch

Table 6.5: Performance comparison of WA + PR at $L = 0.7$ and $n = 32$ [27]

W_a	Retransmission	π_0	π_1	π_2	π_3	π_4	π_5	π_6
0	PR(Sim.)	0.724630	0.251290	0.023879	0.000202	0.000000	0.000000	0.000000
	PR(Ana.)	0.726941	0.251999	0.021060	0.000000	0.000000	0.000000	0.000000
	RR(Sim.)	0.724625	0.199333	0.054979	0.015219	0.004220	0.001177	0.000323
	RR(Ana.)	0.727044	0.198451	0.054168	0.014786	0.004036	0.001102	0.000301
2	PR(Sim.)	0.803690	0.187907	0.008387	0.000017	0.000000	0.000000	0.000000
	PR(Ana.)	0.809070	0.185466	0.005464	0.000000	0.000000	0.000000	0.000000
	RR(Sim.)	0.803699	0.157708	0.030995	0.006100	0.001203	0.000237	0.000047
	RR(Ana.)	0.809065	0.154479	0.029495	0.005632	0.001075	0.000205	0.000039
4	PR(Sim.)	0.847605	0.148550	0.003842	0.000003	0.000000	0.000000	0.000000
	PR(Ana.)	0.855986	0.144013	0.000001	0.000000	0.000000	0.000000	0.000000
	RR(Sim.)	0.847599	0.129161	0.019694	0.003005	0.000459	0.000069	0.000011
	RR(Ana.)	0.854963	0.124001	0.017985	0.002609	0.000378	0.000055	0.000008

at the first transmission; i.e., $\pi_0 \approx 0.72$ and $\pi_0 \approx 0.80$ at $W_a = 0$ and $W_a = 2$, respectively. Under $W_a = 4$, the effective traffic load on each wavelength channel is half of the traffic load; i.e., $L_e = 0.5 \times L$. The case $W_a = 4$ provides the best performance in reducing the volume of traffic that should be retransmitted because using additional wavelength channels reduces traffic load and PLR as a result. One can see that the difference between the performance of PR and RR is reduced when traffic load is decreased or more additional wavelength channels are used on a fiber.

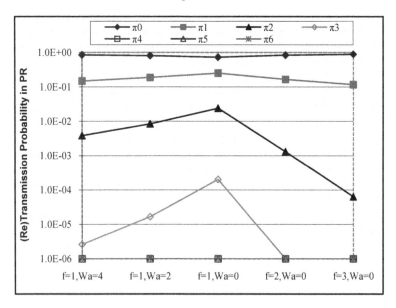

Figure 6.28: Performance comparison of PR under MF and AW contention avoidance schemes [27]

Now, the performance of the single-hop OPS metro network with $n = 32$ and $W = 4$ is compared at traffic load $L = 0.7$ under two network scenarios (see Fig. 6.28 and Fig. 6.29): (1) a single-fiber network with W_a additional wavelength channels on

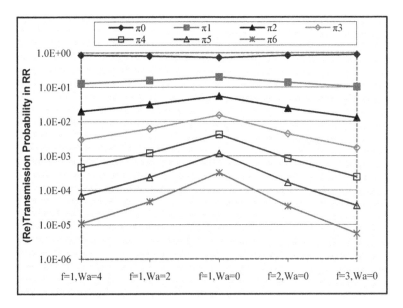

Figure 6.29: Performance comparison of RR under MF and AW contention avoidance schemes [27]

a fiber and (2) a multi-fiber architecture with f fibers on each connection link. Both scenarios use hardware-based contention avoidance techniques (i.e., either using additional wavelengths or using more fibers). Since under PR, there is no observation for more than two retransmissions (i.e., $\pi_3 = 0$) in a multi-fiber architecture, nothing is displayed (due to the logarithmic scale) at the right-hand side of the performance curve of π_3. Since there is no retransmission observed for more than three times under PR, there is also no curve displayed for π_i, where $i > 3$. However, there are observations even up to 14 retransmissions under RR.

In the diagrams, the case ($f = 1, W_a = 0$) at the center of the diagrams denotes to the base network with no contention avoidance. Both diagrams show that by using more contention avoidance hardware, the probability of the arrival of newly generated traffic (i.e., π_0) in the network increases. For example, under PR, π_0 at ($f = 1$, $W_a = 0$) has increased from almost 0.72 to almost 0.83 and 0.88 in a multi-fiber architecture with $f = 2$ and $f = 3$, respectively. The value of π_0 at ($f = 1, W_a = 0$) has also risen to almost 0.80 and 0.85 when using, respectively, $W_a = 2$ and $W_a = 4$ additional wavelengths in a single-fiber network architecture. The diagrams also show that by using more contention avoidance hardware, the probability of retransmission at all levels (i.e., π_i) has decreased. For example, under PR, π_2 at ($f = 1, W_a = 0$) has decreased from almost 0.023 to almost 0.0013 and 0.00006 in a multi-fiber architecture with $f = 2$ and $f = 3$ fibers, respectively. The value of π_2 at ($f = 1, W_a = 0$) has reduced to almost 0.0084 and 0.0038 at $W_a = 2$ and $W_a = 4$ in a single-fiber architecture, respectively. A similar behavior can also be observed under RR. Among the four cases of contention avoidance, a multi-fiber architecture with $f = 3$ fibers provides the highest π_0, and the lowest π_i, for $i > 0$. However, such an architecture needs a core switch of size 96*96, which is a large optical switch. When using additional wavelength channels, we still need a 32*32 core switch, but with a high number of wavelengths on a fiber, say eight wavelengths on each fiber at $W_a = 4$.

6.1.5 Hybrid Reservation/Contention-Based OPS

The DA technique (see Section 2.1.2.2) has better performance results than MP-BvN (see Section 6.1.3.3) under low traffic loads [21] because DA experiences a low number of collisions and loss. Therefore, most of the transmitted traffic successfully passes through the network without requiring further retransmissions. However, re-transmission of dropped traffic may lead to a high traffic transmission in an ingress switch, especially at high loads. This may lead to traffic drop at ingress switches. Unlike DA, there is no traffic loss for MPBvN in a core switch due to reservation. However, the reservation imposes an additional two-way propagation delay on the delay performance of client packets. This is why delay performance at low traffic load is better under DA than MPBvN.

DA can provide much better results than MPBvN even at high traffic loads when a high volume of contention resolution/avoidance techniques are used in the network [21], but at the expense of network cost increase. Otherwise, when a low volume of contention resolution/avoidance techniques is used in the network, MPBvN can provide a better performance than DA at high traffic loads due to the absence of growing volume of retransmitted traffic found in DA.

Algorithm: The HTDM Operation

1: Init system

   Protocol DA:

2: **while** load in the network is low **do**
3:      Run DA in the network
4: **end while**
5: Send a command to all edge switches to switch from DA to MPBvN
6: Goto MPBvN protocol

   Protocol MPBvN:

7: **while** load in the network is high **do**
8:      Run MPBvN in the network
9: **end while**
10: Send a command to all edge switches to switch from MPBvN to DA
11: Goto DA protocol

Figure 6.30: The HTDM protocol. Reproduced from *Computer Networks*, Vol. 54, Akbar Ghaffarpour Rah-bar and Oliver Yang, Agile bandwidth management techniques in slotted all-optical packet interconnection networks, 387–403, copyright 2010, with permission from Elsevier.

Since the same frame-based and time slotted mechanism is utilized for both DA and MPBvN, the same architecture can be used under Hybrid TDM (HTDM). The HTDM combines the good attributes of DA and MPBvN (see Fig. 6.30), where the OPS network works under DA at low traffic loads and under MPBvN at high traffic loads. The core switch decides when to switch between the protocols at frame boundaries and then announces this switching by sending a command on the control channel to all edge switches. Then, each edge switch will have to switch its operation as requested at a synchronized time. For simplicity, it is assumed that the network always starts from DA whenever the HTDM is reset.

6.1.5.1 Switching Protocol from DA to MPBvN   Consider the following definitions required in the HTDM operation.

- $N_{T,i}$: total number of non-empty time slots (including optical packets) transmitted from ingress switch i within a frame.

- U_f: average channel utilization of time slots within a frame.

- U_T: threshold value for the maximum channel utilization of time slots within a frame set by the network manager.

- r_a: average normalized load of the arrival of optical packets from all ingress switches at the core switch.

- r_n: average normalized traffic load of newly generated optical packets from all ingress switches.

- r_T: normalized traffic load of the newly transmitted optical packets when $r_a = U_T$.

- PLR_c: average optical packet loss rate at the core switch when the HTDM system is operating under DA.

- PLR_f: average optical packet loss rate within a frame at the core switch when the HTDM system is operating under DA.

- PLR_T: optical packet loss rate when $r_a = U_T$.

The following procedures are necessary to switch from DA to MPBvN:

- **Switching Criterion:** The core switch performs the algorithm shown in Fig. 6.31 at each frame boundary for a possible switching. When DA is running, we have $r_a = r_n + r_a \times PLR_c$, where the second term represents the normalized feedback traffic load of dropped optical packets due to optical packet loss rate PLR_c at the core switch. The PLR_c is computed by the running average method with smoothing factor σ. By increasing r_n, optical packet loss rate will also increase until the critical point $r_a = r_n + r_a \times PLR_c = 1.0$. Beyond this point, any further increase would cause queues in an ingress switch to grow up. Client packet drop would occur when using a finite buffer in ingress switches. The

Algorithm: Evaluate Switching from DA to MPBvN

1: $PLR_c = \sigma \times PLR_c + (1 - \sigma) \times PLR_f$
2: $C_{U,f} = \sum_{i=1}^{n} \frac{N_{T,i}}{N_s}$
3: **if** $C_{U,f} \geq C_{U,T}$ **then**
4:    $g = g + 1$ {update counter g}
5: **else**
6:    $g = 0$ {reset counter g}
7: **end if**
8: **if** $g \geq S_p$ **then**
9:    $g = 0$ {reset counter g}
10:    $PLR_T = PLR_c$ {calculate PLR_T used to switch from MPBvN to DA}
11:    $r_T = C_{U,T} \times (1 - PLR_T)$ {calculate r_T used to switch from MPBvN to DA}
12:    Send the switching command to all edge switches
13:    Follow the MPBvN operation
14: **end if**
15: END

Figure 6.31: Switching from DA to MPBvN at the core switch. Reproduced from *Computer Networks*, Vol. 54, Akbar Ghaffarpour Rahbar and Oliver Yang, Agile bandwidth management techniques in slotted all-optical packet interconnection networks, 387–403, copyright 2010, with permission from Elsevier.

Algorithm: Switching from DA to MPBvN in Ingress Switch i

1: Receive switching command at time τ_i
2: **for** $j = 1$ to n, where $j \neq i$ **do**
3:    $N_{NG,ij} = N_{B,ij}$
4:    Get $N_{R,ij}$ and $N_{RS,ij}$
5:    $d'_{ij} = N_{NG,ij} + N_{RS,ij} + N_{R,ij}$ {compute the demand of time slots in ingress switch i}
6: **end for**
7: $\gamma_i = \min\left(1, \frac{F \times f \times W}{\sum_{j=1}^{n_e} d'_{ij}}\right)$
8: **for** $j = 1$ to n, where $j \neq i$ **do**
9:    $d_{ij} = \lfloor \gamma_i \times d'_{ij} \rfloor$ {reduce the required number of time slots allocated to each egress switch proportionally by factor γ_i}
10:    $N_{NG,ij} = d'_{ij} - d_{ij}$ {update the number of time slots that cannot be granted}
11: **end for**
12: Send traffic demand to the core switch at the new frame boundary starting from time $\tau_i + 2 \times (D_{P,max} - D_{P,i})$
13: Follow the MPBvN operation
14: END

Figure 6.32: Switching from DA to MPBvN in ingress switch i. Reproduced from *Computer Networks*, Vol. 54, Akbar Ghaffarpour Rahbar and Oliver Yang, Agile bandwidth management techniques in slotted all-optical packet interconnection networks, 387–403, copyright 2010, with permission from Elsevier.

HTDM system issues the switching command from DA to MPBvN to all edge switches before reaching this critical point, handled by threshold U_T. At the critical point, we have $U_T = r_T + U_T \times PLR_T$. Before switching to MPBvN, the core switch calculates r_T, which will be used to switch from MPBvN to DA later on.

The system will switch from DA to MPBvN if the average channel utilization remains greater than or equal to threshold U_T; i.e., $U_f \geq U_T$, for at least a stability period S_p (handled by counter g). Using S_p avoids false switching from DA to MPBvN due to traffic fluctuations. For example, $S_p = 100,000$ means that the condition $U_f \geq U_T$ must be held for a period of at least 100,000 consecutive frame times.

- **Switching Protocol in an Ingress Switch:** Upon the arrival of a switching command, ingress switch i runs the procedure depicted in Fig. 6.32. In this procedure, $N_{R,ij}$ is defined to be the number of optical packets that must be retransmitted from ingress switch i to egress switch j when DA system has already switched from DA to MPBvN. Define $N_{B,ij}$ to be the number of time slots required to carry the traffic left in the buffers relevant to ingress switch i and egress switch j under DA, as a non-granted traffic under MPBvN. Define $N_{RS,ij}$ to be the number of required time slots to carry the traffic from ingress switch i to egress switch j. Define $N_{NG,ij}$ to be the number of time slots that are not granted between ingress switch i and egress switch j during the previous frame. Define d'_{ij} to be the demand of required time slots (as an integer number) between ingress switch i and egress switch j. Finally, d_{ij} is defined to be the number of required time slots between ingress switch i and egress switch j that OBM of ingress switch i can report to the core switch. When the system has already switched from DA to MPBvN, there may be a number of optical packets left by DA that must be retransmitted under MPBvN. Therefore, it is necessary to account $N_{R,ij}$ in d'_{ij} right after switching. Next, traffic demands are sent to the core switch from a new frame boundary. Then, MPBvN is followed by each edge switch and the core switch in the network.

6.1.5.2 Switching Protocol from MPBvN to DA   The following considerations and procedures are required to switch from MPBvN to DA:

- **Switching Criterion:** The core switch performs the algorithm shown in Fig. 6.33 at each frame boundary for a possible switching from MPBvN to DA. In the pseudo code, N_s is total number of optical packets that can be carried in one frame from all ingress switches within the frame period (i.e., $N_s = n \times F \times f \times W$). In addition, r_c is normalized traffic load at the core switch, where $0 \leq r_c \leq 1$. Besides, a_{ij} is the number of optical packets that ingress switch i is allowed to send to egress switch j within a frame period. Here, the core switch will request the edge switches to switch from the MPBvN to DA operation, provided that average load r_c at the core switch remains less than r_T (obtained above when switching from DA to MPBvN) for a stability period S_p. This is controlled using counter g in the pseudo code. By this way, false switching due to traffic fluctuations can be avoided.

Algorithm: Evaluate Switching from MPBvN to DA

1: $r_c = \frac{1}{N_s} \sum_{i=1}^{n} \sum_{j=1}^{n} a_{ij}$
2: **if** $r_c < r_T$ **then**
3:     $g = g + 1$ {update counter g}
4: **else**
5:     $g = 0$ {reset counter g}
6: **end if**
7: **if** $g \geq S_p$ **then**
8:     $g = 0$ {reset counter g}
9:     Send the switching command to all edge switches
10:    Follow the DA operation
11: **end if**
12: END

Figure 6.33: Evaluation of switching from MPBvN to DA at the core switch. Reproduced from *Computer Networks*, Vol. 54, Akbar Ghaffarpour Rahbar and Oliver Yang, Agile bandwidth management techniques in slotted all-optical packet interconnection networks, 387–403, copyright 2010, with permission from Elsevier.

- **Switching Protocol in an Ingress Switch:** After receiving the switching command by an ingress switch, the ingress switch sends its traffic based on DA at the first available frame boundary.

6.1.5.3 HTDM Parameters   A network manager should choose S_p and U_T parameters in a heuristic manner:

- **Choice of S_p:** Under bursty traffic, parameter S_p should be set in a way to avoid HTDM from oscillation between DA and MPBvN. Note that a small S_p may lead to the oscillation whenever traffic is bursty, whereas a high S_p may lead to delay in switching between DA and MPBvN. Network manager sets S_p in a way that $70 \leq S_p \times F \times S_T \leq 400$ s to avoid oscillation and provide an almost on time switching. If traffic is less bursty, a small value should be assigned to S_p. Otherwise, a large value must be assigned to S_p.

- **Choice of U_T:** Choosing a value very close to 1.0 may prevent switching from DA to MPBvN at all. For example, in a network with ten edge switches and $f \times W = 16$ wavelengths, there are 160 time slots arriving at the core switch at each time slot. Now consider the time slots coming from the first nine edge switches are all fully occupied. However, one-quarter of the time slots coming from the last edge switch is full. In this scenario, the average channel utilization is $U_f = \frac{148}{160} = 0.925$. Now, if one chooses $U_T = 0.94$, the system will never switch from DA to MPBvN. However, this system must switch from DA to MPBvN in order to obtain a better performance. On the other hand, picking a small value for U_T may avoid using the benefits of DA. It is suggested using $U_T \in [0.80, 0.95]$. When more contention resolution/avoidance techniques are used in network, U_T should be set to a small value because network would never reach high channel utilization. In this case, however, switching from DA to MPBvN should not happen since DA may provide a better performance than MPBvN.

6.1.5.4 Performance Comparison Here, the performance of HTDM is evaluated, where DA follows the mechanism stated in Section 6.1.4.1 and MPBvN follows the mechanism stated in Section 6.1.3.3. There are $M = 3$ classes of traffic: HP, MP, and LP. In a network that uses relatively more contention resolution techniques, HTDM may not be appropriate. Otherwise, HTDM can significantly improve the network performance .

Define $L_i(t)$ to be the normalized (with respect to available bandwidth) load of client packets arrival in ingress switch i at time t, where $A_{min} \leq L_i(t) \leq A_{max}$ in which A_{min} is the lowest level of normalized traffic load during low traffic load times (say mid-night time) and A_{max} is the highest level of normalized traffic load for peak times within a day in an ingress switch. To model the diurnal pattern resulted from human activity in Internet traffic [29, 30], we have $L_i(t) = \frac{A_{max}+A_{min}}{2} + \frac{A_{max}-A_{min}}{2} \times$ $sin\left(2 \times \pi \times \frac{t}{T} + \rho_i\right)$, where T is a 24-hour period, and $-\frac{\pi}{6} \leq \rho_i \leq +\frac{\pi}{6}$ is the phase representing the time zone of ingress switch i. The diurnal pattern is divided into two congested and non-congested intervals. For example for $\rho_i = \frac{\pi}{6}$, $L_i(t)$ reaches its maximum and minimum values at time $t = 4$ and $t = 16$, respectively. Here, time $t = 0$ can be thought of as 7:00 am when daily work starts. For $n = 8$, the following phases are used: $[0, \frac{\pi}{8}, -\frac{\pi}{8}, \frac{\pi}{12}, -\frac{\pi}{12}, \frac{\pi}{6}, -\frac{\pi}{6}, \frac{\pi}{15}]$.

The simulations are run for one single-star with $n = 8$ edge switches, $S_T = 9$ μs, $S_O = 1$ μs, and $B_C = 10$ Gbits/s. Other parameters are: $F = 100$ (i.e., a 1 ms frame), $f = 2$, $W = 2$, $\sigma = 0.5$, $A_{min} = 0.2$, $A_{max} = 0.8$, $U_T = 0.9$, $V_{HP} = 0.6$, $V_{MP} = 0.3$, and $V_{LP} = 0.1$ (where the last three significance parameters are used under DA in core switches as stated in Section 4.2.8). Four shared-per-node WCs are used under DA in the core switch.

This traffic at any load includes 25%, 35%, and 40% of HP, MP, and LP classes, respectively. The length of client packets follows 46% of size 40 bytes, 18% of 552 bytes, 18% of 576 bytes, and 18% of 1500 bytes. Packet inter-arrival times of each class is based on Pareto distribution to model the burstiness of Internet traffic. Traffic generated by each ingress switch follows the aforementioned sinusoidal pattern and is symmetric to any egress switch.

Define end-to-end (delay referred to as "delay" from now on) to be the interval between the arrival-time of a client packet in an ingress switch until its successful arrival at its desired egress switch. When generating daily traffic, traffic load varies 120 times in a 24-hour period. Therefore, each traffic load is kept constant for 12 minutes. In the following diagrams, delay and traffic loss performances are displayed in 12-minute intervals within a 24-hour period starting from $t = 1$. Each point obtained at a 12-minute interval in the diagrams is computed from the averaging of six replications for identical traffic conditions.

Figure 6.34 illustrates end-to-end delay for HP, MP, LP traffic and loss for LP traffic in ingress switches under the DA, MPBvN, and HTDM protocols. Here, HTDM(1) and HTDM(2) represent the performance evaluation of HTDM at $S_p = 72,000$ (i.e., 72 s), $S_p = 216,000$, respectively. In this simulation scenario, the dis-

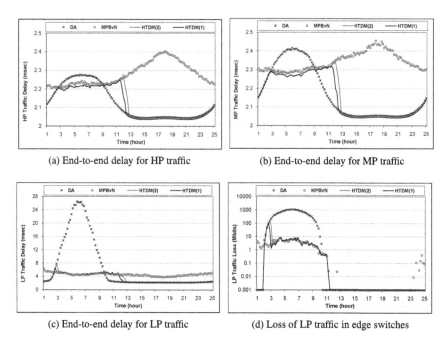

(a) End-to-end delay for HP traffic

(b) End-to-end delay for MP traffic

(c) End-to-end delay for LP traffic

(d) Loss of LP traffic in edge switches

Figure 6.34: HTDM performance evaluation results including HP, MP, LP end-to-end delays and LP traffic loss. Reproduced from *Computer Networks*, Vol. 54, Akbar Ghaffarpour Rahbar and Oliver Yang, Agile bandwidth management techniques in slotted all-optical packet interconnection networks, 387–403, copyright 2010, with permission from Elsevier.

tances of all edge switches from the core switch are 200 km. Under DA, the HP delay is an increasing function of traffic load (see Fig. 6.34a). On the other hand, MPBvN has a higher HP delay at the non-congested interval than the congested interval because the number of required and granted time slots within the frame to carry the MPBvN traffic at the non-congested interval is less than the congested interval. Hence, client packets must wait for a long time before transmission. The HP delay is higher for DA than for MPBvN at the congested interval since DA experiences a high PLR at the congested interval, and therefore the number of retransmissions goes up. The retransmission of lost optical packets reduces the available bandwidth for new traffic, thus leading to a high waiting time and even traffic drop in ingress switch buffers. Under the same protocol, Fig. 6.34b shows that the HP delay is less than the MP delay. This differentiation is provided by the operation of the packet schedulers used in each ingress switch. At a low traffic load, the difference between HP and MP delays is small. The LP delay in Fig. 6.34c is higher for DA than for MPBvN at the congested interval due to the retransmission issue. However, DA outperforms MPBvN in the non-congested interval.

Recall that the overlaid star OPS network under DA, MPBvN, and HTDM guarantees to deliver a transmitted optical packet to its destination. Therefore, traffic loss occurs only in ingress switches. Figure 6.34d shows the network-wide loss for LP traffic in electronic buffers of all ingress switches, where no traffic loss is observed for HP and MP client packets because of using the Output-Controlled Grant-based Round Robin (OCGRR) packet scheduler in ingress switches. The OCGRR always starts scheduling traffic from the HP traffic, then the MP traffic, and then the LP traffic [31]. Therefore, only the LP traffic may be under the risk of traffic drop and delay. Since there is a low traffic arrival within the non-congested interval, there is no loss under any protocol in the interval [12.25, 23.25] hour. The LP traffic loss is higher under DA than MPBvN at the congested interval because of the retransmission of optical packets lost at the core switch. Unlike DA, MPBvN experiences LP traffic drop even at mid-range traffic loads.

In summary, HTDM(1) provides a better performance in terms of delay and loss for the HP, MP, and LP traffic in all traffic loads compared with DA, MPBvN, and HTDM(2), except around time intervals in which switching happens. The sensitivity of switching between DA and MPBvN is controlled by S_p in HTDM. Note that reducing S_p too much may lead to the network oscillation between DA and MPBvN. In the diagrams, HTDM(1) and HTDM(2) switch from DA to MPBvN at 2.62 hour and 2.85 hour, respectively. Then, HTDM(1) and HTDM(2) switch from MPBvN to DA at 11.88 hour and 12.38 hour, respectively.

Additional performance evaluations reveal that in a network that uses more contention resolution techniques (i.e., an expensive network), HTDM cannot provide good results because DA can provide better performances in this case. However, when a low number of contention resolution techniques are used (i.e., an inexpensive network), HTDM is more superior than both DA and MPBvN since it can achieve a better performance under both low and high traffic loads most of the time.

6.2 OPS in Ring Topology

In this section, some recent MAC protocols developed for improving the operation of OPS networks under the ring topology are studied. An interested reader is referred to [32] for various issues of a WDM optical packet ring such as design, packet scheduling, delivered QoS, and efficiency of optical transport.

Two common ring architectures could be used for OPS ring networks:

- **The ECOFRAME Architecture**: ECOFRAME is a slotted and unidirectional WDM metro ring network, where its nodes are connected to the ring via Packet Optical Add and Drop Multiplexers (POADM) [36, 37]. As depicted in Fig. 6.35, each node of this ring has one tunable transmitter and one or multiple receivers. The adaptation layer displayed in the figure is responsible for adapting the size of the upper layer packets into fixed-size packets by means of concatenation and fragmentation mechanisms. It is assumed that transit traffic passes transparently through each node, and the length of packets is fixed.

Data packets are carried through data channels, while control packets are carried through a control channel. Each ECOFRAME interface is associated with both data channels and the control channel. Control packets inform the ring nodes for the occupancy of each time slot on each wavelength channel. Based on

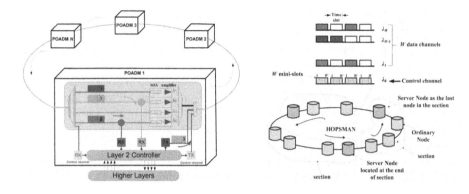

Figure 6.35: The ECOFRAME architecture [33]   Figure 6.36: The HOPSMAN architecture [34, 35]

such information, a node can identify the optical packets to receive and drop. Once an optical packet is dropped from a time slot on a given wavelength, the time slot can be re-used for transmitting a locally generated optical packet to a ring node receiving traffic on that wavelength (e.g., the optical packet added on wavelength λ_2 destined to node 3 displayed in the figure). Since there is one fully tunable transmitter, a given node can transmit at most one optical packet per time slot. A time slot can be either busy or free. If a node has multiple receivers, it can receive more than one optical packet in each time slot.

A bypassing optical packet (e.g., the optical packet destined to node 3 on wavelength λ_1) can be re-timed and re-amplified all-optically by placing a fiber delay line and a semiconductor optical amplifier for each wavelength channel in Fig. 6.35.

- **The HOPSMAN Architecture**: High-performance Optical Packet-Switched MAN (HOPSMAN) is a WDM, slotted and unidirectional ring network (see Fig. 6.36) with W 10 Gbits/s data channels (λ_1 to λ_W) and one 2.5-Gbits/s control channel [34, 35]. Within a time-slot period, the control packet has W mini-slots, where each of these mini-slots carries the status of a wavelength in that time slot. There are two types of nodes as ordinary nodes and server nodes (at least one). Both types of nodes have one fixed transmitter and one fixed receiver to send control packets on the control channel and receive control packets from this channel, respectively. Each ordinary node has a tunable transmitter and a tunable receiver to send/receive on/from a data channel. On the other hand, a server node has multiple receiver and transmitter pairs and one additional component as optical slot eraser, where bandwidth efficiency increases with erasing optical slots.

The MAC protocols in OPS rings can be divided into three groups: reservation-based, contention-based, and hybrid reservation-based and contention-based.

6.2.1 Reservation-Based OPS

Reservation-based algorithms reserve time slots according to the bandwidth requests of nodes. Some of these algorithms and their procedures are detailed in the following.

6.2.1.1 Local Cyclic Reservation (LCR) LCR is a fair, simple, and proactive protocol, which is easily compatible with the changes of traffic conditions [38]. This protocol tries to provide node fairness. Consider a set of nodes (more than one node) that compete to share the bandwidth of a link. Node fairness policy under LCR gives all the nodes the same amount of bandwidth. LCR has three main steps that can be run both in slotted and asynchronous OPS:

- Cyclic Reservation: This step collects and distributes bandwidth reservation requests. Each node in each cycle declares the amount of required bandwidth on each link on the ring and reserves its requests. A reservation request shows the capacity a node needs over a certain link in the next fairness cycle to cope with its demands over that link. These requests are placed in traffic matrix r that shows how much bandwidth each node has requested. The traffic matrix is carried by control packet. As the control packet circulates the ring, each node inserts its request over each link by updating the corresponding elements in matrix r. The time between two sequential departures of the control packet from the same node is called fairness period. Note that the demand of each node on a certain link is the traffic accumulated on that node; i.e., the sum of that node's requests on that link and downstream links.

- Fair Rate Calculation: This step determines how much traffic a node can transmit over each link during a fairness cycle. A node calculates its own per link fair rate for the fairness cycle c at the end of the fairness cycle $c - 1$. The algorithm for the computation of the fair rates starts from the heaviest bottleneck link towards the lightest bottleneck link. This is because solving a heavy bottleneck link may imply in solving a lighter bottleneck link. Note that the bottleneck link is the link which carries larger demands, and the traffic on it should be normalized. To solve the bottleneck of a given link, LCR evaluates the requests made by the nodes over that link, one after another, stating from the lightest request to the heaviest request. At the end of this step, each node obtains a transmission rate as a credit over every link in the ring.

- Fairness Implementation: This step assures that a transmission occurs only if it is in accordance with the corresponding calculated fair rate. LCR must assign the calculated fair rates to each node. At the beginning of the next fairness cycle, the ith node updates its transmission credit over link k. Given a packet of size s destined to node $k + 1$, node i can transmit that packet only if transmission credit over link k is greater than or equal to s. If transmission is granted, then the fairness enforcement mechanism subtracts s from transmission credit over link k to account for the transmission.

6.2.1.2 Local Cyclic Reservation with Source–Destination (LCR-SD) All of operations and phases of LCR-SD are similar to LCR, but with considering flow fairness instead of node fairness [39]. Flow fairness is similar to node fairness, but with this difference that the same amount of bandwidth is given to all flows competing to share a link under flow fairness. Three main steps of LCR are executed under LCR-SD, but each element of request matrix pertains to the amount of bandwidth requested from each flow. Similar to LCR, LCR-SD is also simple and easily compatible with the changes of traffic conditions, but this method provides higher throughput and fairness in comparison with LCR.

6.2.2 Contention-Based OPS

Unlike reservation-based mechanisms, there is no bandwidth reservation for ring nodes under contention-based mechanisms. In the following, some contention-based mechanisms proposed for ring networks are detailed.

6.2.2.1 Label-Based MAC in Packet-Based ECOFRAME This MAC protocol has been designed for the ECOFRAME ring architecture [40]. Recall that a separate transmitter exists for sending control packets on the control channel in ECOFRAME. There is also one tunable transmitter for all data channels so that a given node can transmit at most one optical packet per time slot.

The basic information units under ECOFRAME are Service Data Unit (SDU) and Protocol Data Unit (PDU). A variable length SDU encapsulates client data, whereas a fixed length PDU can include one or more SDUs. The header of a PDU contains three fields as synchronizing and framing information, length of PDU, and Cyclic Redundancy Check (CRC) field to prevent errors in a transmitted PDU. On the other hand, the header of a SDU contains five fields as Segmentation and Reassembly (SAR) information, type of data from the quality of service perspective, length of SDU, label of SDU to indicate which flow the SDU belongs to, and client data.

A control packet carried on the control channel contains some information about the state of data channels and PDUs carried on data channels, such as whether time slot is free or busy, how to behave with the PDU that a node will receive in a time slot, and so on. Note that a node may behave differently according to the PDU type; it can receive packet, receive and erase, or bypass transparently.

Each Control packet PDU (CPDU) contains different fields such as framing and synchronizing information, length of CPDU, cyclic redundancy check, global information, stuffing bits to fill the remaining space of the CPDU according to the CPDU length value, and 40 control fields for at most 40 wavelengths in ECOFRAME, where each of them is relevant to one of the data channels. A control field for wavelength channel λ_i in a given time slot that carries an optical packet includes the following fields: occupancy flag to indicate whether λ_i is busy or free at this time slot, service type indicator to show how the node will behave with the optical packet carried on λ_i, type of the optical packet and traffic to show whether the optical packet is a guaranteed packet or best effort and whether it is multicast or unicast traffic, reservation flag, source and destination addresses for the optical packet, and a stack of labels.

When a certain node receives a CPDU for a given time slot, it can detect which PDU is for it and how it should behave with the PDU. After receiving a CPDU, the node checks whether the destination address field in CPDU matches with the node address. If yes, the node receives the PDU, removes it from the network, and sends it to the relevant client layer. If the PDU traffic type is multicast, the node cannot remove the PDU.

Guaranteed SDUs are inserted into a PDU as soon as a time slot in an appropriate wavelength becomes free. Because of fixed length PDUs, if there is not any guaranteed SDU for insertion, Best Effort SDUs can be inserted into the PDU instead. In this mechanism, transit traffic has always higher priority even if the insertion packet is guaranteed and transit packet is best effort. In other words, a BE traffic cannot be removed from the network.

6.2.2.2 Removing of Overdue Blocks (ROB) The ROB is used for supporting multimedia traffic in the ECOFRAME architecture [41]. Recall that multimedia traffic should be received at its destination on time, but this traffic does not have delivery guarantee. The ROB mechanism removes those multimedia (higher priority) packets that have waited for a long time. Here, there are $M = 3$ classes of priority, where the highest-priority class does not have delivery guarantee since this class is mainly used for multimedia traffic. Transportation unit in the network is a frame which is carried in a time slot with constant capacity.

The Add/Drop mechanism is used in each node that makes it possible to remove a packet from a frame if it is assigned to this node or if the quality management mechanism has requested this removal; to add a packet to the empty frame if the node has traffic to send; or to pass the optical packet in the frame unchanged if the node is an intermediary node to pass transit traffic between other nodes. Optical packets sent over the ring are switched all-optically.

For every class c, there is a separate buffer that can save K blocks of length b. Data received from customers are divided into blocks of constant length b in each node. Each block has QoS class assigned to it and is stored in the buffer dedicated to its class, or lost when the buffer is overflowing.

Every frame can carry p blocks of data. If there are p waiting blocks of the same class in a node, an optical packet is formed and sent inside the next possible frame. If there are blocks waiting for a longer time than a certain threshold in the buffer of the class, an optical packet can still be formed, despite not being filled in full. Note that an optical packet can contain only blocks of the same class. This facilitates QoS management in transit nodes.

Optical packets with lower classes can only be sent if there are no optical packets from higher classes. In other words, blocks from class M can be sent if there is no traffic waiting in buffers of classes 1 to $M - 1$. Additionally, an optical packet of the highest- priority class can be inserted in place of a lowest-priority class optical packet, and therefore the lowest-priority class optical packet is lost. This is the preemption of low-priority traffic with high-priority traffic. This is in accordance with the lack of delivery guarantee for the lowest-priority class traffic.

To ensure delivery time guarantee for the highest-priority class blocks which are suitable for multimedia traffic, the mechanism of removing of overdue blocks is used. For every block of a class in a buffer, a counter is used to keep the time of stay of the block in the buffer. A block which would stay in buffer for a long time (i.e., the counter value is bigger than a threshold value), is removed from the queue. This is because according to the delivery time guarantee rules, such a block is not suitable to be delivered to the receiver.

Observations show that probability of packet loss rate, average waiting time, and average number of blocks in buffers for all types of classes under the ROB mechanism decrease compared with simple QoS methods (i.e., without preemption and removing of overdue blocks).

6.2.2.3 CSMA/CA-Based MAC Protocol for Non-Slotted Ring Ne tworks   In this section, the MAC protocol proposed in [42, 43] for WDM ring OPS networks is explained that supports variable size and DiffServ packets, called CSMA/CA-based MAC protocol. One tunable transmitter and W fixed receivers are used in each node to receive optical packets from any data channel. This MAC protocol uses the CSMA/CA mechanism and then sends optical packets to the ring. It is a fully distributed protocol and does not use a control channel to coordinate operations among nodes. It is not slotted and does not require any segmentation of optical packets as it supports variable-size client packets.

The network supports three traffic classes as EF, AF, and BE. There are three kinds of queues for each class based on IP packet sizes as Q_1, Q_2 and Q_3 are for 40, 41 to 552, and 553 to 1500 bytes IP packet sizes, respectively. The buffer selector unit used in each node pre-classifies an arriving IP packet based on its size to one of these three kinds of queues and saves it either in Q_1, Q_2 or Q_3.

For an optical packet passing a node on a wavelength channel, a splitter is used to tap off a small portion of the optical power to detect the optical packet destination address. If the destination address located in the optical packet header matches the node address, the optical packet data are dropped in the relevant egress switch connected to the node. Meanwhile, the MAC controller is signaled to activate the switch for the wavelength channel to remove the received optical packet. If the destination address is irrelevant to the node address, the detected optical packet is ignored and bypassed to the next node.

Note that an optical receiver adjusted for a given wavelength channel in a node is responsible for both checking the destination address of an incoming optical packet and detecting the availability of the wavelength channel and computing Available Channel Length (ACL). This can be performed by the carrier-sense module depicted in Fig. 6.37. Each incoming optical packet on every wavelength channel is delayed using FDL, where the length of the FDL is equal to the length of maximum IP packet (i.e., 1500 bytes). After determining the available length on a given wavelength channel (i.e., ACL), the node selects an appropriate IP packet with size of less than ACL (that can be transmitted in the free space of the wavelength channel) from the buffers Q_1 to Q_3 of a class (starting from EF class, then AF class, and finally BE class) according to the size of packets that are placed at the head of each class queue.

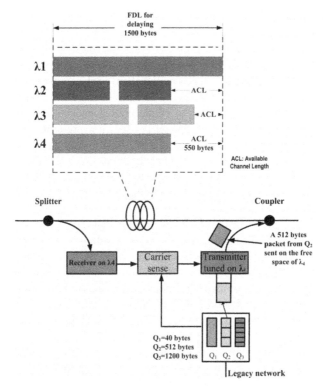

Figure 6.37: Node architecture in CSMA/CA-based MAC for non-slotted ring networks [42]

This mechanism is used to prevent collision of the incoming optical packet with the new inserted optical packet. In Fig. 6.37, assume we have ACL = 550 bytes free on wavelength λ_4. Then, an IP packet is extracted from buffer Q_2 with size 512 bytes and is then converted to an optical packet and sent on wavelength λ_4 by the tunable transmitter. Clearly, this MAC protocol can avoid collision of optical packets sent to the network with transit optical packets and can reuse wavelengths.

6.2.3 Hybrid Reservation/Contention-Based OPS

Hybrid protocols combine reservation-based and contention-based mechanisms in OPS ring networks as detailed in the following.

6.2.3.1 SWING MAC Protocol   The Simple Wdm rING (SWING) protocol has been proposed for slotted OPS in WDM ring networks, where both opportunistic access and reservation mechanisms are used [44]. The ring has a number of data wavelength channels and one control channel for transmission of control packets. Each node can access to all data channels using fixed transmitters. In each time slot, a node can transmit more than of one optical packet. The control packet carries the following information for each time slot in the control channel: Busy, Free, Reserved, and Unreserved. Note that the control packet can determine which node has reserved a time slot.

In an n-node ring network, each node keeps two buffers (one buffer for saving premium client packets and the other one for saving BE client packets) to each one of the remaining $n-1$ nodes. Therefore, the client packets going to the same egress node are saved in their relevant buffers. The SWING uses the packet aggregation mechanism (see Section 2.1.2.1.1) to create optical packets and then sends them when appropriate. In creating an optical packet, first premium client packets are filled, and then BE packets are filled if there is still enough room in the optical packet. Hence, an optical packet may carry both premium and BE packets. All the created optical packets are treated equally in the network as classless optical packets. In short, all scheduling operations are performed in the electronic domain in ingress nodes, and the optical layer only transmits classless optical packets.

One of the key properties of this network is spatial reuse of time slots. When an optical packet reaches its destination, the corresponding time slot must be released immediately. By doing so, the upstream nodes, even the destination node, can reuse this time slot. Another way of using spatial reuse is to use reserved time slots in non-overlapping sections of the ring. For example, in a network with 8 nodes which are sequentially numbered, a time slot can be used between node 2 to node 4 as one section and node 5 to node 8 as another section on the ring, where these sections do not overlap.

The SWING MAC protocol is a distributed protocol that combines reservation scheme with the opportunistic access scheme. The opportunistic scheme avoids wasting of network capacity and the reservation scheme provides fairness among nodes. Under the opportunistic access, each node tries to use any opportunity to send its optical packet in a time slot if the time slot is not busy. The SWING efficiently uses the opportunistic access in the congestion cases and uses the distributed reservation mechanism to establish fairness among nodes. Using SWING, there is no significant loss in network.

The whole ring includes S time slots as a cycle. The first time slot is used to synchronize the reservation mechanism. Unlike the opportunistic access used under normal traffic, the reservation is used when the traffic is bursty. If a node wants to reserve a time slot for cycle c, this reservation is performed in cycle $c-1$ without determining the destination of the optical packet that will be sent in the reserved time slot. Then, right before transmission, the node determines where to send its optical packet in the reserved time slot. The objective of reservation is to establish fairness among nodes that are competing to access to time slots under bursty traffic.

Time-slot reservation in SWING is as follows. Define $R_{i,j}$ to be a counter maintained by node i that shows the number of time slots reserved by node j since crossing the synchronization time slot in the current cycle, $X_{i,j}$ to be the number of aggregated packets in node i waiting to be sent to node j, and X_i to be the total number of aggregated packets waiting in node i. Node i tries to reserve a time slot on a given wavelength channel when it has more waiting aggregated packets than the number of time slots pending for reservation; i.e., $X_i > R_{i,i}$. However, the reservation of a time slot is not definitive and it can be preempted by a downstream node. The reservation of a given time-slot reserved by node j can be preempted by node i, provided that node i has outstanding aggregated packets (i.e., $X_i > R_{i,i}$) and smaller number of pending reservations than node j (i.e., $R_{i,i} < R_{i,j}$). By this mechanism, the node that suffers from congestion can share fairly the bandwidth of the ring.

Under SWING, a node can only transmit an optical packet in a given time slot if:

- The time slot at the previous cycle is reserved by this node and has not been preempted by other nodes.

- The time slot is reserved by other nodes, but spatial reuse is possible because of non-overlapping section.

- The time slot is not busy and not reserved by other nodes.

- Recall that when an optical packet reaches its destination, the destination node should release the time-slot reservation. After getting the optical packet and freeing up the time slot, this node can immediately use this time slot if none of the ring nodes has reserved it yet.

6.2.3.2 PQOC MAC Protocol   The MAC protocol proposed for the HOPSMAN network (see Section 6.2) is called Probabilistic Quota plus Credit (PQOC) [34, 35]. The PQOC is a quota-based protocol based on probabilities, where each network node can access to time slots with a probabilistic quota and credit. As shown in Fig. 6.38, the entire WDM ring is divided into several cycles, where each cycle has a pre-defined fixed number of time slots. Each mini-slot of a control packet carries the status of the corresponding wavelength in that time slot. This status has four modes as Busy, Busy/Read (BREAD), Idle, and Idle/Marked (IMRKD).

There are s server nodes and the whole ring is divided into s sections, where each section has one server node and several ordinary nodes. There are n_k nodes in section k, and therefore we have $n = \sum_{k=1}^{s} n_k$ nodes totally in the ring. Figure 6.36 shows $s = 3$ sections, each with one sever node. The server node of each section is placed at the last point of the section in the direction of packet transmission on the ring.

The PQOC fairly allocates a transmission quota in each cycle to each node. The value of a quota shows the number of allowed time slots that the node can use. Without using a quota, an unfairness issues may happen so that when the traffic load is bursty, the upstream nodes can access to empty time slots before downstream nodes. On the other hand, each ordinary node can send in each time slot at most one optical packet since these nodes have only one tunable transmitter. This limitation is called as "Vertical Access Constraint". There are several scenarios for transmission of traffic in a time slot:

- If the time slot is empty, the permission of accessing to this time slot is granted according to the probability defined by the quota divided by the cycle length. When there is at least one quota, the node can transmit one optical packet. The node can transmit an optical packet and update the status of the corresponding wavelength mini-slot in the control packet from Idle to Busy.

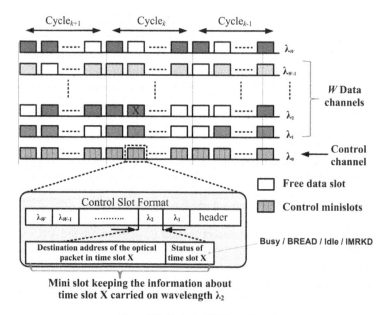

Figure 6.38: Cycles in HOPSMAN [34, 35]

- If the permission access is granted but the time slot is not empty, the node can transmit one optical packet in the next idle and free time slot unconditionally.

- If the node has fewer optical packets than its quota, the number of these missed free time slots will be accounted as a credit for that node. With this credit, the node has more opportunity than its basic quota to transmit optical packets in the next coming cycles. According to its credit amount, the node marks the time slots as IMRKD. Therefore, this node can access to the same number of time slots marked as IMRKD in the next cycles in order to send its optical packets. After sending its optical packets, the node updates the status of the corresponding mini-slot from IMRKD to Busy. The rationale behind this mechanism is to regulate a fair use of the unused remaining bandwidth, especially in the metro environment with bursty traffic. In short, the remaining quota of the node that has failed to use becomes a credit for that node in the next incoming cycles. This can be thought as a kind of time-slot reservation.

- When a destination node receives an optical packet, it updates the status of the corresponding time slot in the relevant control packet from Busy to Busy/Read. By doing so, the next downstream server node will be able to update the corresponding field in the control packet from Busy/Read to Idle. Therefore, the downstream nodes can use this free time slot.

- A time slot passing through a server node is considered as either an available bandwidth if the time slot is empty (Idle) or erased (Busy/Read)) or as a used bandwidth if the time slot is non-empty and cannot be erased since it has not been read yet. Thus, the quota for a node can be computed as the mean value of total amount of available bandwidth observed by a section divided by the total number of nodes in that section. Note that in the Busy/Read case, the server node can erase the time slot and update its status to Idle because the corresponding packet of this time slot has already been reached its destination and read.

Performance evaluation results show that HOPSMAN can achieve 100% throughput when there are only two server nodes in the network. Furthermore, HOPSMAN with PQOC can achieve highly efficient and fair bandwidth allocation under various traffic loads and burstiness.

6.3 Summary

In this chapter, the architectures and recent MAC protocols developed for overlaid star networks and ring networks in metro environments have been studied. The reservation-based, contention-based, and hybrid reservation-based/contention-based mechanisms for traffic transmission in overlaid star OPS have been studied. The similar mechanisms have also been studied for MAC protocols in OPS ring networks. The reservation-based mechanisms are suitable when traffic load is high and bursty. These schemes can provide fairness for nodes as well. On the other hand, contention-based mechanisms are suitable for the case that traffic load is low in which the network traffic can experience the lowest delay. A hybrid contention/reservation mechanism combines the advantages of both reservation-based and contention-based mechanisms, but at the expense of increasing the complexity of core switches.

REFERENCES

1. C. Papazoglou, G. Papadimitriou, and A. Pomportsis. Design alternatives for optical-packet-interconnection network architectures. *OSA Journal of Optical Networking*, 3(11):810–825, 2004.

2. F. J. Blouin, A. W. Lee, A. J. M. Lee, and M. Beshai. Comparison of two optical core networks. *Journal of Optical Networking*, 1(1):56–65, 2002.

3. M. Maier and M. Reisslein. AWG-based metro WDM networking. *IEEE Communications Magazine*, 42(11):S19–S26, 2004.

4. M. Jin and O. W. W. Yang. APOSN: operation, modeling and performance evaluation. *Computer Networks*, 51(6):1643–1659, 2007.

5. I. Khazali and R. Vickers. The agile all-photonic network: architectures, algorithms, and protocols. In *IEEE 5th International Multi-Conference on Systems, Signals and Devices*, 2008.

6. M. Jin and O. Yang. A TDM solution for all-photonic overlaid-star networks. In *Information Sciences and Systems*, pages 1691–1695, Princeton, NJ, 2006.

7. Z. Liu and O. W. W. Yang. Terminal-pair reliability analysis of overlaid-star networks. In *International Conference on Computing, Communications, and Control Technologies (CCCT)*, pages 406–411, Austin, TX, Aug. 2004.

8. C. Fan, M. Maier, and M. Reisslein. The AWG–PSC network: a performance enhanced single-hop WDM network with heterogeneous protection. *IEEE Journal of Lightwave Technology*, 22(5):1242–1262, 2004.

9. L. Mason, A. Vinokurov, N. Zhao, and D. Plant. Topological design and dimensioning of agile all-photonic networks. *Computer Networks*, 50(2):268–287, Feb. 2006.

10. X. Liu, A. Vinokurov, and L. G. Mason. Performance comparison of OTDM and OBS scheduling for agile all-photonic network. In *IFIP 2005 Conference on Metropolitan Area Networks*, Viet Nam, 2005.

11. A. G. Rahbar and O. Yang. Agile bandwidth management techniques in slotted all-optical packet interconnection networks. *Elsevier Computer Networks*, 54(3):387–403, Feb. 2010.

12. I. Keslassy, M. Kodialam, T. Lakshman, and D. Stiliadis. On guaranteed smooth scheduling for input-queued switches. *IEEE/ACM Transactions on Networking*, 13(6):1364–1375, Dec. 2005.

13. C. Peng, G. V. Bochmann, and T. J. Hall. Quick Birkhoff–von Neumann decomposition algorithm for agile all-photonic network cores. In *IEEE ICC*, Istanbul, Turkey, 2006.

14. N. Saberi. *Photonic Networks: bandwidth Allocation and Scheduling*. LAP LAMBERT Academic Publishing, 2011.

15. C. S. Chang, D. S. Lee, and Y. J. Shih. Mailbox switch: a scalable two-stage switch architecture for conflict resolution of ordered packets. In *IEEE Infocom*, pages 1995–2006, Hong Kong, 2004.

16. D. Careglio, A. Rafel, J. S. Paretat, S. Spadaro, A. Hill, and G. Junyent. Quality of service strategy in an optical packet network with multi-class frame-based scheduling. In *IEEE International Workshop on High Performance Switching and Routing (HPSR 2003)*, pages 129–134, Torino, Italy, 2003.

17. C. S. Chang, W. J. Chen, and H. Y. Huang. Birkhoff–von Neumann input buffered crossbar switches. In *IEEE Infocom*, pages 1614–1623, Tel Aviv, 2000.

18. S. Chang, W. J. Chen, and H. Y. Huang. On service guarantees for input-buffered crossbar switches: a capacity decomposition approach by Birkhoff and Von Neumann. In *Seventh International Workshop on Quality of Service*, pages 79–86, 1999.

19. N. McKeown. The iSLIP scheduling algorithm for input-queue switches. *IEEE Transactions on Networking*, 7:188–201, 1999.

20. A. G. Rahbar. EBvN: efficient BvN in multi-fiber/multi-wavelength overlaid-star optical networks. *Springer Journal of Annals of Telecommunications*, 67(11-12):575–588, Nov. 2012.

21. A. G. P. Rahbar and O. Yang. An integrated TDM architecture for AAPN networks. In *SPIE Photonics North*, volume 5970, Toronto, Canada, Sept. 2005.

22. C. S. Chang, W. J. Chen, and H. Y. Huang. Birkhoff–von Neumann input-buffered crossbar switches for guaranteed-rate services. *IEEE Transactions on Communications*, 49:1145–1147, 2001.

23. T. H. Cormen, C. E. Leiserson, R. L. Rivest, and C. Stein. *Introduction to Algorithms*. Second edition, MIT Press and McGraw-Hill, 2001.

24. A. G. Rahbar and O. Yang. Reducing loss rate in slotted optical networks: a lower bound analysis. In *IEEE ICC*, pages 2770–2775, Istanbul, Turkey, 2006.

25. A. G. Rahbar and O. Yang. CST: a new contention reduction scheme in slotted all-optical packet switched networks. *Performance Evaluation*, 67(5):361–375, May 2010.

26. A. G. Rahbar. PTES: a new packet transmission technique in bufferless slotted OPS networks. *IEEE/OSA Journal of Optical Communications and Networking*, 4(6):490–502, June 2012.

27. A. G. P. Rahbar and O. Yang. Contention resolution by retransmission in single-hop OPS metro networks. *Journal of Networks (JNW)*, 2(4):20–27, Aug. 2007.

28. A. G. Rahbar and O. Yang. Prioritized retransmission in slotted all-optical packet-switched networks. *Journal of Optical Networking*, 5(12):1056–1070, Dec. 2006.

29. D. Medhi. QoS routing computation with path caching: a framework and network performance. *IEEE Communications Magazine*, 40(12):106–113, Dec. 2002.

30. S. Floyd and V. Paxson. Difficulties in simulating the Internet. *IEEE/ACM Transactions on Networking*, 9(4):392–403, Aug. 2001.

31. A. G. Rahbar and O. Yang. OCGRR: a new scheduling algorithm for differentiated services networks. *IEEE Transactions on Parallel and Distributed Systems*, 18(5):697–710, May 2007.

32. B. Uscumlic. *Optical Packet Ring Engineering: design and Performance Evaluation*. LAP LAMBERT Academic Publishing, 2011.

33. D. Chiaroni. Optical packet add/drop multiplexers for packet ring networks. In *ECOC*, Brussels, Belgium, Sept. 2008.

34. M. C. Yuang, I-F. Chao, and B. C. Lo. HOPSMAN: an experimental optical packet-switched metro WDM ring network with high-performance medium access control. *IEEE Optical Communications and Networking*, 2(2):91–101, 2010.

35. M. C. Yuang, I-F. Chao, B. C. Lo, P. L. Tien, J. J. Chen, and C. Wei. HOPSMAN: an experimental testbed system for a 10-Gb/s optical packet-switched WDM metro ring network. *IEEE Communications Magazine*, 46(7):158–166, 2008.

36. B. Uscumlic, I. Cerutti, A. Gravey, P. Gravey, D. Barth, M. Morvan, and P. Castoldi. Optimal dimensioning of the WDM unidirectional ECOFRAME optical packet ring. *Photonic Network Communications*, 22:254–265, 2011.

37. B. Uscumlic, A. Gravey, P. Gravey, and M. Morvan. Scheduling issues in ECOFRAME optical packet switched ring. *Telfor Journal*, 3:49–53, 2011.

38. M. R. Salvador, M. M. Uesono, and N. L. da.Fonseca. A local fairness protocol for optical packet-switched WDM ring networks. In *IEEE International Conference on Communications*, pages 1524–1528, Seoul, Korea, May 2005.

39. M. R. Salvador, M. M. Uesono, and N. L. da Fonseca. A packet ring fairness protocol and its impact on TCP fairness. In *IEEE Global Telecommunications Conference (GLOBECOM)*, pages 1044–1048, ST.Louis, MI, 2005.

40. L. Sadeghioon, A. Gravey, and P. Gravey. A label based MAC for OPS multi-ring. In *The 15th International Conference on Optical Network Design and Modeling (ONDM)*, Bologna, Italy, Feb. 2011.

41. M. Nowak and P. Pecka. QoS management for multimedia traffic in synchronous slotted-ring OPS networks. In *HET-NETs*, 2010.

42. W. P. Chen, W. S. Hwang, T. C. Hsin, and C. T. Sung. A novel MAC protocol of IP over metro ring network for end-to-end QoS. In *IEEE International Conference on System & Signals*, pages 1184–1189, 2005.

43. W. P. Chen and W. S. Hwang. A packet pre-classification CSMA/CA MAC protocol for IP over WDM ring networks. In *The 8th ICCS IEEE International Conference on Communication Systems*, pages 1217–1221, Singapore, Nov. 2002.

44. T. Bonald, S. Oueslati, J. Roberts, and C. Roger. SWING: traffic capacity of a simple WDM ring network. In *21st International Conference in Teletraffic Congress, ITC*, Paris, Sept. 2009.

Index

Quality of Service in Optical Packet Switched Networks, First Edition.
By Akbar Ghaffarpour Rahbar Copyright © 2015 IEEE. Published by John Wiley & Sons, Inc.

IEEE Press Series on
Information and Communication Networks Security (ICNS)

Series Editor, **Stamatios Kartalopoulos, PhD**

Mission Statement:
This series provides high quality technical books on Information and Communication Networks Security Theory and Technology. The series is interested in the security aspects of all types of communication networks (wireless, wired, optical, quantum, chaotic, hierarchical, non-hierarchical, IP, ad-hoc, cloud, and so on), and in the security of information transported through their nodes and across them. Our security interests are on all levels of the OSI model, from the network physical layer to the application layer, on the node level and end-to-end, and on all levels of mathematical and technical complexity. Books are intended for professionals, researchers, and students, as well as for private, academic and government organizations.